PRINCIPLES OF ELECTRICAL MEASUREMENT

Sensors Series
Senior Series Editor: **B E Jones**
Series Co-Editor: **W B Spillman, Jr**

Novel Sensors and Sensing
R G Jackson

Hall Effect Devices, Second Edition
R S Popović

Sensors and their Applications XII
Edited by S J Prosser and E Lewis

Sensors and their Applications XI
Edited by K T V Grattan and S H Khan

Thin Film Magnetoresistive Sensors
S Tumanski

Electronic Noses and Olfaction 2000
Edited by J W Gardner and K C Persaud

Sensors and their Applications X
Edited by N M White and A T Augousti

Sensor Materials
P T Moseley and J Crocker

Biosensors: Microelectrochemical Devices
M Lambrecht and W Sansen

Current Advances in Sensors
Edited by B E Jones

Series in Sensors

PRINCIPLES OF
ELECTRICAL MEASUREMENT

S Tumanski

Warsaw University of Technology
Warsaw, Poland

Taylor & Francis
Taylor & Francis Group
New York   London

Published in 2006 by
CRC Press
Taylor & Francis Group
6000 Broken Sound Parkway NW, Suite 300
Boca Raton, FL 33487-2742

International Standard Book Number-10: 0-7503-1038-3 (Hardcover)
International Standard Book Number-13: 978-0-7503-1038-3 (Hardcover)
Library of Congress Card Number 2005054928

Library of Congress Cataloging-in-Publication Data

Tumanski, Slawomir.
    Principles of electrical measurement / by Slawomir Tumanski.
        p. cm.-- (Series in sensors)
    Includes bibliographical references and index.
    ISBN 0-7503-1038-3
    1. Electric measurements. 2. Electronic measurements. 3. Signal processing. I. Title. II. Sensors series.

TK275.T75 2005
621.37--dc22   2005054928

Contents

Preface   ix

1. Introduction to Measurements   1

2. Fundamentals of Electrical Measurements   13
2.1. Main Terms and Definitions   13
2.1.1. Basic terms of measurement technique   13
2.1.2. The main methods of measurements   18
2.2. Uncertainty of Measurements   26
2.2.1. Errors, uncertainty, and reliability of signal processing   26
2.2.2. Basic statistical terms and concepts   34
2.2.3. Methods of evaluation and correction of the uncertainty
     related to limited accuracy of measuring devices   40
2.2.4. The estimation of uncertainty in measurements   52
2.3. Standards of Electrical Quantities   57
2.3.1. Standards, etalons, calibration and validation   57
2.3.2. The standards of electrical quantities referred to
     the physical phenomena and laws   58
2.3.3. Material standards of electrical quantities   63
2.3.4. The reference multimeters and calibrators   69
References   71

3. Classic Electrical Measurements   73
3.1. Indicating Measuring Instruments   73
3.1.1 Electromechanical instruments versus digital
     measuring systems   73
3.1.2. The moving coil meters   74
3.1.3. The moving iron meters   81
3.1.4. Electrodynamic meters – wattmeters   82
3.1.5. Induction type watt-hour meters   86
3.2. Recording and Displaying Measuring Instruments   88
3.2.1. Fundamentals of oscilloscopes   88
3.2.2. Recorders and data storage devices   93
3.3. Bridge Circuits   94
3.3.1. Balanced and unbalanced bridge circuits   94
3.3.2. Null-type DC bridge circuits   96

3.3.3. The AC bridge circuits   99
3.3.4. The transformer bridge circuits   104
3.3.5. The unbalanced bridge circuits   107
3.3.6. The alternatives for bridge circuits – Anderson Loop   112
3.4. Potentiometers and Comparators   **114**
References   118

4. Processing of the Analogue Measurement Signals   **121**
4.1. Signal Conditioning   **121**
4.1.1. Analogue measurement signals   121
4.1.2. Conditioning of resistance, capacitance and inductance 126
4.1.3. AC/DC conversion   131
4.1.4. Voltage to frequency conversion   141
4.2. Amplification of the Signals   **143**
4.2.1. Differential, operational and instrumentation amplifiers 143
4.2.2. Isolation amplifiers   147
4.2.3. Amplifiers of very small DC signals   150
4.2.4. Amplifiers of very small AC signals   154
4.2.5. Amplifiers of very large input resistance (electrometers) 159
4.2.6. The function amplifiers   161
4.3. Negative Feedback in the Measuring Technique   **169**
4.4. The Improvement of the Quality of the Analogue Signals **179**
4.4.1. The noises and interferences of the analogue signals   179
4.4.2. The connection of the measuring signal to the amplifier 184
4.4.3. The analogue filtering of the signals   191
References   201

5. Digital Processing of the Measurement Signals   **205**
5.1. Analogue-to-Digital Converters   **205**
5.1.1. Sampling, quantization and coding of signals   205
5.1.2. Analogue-to-digital converters ADC   218
5.1.3. The main specifications of analogue-to-digital converters 234
5.2. Digital-to-Analogue Converters   **238**
5.2.1. The reconstruction of the analogue signal   238
5.2.2. The digital-to-analogue converters DAC   242
5.2.3. The main specifications of digital-to-analogue converters 247
5.3. Methods and Tools of Digital Signal Processing   **249**
5.3.1. The main terms of digital signal processing   249
5.3.2. The Discrete Fourier Transform DFT and Fast Fourier
       Transform FFT   259
5.3.3. Short-time Fourier Transform and Wavelet transform   268
5.3.4. Digital filters   275

5.4. Examples of Application of Digital Signal Processing in Measurements **287**
    5.4.1. The spectral analysis 287
    5.4.2. Digital signal synthesis 297
    5.4.3. Improvement of the signal quality and the signal recovery 303
5.5. Digital Measuring Instruments **312**
    5.5.1. Digital multimeters and frequency meters 312
    5.5.2. Digital oscilloscopes 318
    5.5.3. Digital measurement of power and energy 323
5.6. Intelligent Data Analysis **326**
    5.6.1 The artificial intelligence in measurements 326
    5.6.2. The adaptive filters 327
    5.6.3. Artificial neural networks 331
    5.6.4. Fuzzy Logic 340
References 344

6. Computer Measuring Systems **349**
6.1 Introduction **349**
6.2. Input Circuits of the Measuring Systems **353**
    6.2.1. Circuits for data conditioning and acquisition 353
    6.2.2. The sensors with built-in interface – intelligent sensors 354
    6.2.3. Analogue and digital transmitters 356
    6.2.4. Data loggers 357
    6.2.5. IEEE P1451 standard – smart sensors 359
6.3. Data Acquisition Circuits – DAQ **362**
    6.3.1. Plug-in data acquisition board 362
    6.3.2. External data acquisition board 365
6.4. Data Communication in Computer Measuring Systems **367**
    6.4.1. Interfaces, buses and connectors 367
    6.4.2. Serial interfaces: RS-232C and RS-485 368
    6.4.3. Serial interfaces: USB and FireWire 373
    6.4.4. Parallel GPIB interface (IEEE-488/IEC-625) 377
    6.4.5. Wireless interfaces: IrDA, Bluetooth and WUSB 382
    6.4.6. Mobile telephony systems GSM and UMTS as a tool
        for data transfer 385
    6.4.7. Radio data acquisition and transfer 389
    6.4.8. Computer systems using Ethernet and Internet 392
    6.4.9. Dedicated interfaces: CAN, I^2C, MicroLAN, SDI-12 396
    6.4.10. HART interface and the 4 – 20 mA standard 400
    6.4.11. Industrial communication standards – Fieldbus, Profibus,
        SCADA 401
    6.4.12. Modular systems – VXI, PXI 406
    6.4.13. Standard command for measuring devices – SCPI 408

6.5. Measuring Systems Basing on the Signal Processors   **410**
   6.5.1. Microcontrollers and signal processors
      in measuring technique   410
   6.5.2. Microinterfaces – SPI and UART   418
6.6. Virtual Measuring Systems   **421**
   6.6.1. What is the virtual measuring device?   421
   6.6.2. TestPoint   424
   6.6.3. Agilent VEE Pro   428
   6.6.4. LabVIEW of National Instruments   431
6.7. The Examples of Computer Measuring Systems   **438**
   6.7.1. The measuring system for testing of magnetic materials   438
   6.7.2. The arbitrary wave excitation systems   442
   6.7.3. The scanning device for magnetic field imaging   449
References   452

Symbols used in the Book   455
Abbreviations used in the Book   457
Index   461

Preface

In libraries and bookshops we can find various books on electrical measurements[1]. Most of them describe various aspects of electrical measurements: digital or analogue techniques, sensors, data acquisition, data conversion, etc. However, it can be difficult to find a book that includes a complete guide on the techniques used in taking electrical measurements. The reason for this is rather obvious –modern measuring requires knowledge of many interdisciplinary topics such as computer techniques, electronics, signal processing, micro- and nanotechnology, artificial intelligence methods, etc. It is practically impossible for one author to know and explain all these subjects. Therefore, there are frequently available books called "*Handbook of...*" written by dozens of co-authors. Unfortunately, such books are mainly more conglomerates of many encyclopaedia entries of unequal levels than comprehensive and compact knowledgeable books.

The other aspect of this problem is that the progress in measuring techniques is very fast, with every year bringing new developments. It is really difficult to catch the state of the art in measurements. It is much easier to gather knowledge on a particular subject in the form of a monograph focused on a special problem. But on the other hand, students and industry engineers look for comprehensive books that are easy to understand and most of all include recent developments, such the computer measuring systems or virtual measuring methods. I lecture on electrical measurements to students of electrical engineering, robotics and informatics. To tell the truth I could not find a suitable book on the whole subject and therefore I decided to write one myself. Last year I "tested" this book on students and the results were quite promising. Most of the students understood the electrical measurements and what most importantly, they found that this subject was interesting, and even fascinating.

Let us look at modern measurement techniques, the present state and the future perspectives. There is no doubt that the future is reserved for computer measuring systems. It is no wonder that today, when a simple electric shaver is supported by a microcontroller that the measuring instruments are also

---

[1] A non-exhaustive list of market available books on measurements is included at the end of this preface.

computerized. Recently, computer measuring systems have become main tools and the subject of research. The result is that many important topics, discussed in this book as "*Classic Electrical Measurements*" are today on the periphery of interest. However knowledge of these subjects is important to understand the principles of modern measuring instruments.

Other consequence of the development of computer and microelectronics supported measuring systems is that they are now also available to non-specialists. Today, what was reserved exclusively in past, measuring devices as high quality analogue to digital converters or amplifiers, are now available to all at modest prices. User friendly software such as LabVIEW helps in the design of sophisticated measuring instruments. So-called intelligent sensors are today designed in "plug and play" technology, ready to connect into worldwide computer networks. Thus currently, the measurement technique is open to everyone (including persons far from electrical engineering) and it is important to show them, how to perform the measurements correctly. This brings us to the fundamental question: which knowledge about measurements is indispensable?

After discussing with many university colleagues, practicing industry engineers and of course students, the proposal of contents for such indispensable subject was formulated. But it appeared that to present such subjects more than a thousand pages book was advisable. Therefore, the whole programme was divided into two clearly separated parts: "*Principles of Electrical Measurements*" and "*Application of Electrical Measurements in Science, Industry and Everyday Life*". This first part is presented in this book.

I understand the "*Principles of Electrical Measurements*" as the whole knowledge, common for all types of electrical measurements. These common subjects include most of all *signal processing techniques* (digital and as well analogue), *classic measurement techniques, methods of estimation of accuracy and uncertainty of measurement results, data acquisition and signal conditioning, application of computers and digital signal processors in measurement* and *virtual measurements techniques*. When such subjects are understood (for example, after reading this book, I hope) it should be more easy to adapt to the more practical subjects: "*Application of Electrical Measurements*" – *sensors, measurements of electrical and non-electrical quantities, non-destructive testing and material evaluation, design of measuring instruments, etc.*).

This book is divided into three main parts. In the first one (Chapters two and three) the fundamentals and classic electrical methods are described (main terms and methods, standards and measurement uncertainty). The second part (Chapters four and five) are devoted to signal processing – analogue and digital. And the last part (Chapter six) informs about computer measuring systems. Taking into account the state of the arts techniques and perspectives of electrical measurements presented above, we understand why the "classical part" occupies only about quarter of the book while the "digital signal processing and computer measuring systems" fill more than half of it.

This book is addressed mainly to students, but the proposed material should be also useful for practicing engineers. As was earlier mentioned, this book was "tested" on several groups of students of Warsaw University of Technology. I would like to thank many colleagues from that University for valuable discussions and remarks. I would especially like to thank professors Jerzy Barzykowski, Marek Stabrowski, Zygmunt Warsza, Dr Stan Zurek (from Cardiff University) and Ph.D. student Slawomir Baranowski.

Slawomir Tumanski

Most important books related to Electrical Measurements

Analog Devices 2004 *Data Conversion Handbook*, Newnes
Anderson N.A. 1997 *Instrumentation for Process Measurement and Control*, CRC Press
Austerliz H. 2002 *Data Acquisition Techniques using PCs*, Academic Press
Baican R., Nesculescu D.S. 2000 *Applied Virtual Instrumentation*, Computational Mechanics
Battigha N.E. 2003 *The condensed Handbook of Measurement and Control*, ISA Instrumentation
Bentley J.P. 2004 *Principles of Measurement Systems*, Prentice Hall
Bolton W. 2001 *Newnes Instrumentation and Measurement Pocket Book*, Newnes
Boyes W. 2002 *Instrumentation Reference Book*, Butterworth-Heinemann
Brignel J. White N. 1996 *Intelligent Sensor System*, IOP Publ.
Dally J.W., Riley W.F., McConnell K.G. 1993 *Instrumentation for Engineering Measurements*, John Wiley & Sons
Doebelin E.O. 2003 *Measurement Systems*, McGraw-Hill
Dunn W.C. 2005 *Introduction to Instrumentation, Sensors, and Process Control*, Artech House
Dyer S.A. 2001 *Wiley Survey of Instrumentation and Measurements*, IEEE Computer Society
Elgar P. 1998 *Sensors for Measurement and Control*, Prentice Hall
Eren H. 2003 *Electronic Portable Instruments: Design and Application*, CRC Press
Fraden J. 2003 *Handbook of Modern Sensors*, Springer
Frank R. 2000 *Understanding Smart Sensors*, Artech
Gardner J.W., Varadan V.K., Awadelkavim O.A. 2001 *Microsensors, MEMS and Smart Devices*, Wiley & Sons
Hughes T.A. 2002 *Measurement and Control Basic*, ISA-Instrumentation

James K. 2000 *PC Interfacing and Data Acquisition: Techniques for Measurements, Instrumentation and Control*, Newnes

Kester W. 2005 *Data Conversion Handbook*, Butterworth-Heinemann

Kester 2003 *Mixed Signals and DSP Design Techniques*, Newnes

Klaasen K.B. 1996 *Electronic Measurement and Instrumentation*, Cambridge University Press

Kularatna N. 2002 *Digital and Analogue Instrumentation Testing and Measurement*, IEE

Liptak B.G. 2003 *Instrument Engineering Handbook: Process Measurement and Analysis*, CRC Press

Morris A.S 2001 *Measurements and Instrumentation Principle*, Butterworth-Heineman

Morris A.S 1996 *The Essence of Measurement*, Prentice Hall

Nawrocki W. 2005 *Measurement Systems and Sensors*, Artech

Northrop R.B. 1997 *Introduction to Instrumentation and Measurements*, CRC Press

Pallas-Areny R., Webster J.G. 1991 *Sensors and Signal Conditioning*, John Wiley & Sons

Park J., Mackay S. 2003 *Practical Data Acquisition for Instrumentation and Control*, Newnes

Paton B.E. 1998 *Sensors, Transducers, LabVIEW*, Prentice Hall

Potter R.W. 1999 *The Art of Measurement*, Prentice Hall

Putten van A.F. 2003 *Electronic Measurement Systems: Theory and Practice*, IOP Publ.

Ramsey D.C. 1996 *Principles of Engineering Instrumentation*, Butterworth-Heinemann

Rathore T.S. 2004 *Digital Measurement Techniques*, CRC Press

Romberg T.M., Ledwige T.J., Black J.L. 1996 *Signal Processing for Industrial Diagnostics*, John Wiley & Sons

Schnell L. 1993 *Technology of Electrical Measurements*, John Wiley & Sons

Sinclair I. 2001 *Sensors and Transducers*, Newnes

Swanson D.C. 2000 *Signal Processing for Intelligent Sensor Systems*, Marcel Dekker

Sydenham P.H. (Ed) 2005 *Handbook of Measuring System Design,* John Wiley & Sons

Taylor H.R. 1997 *Data Acquisition for Sensor Systems*, Springer

Tran Tien Lang 1987 *Electronics of Measuring Systems*, John Wiley & Sons

Turner J.D., Hill M. 1999 *Instrumentation for Engineers and Scientists*, Oxford University Press

Webster J.G. 1998 *Measurements, Instrumentation and Sensors Handbook*, CRC Press

Webster J.G (Ed) 2004 *Electrical measurement, Signal Processing and Display*, CRC Press

Wilson J.S. 2004, *Sensor Technology Handbook*, Newnes

1

Introduction to Measurements

The main person of the Molier's comedy "The Bourgeois Gentleman[1]" Monsieur Jourdain states with amazement "*By my faith! For more than forty years I have been speaking prose without knowing about it...*". Probably many of the readers would be also surprised by the information that they perform measurements almost all the time and everywhere without knowing about it. When we say "*it is cold today*" we describe the result of a measurement carried out by our senses (receptors). Such measurement is performed in a subjective way - another person could state in the same conditions that it is not cold. But generally we estimate the temperature by comparison with the temperature memorized as a reference one. Thus we performed the measurement.

Furthermore, when we say "*I do not feel well today*" we describe the results of the analysis of the state of our organism. Our receptors tested the parameters: blood pressure, body temperature, pulse, level of adrenaline, etc. as incorrect. The measuring system in our body operates very similarly to a computer measuring systems used for instance in the industry. The receptors (the sensors) determine the value of many quantities: light, sound, smell, temperature, etc. The results of the sensing are transmitted to the brain as the electrical signals by the interface consisting of billions of nervous fibers[2]. Our brain acts as a central computer unit - it controls the measuring system and processes all incoming signals. It is worth noting that the human organism is a very excellent temperature conditioner – it stabilizes the temperature of the body at *36.6 °C* with the precision of *0.1 °C*.

---

[1] or "The Would-Be Gentleman" or "The Middle-Class Gentlemen"
[2] This current is very small, about *100 pA*, but we are able to measure such currents using the SQUID superconducting method – this way we have been registered the current variations during the reading of various letters.

The Oxford Dictionary explains the term *measure* as *"ascertain the size, amount or degree of (something) by using an instrument or device marked in standard units or by comparing it with an object of known size"* (from the Latin *mensurare* – to measure)[1].

For people working professionally in the measurement field this explanation is unacceptably incomplete. It contains two important terms, namely *ascertainment* or better **(1) estimation** and **(2) standard unit**. But there is a lack of a third, absolutely indispensable term – the accuracy of estimation, or better **(3) uncertainty** of estimation. Without the knowledge of the uncertainty of estimation the whole measurement process is worthless. More exact discussion of the main terms of measurement is presented in the next Chapter. However, in this Chapter we should assume the following intuitive definition: *measurement is the estimation of the quantity of certain value (with known uncertainty) by comparison with the standard unit*. This simplified definition given above emphasizes the important aspect of the measurement process – this action is always present in our lives.

Practically almost whole activity of our lives is related to measurements, because we constantly compare various objects, evaluate their properties, determine their quantities. We persistently discover surrounding us world. Where is the limitations of the term *"measurements"* in the sense of the title of this book? Consider following examples.

We pay in the supermarket with cash for the shopping. Is it the measurement? Theoretically all elements of given above definition are present. In the case of cash payment we estimate the value of the amount; there is a standard unit (quant) of amount – for example one cent or one penny. If we are absentminded or with poor eyesight our counting of money is with certain level of uncertainty.

The payment can be realized in traditional way. But it is forecasted that in the future the supermarket cashiers will be not necessary. All products can be marked (by for example the magnetic code signature) and the sensor in the gate can detect all items. The computer system determines the cost and withdraws necessary amount of money directly from our bank account. The reliability and accuracy of such system strongly depends on the quality of magnetic field sensors and magnetic signature detection.

And other situation. We choose the color for painting of the walls. Typically such choice is very subjective. But the colors are very precise described as the length of the light wave. In the case of mixture of colors (it can be for example *RGB* mixture – red, green and blue or *CMYK* mixture – cyan, magenta, yellow and black) we can precise describe the percents of

---

[1] The most of terms related to measurements are defined by "International Vocabulary of Basic and General Terms in Metrology – ISO VIM", International Organization for Standardization ISO, Geneva, 1993 (revised edition 2004).

every components. Moreover exist special measuring instruments for determination of color. We can describe the color with various precision, even we can use the fuzzy logic system for not precise color describing.

And other situation, seemingly far from the measurements - the rock concert. The singer produces the air pressure variations, which are sensed by the microphone (the transducer converting the air pressure into the electrical signals), next the electrical signals representing the sound (characterized by the frequency and the magnitude) are processed and converted back to the sounds, which we can hear. The recorded sound (electrical signals) we can further use for analysis of the acoustic characteristic of the concert hall.

We see that the distinguish of the everyday life activities and the measurement technique is very fluent and relative (depending on the purpose of this activity).

The difference between a measurement and an everyday routine activity lies in the goal of these actions. *The measurement is the process of gathering* **information** *from the physical world* (Sydenham *et al* 1989). This aspect of a measurement process is very important. Of course most of measurements serve simple practical purposes. For example, when a shop assistant weights our goods it helps us in assessment of the quantity (and price) of the shopping. When we look at the thermometer it helps us in decision what to wear. The sensors in factory help in control of the technological processes of manufacturing. But looking wider – the importance of measurements has crucial significance for human civilization. From beginning of our civilization people tried to understand and comprehend the surrounding world. And the *science of measurements (metrology)* offers still better tools and methods for these purposes. No wonder that such large number of the Nobel prizes were awarded for the measuring achievements (for example for accurate measurement of the resistance by means of the quant Hall effect – 1985, for the scanning tunneling microscope – 1986, for the cesium atomic clock – 1989 or for the magnetic resonance imaging – 2003).

It is also the formal aspect of the definition of measurement. It is called *traceability of measurements*. This term means that all results of measurement are traceable to the standards and standardized units. The standards are arranged in the form of the pyramid. On the top of this pyramid are the international standards (under supervision by the *Bureau International Poids at Mesures BIPM – Paris*). From this standards are traced back the National Standards, from that the standards in Accredited Laboratories and at the end is our measuring device. Similarly on the top of other pyramid there are seven main units of *SI system* (*System Internationale*). From this units are traced back all *derived units* of various quantities. All quantities and their units are collected in the ISO (*International Standard Organization*) standard.

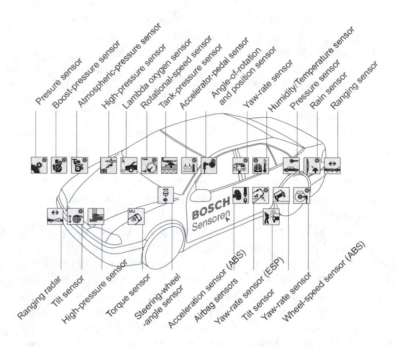

Figure 1.1 The typical Bosch sensors used for automotive application (from Bosch –
Automotive Sensors 2002) (permission of Robert Bosch GmbH)

At present, the measuring devices are almost everywhere. Let us look at
the cars. Some time ago a typical car was equipped with only several
measuring instruments – for detecting the fuel level, speed of the vehicle,
temperature of the engine. Today, dozens (or often hundreds) of various
sensors are installed in any new car (Fig. 1.1) – from sensors important for
the security (testing the rotational speed of each wheel in the ABS system),
through swanky sensors memorizing the positioning of the seats (Robert
Bosch 2002, Jiri 2003). The action of the air-bags is controlled by the stress
sensors. Often the windscreen wipers are controlled accordingly to the
intensity of the rainfall. Many drivers do not know how to reverse without
ultrasonic detectors of the evidence of barriers. It is not a surprise when the
car is equipped with the satellite GPS system (*Global Positioning System*).
The number of sensors is so large that there was a need for a special interface
CAN (*Controller Area Network*) designed by Bosch for connecting of the
intelligent sensors in automotive applications. Modern sensors (so called
intelligent or smart sensors) are equipped with suitable interfaces (Ethernet,
CAN, RS-232 interfaces) and it is possible to connect them directly to the
network system. There are also available special microcontrollers equipped
with CAN output.

The modern cars are additionally equipped with hundreds of sensors for the control of the engine performance. Starting from 1996 practically all cars are equipped with OBDII (*On Board Diagnostics*) system (Cox 2005, David P. 2002, David P. 2004, Delmar A. 2005). Fig. 1.2 presents the example of the operator interface used in OBDII system.

At present the cars are tested and diagnosed continuously. When something wrong appears then special lamp indicates it to the driver that it is necessary to go as soon as possible to the service station. In the station a computer system is connected to the special standardized socket and it is possible to test practically all elements of the car. Moreover, some of the manufacturers equip the cars with the consumer versions of such systems. The OBDII helps drivers to connect the computer, or especially designed palm-top unit, to the car – even on the road. Probably in the nearest future such systems will be introduced to the typical cars.

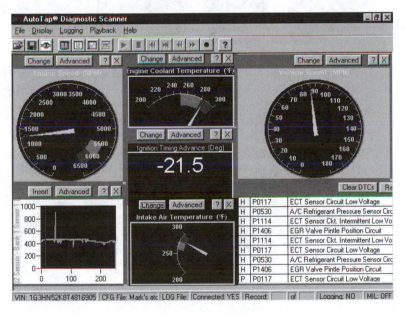

Figure 1.2. The example of *On-board diagnostic* system operator screen (from www.autotap.com) (permission of Autotap)

Recently the measuring techniques changed significantly. Due to the development of informatics, microelectronics and mechatronics we can observe the real revolution in measurements. Generally measuring devices are substituted by more flexible and universal computer measuring systems. The widespread of computer systems stimulated the development of sensor technology, interface systems, signal processing techniques, digital signal processors, measuring software (virtual instruments) and intelligent data analysis methods. Many of measuring devices disappeared from the market. In common applications only several devices remained as "measuring devices", the examples being: digital multimeter, digital oscilloscope and arbitrary wave generator. Using these three devices and computer unit it is possible to design many various measuring systems. But is too simplified thinking that the modern measuring technique means only that the analogue measurements are substituted by the digital ones and the human activity is substituted by the computer. The whole philosophy of measurements has been changed – many traditional methods disappeared and many new methods are being developed.

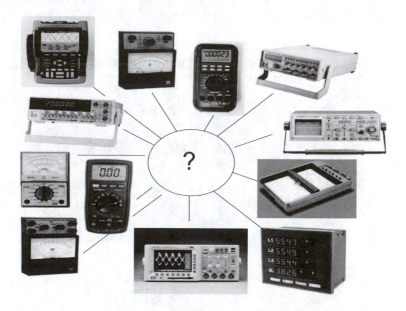

Figure 1.3. The structure of "traditional" measuring system

Figure 1.3 presents the structure of traditional measuring system used some time ago. Properties of the investigated object (for example technological process or physical phenomena) were determined by various measuring devices, sensors, indicating instruments, bridge circuits, etc. placed usually directly near the tested object. Such arrangement of the devices was caused by the fact that most of them did not have output interfaces. There were a lot of such instruments, because each of them fulfilled various functions (ammeter, voltmeter, power meter, etc.) and often each instrument enabled the measurement of different signals (moving-coil device for DC values, moving-iron device for AC measurement, electrodynamic device for power measurement etc.). Thus typical researcher was surrounded by many instruments, like a pilot in the jet cockpit. Usually, the experiment required the activity and presence of a researcher (for example for balancing the bridge circuit or for changing the range of an instrument). Even when digital instruments equipped with output interface appeared on the market, such interfaces was utilized rather infrequently. Only in industrial environment, where the presence of a researcher would be an obstacle the method of transmission of signals was introduced long time ago (sometimes as the non-electric pneumatic signals).

| object | sensors transducers | signal conditioning | data acquisition | data processing | data transmission |

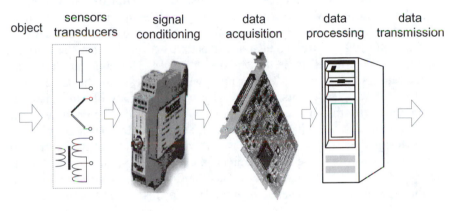

Figure 1.4. The example of the structure of computer measuring system

Figure 1.4 presents an example of the structure of a modern measuring system. Properties of investigated object (electrical and non-electrical ones) are determined by application of various sensors, which convert the measured values into electrical signals (e.g. thermocouples for temperature measurements, Hall sensors for magnetic field and current measurements, strain gauge sensors for stress measurements). The sensors can be very simple – for example displacement sensor in form of the capacitor with one moving electrode or thermistor changing its resistance with the temperature. But the sensors can also be very sophisticated. Due the progress of the microelectronics they can be integrated with electronics – amplifier, correction circuits, analogue-to-digital converters and even microcontrollers. Recently, quite often the so-called intelligent or smart sensors equipped with the output interface (USB[1], RS232c[2] or Ethernet[3]) are utilized. Last years

---

[1] USB – Universal Serial Bus.
[2] RS232c – Recommended Standard-232.
[3] Ethernet - the most popular communication system for Local Area Networks (LAN).

was developed standard IEEE P1451 introducing "plug and play" technology to sensors and helping in easy transferring the measured data by Internet.

It is important to establish such measuring conditions that the sensor (and generally the measuring device) do not influence or disturb the measured environment. It means for instance that the temperature or magnetic field sensor should be so small that it would not influence the distribution of the temperature or the magnetic field measured. The best situation is when the sensor does not take the energy from the investigated environment. Such case is for example when the voltmeter exhibits very large (the best infinitely large) input resistance.

Because there are a lot of various sensors with a lot of various output signals it is necessary to convert these output values into more standardized signals, which are more convenient for further processing (Pallas-Areny, Webster J.G 1991). Often voltage or current are accepted as the standardized output signals – for example *0 – 5 V* or *0 – 20 mA*. The same output signals of the sensors facilitate their further processing – we can use the same output devices for various sensors. That is why various signals of the sensor are transformed to the standardized form with an aid of so-called *signal conditioning devices* (Fig.1.5). Some of the sensors provide directly output voltage signal depending on the measured value. But most of the sensors are parametric (passive) type – they convert the measured value into the change of impedance, often the resistance. Thus the first step in signal conditioning is the conversion of the change of impedance or resistance to the change in voltage.

Figure 1.5. The example of the signal conditioning units for inductive sensors of MacroSensors (Macrosensors 2005) (permission of MacroSensors)

Analogue signal processing is usually the first step in the signal conditioning circuit (Pallas-Areny, Webster 1999). Often the designers fascinated by the possibilities of digital signal processing and software flexibility underestimate this process. Among various capabilities of analogue techniques mainly the amplification methods should be appreciated. These methods are especially important, when the output signal of a sensor is rather small – typical analogue-to-digital converters require the voltage signal in the range *0 – 5 V*. It will be shown in Chapter 4 that also other features of analogue techniques can be very useful in obtaining a measuring signal of good quality, for example if such signal is disturbed by noises and interferences.

Figure 1.6. An example of a data acquisition board with PCI interface

All parts of the measuring system should be connected to each other. In the connection important role play standardized connection/transmission systems called *interfaces*. They can be typical computer interfaces, as RS232 or USB. Especially important is the parallel GPIB interface (*General Purpose Interface Bus*) designed for measuring purposes. Many measuring instruments utilize the GPIB interface as the standard input/output circuit and method of connection with other instruments or computer.

When we have connected all the parts of the typical measuring system we may have some troubles with the design of the program. Some time ago the software was a knowledge reserved only for specialists. But also in this area a real revolution happened. Several computer companies proposed *"user friendly"* software enabling the design to be made directly by end-users of the measuring instruments and even the whole measuring systems. The ease of use can be so "easy" that even non-experienced in programming user can design fully functional measuring system (of course after short introduction to the subject). Some of the software have simple graphical programming language – for instance the *TestPoint* of *Capital Equipment Corp.* permits the programming only with mouse without using the keyboard at all. The most popular software of such type is *LabVIEW* proposed by *National Instruments* (Chugani 1998, Tlaczala 2005). Using the measuring software it is possible to "construct" multimeters, oscilloscopes, spectrum analyzers or other popular measuring instruments having only the computer with the data acquisition board. Because the measuring device is inside the computer and it is represented by artificial graphical elements: indicators, switches, graphs, etc. such design is often called as a "virtual instrument". An example of a virtual instrument designed for students in Laboratory of Physics of Warsaw University of Technology is presented in Fig. 1.7.

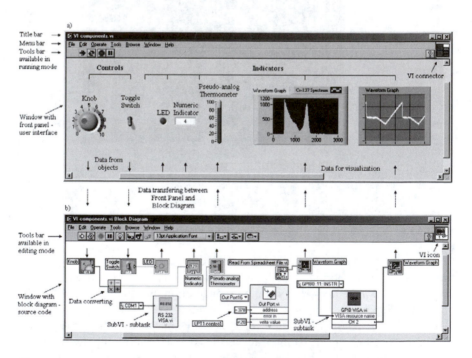

Figure 1.7. The example of virtual measuring device (Tlaczala 2005)

To carry out the measurements today is as easy as never before. The knowledge reserved for specialist is currently available for non-professional users. Many manufacturers offer the measuring equipment resembling popular "auto photo-camera" – it is sufficient just to press a button. For instance, most of modern oscilloscopes are equipped with the button "*Auto Scale*". This simplicity is misleading and even dangerous, because it does not require thinking from lazy researchers. It is very important to perform the measurements consciously, with understanding of the principles of used methods, its limitations and uncertainties. If we assume the incorrect model of investigated object, if we use incorrect methods or if we do not take into account the uncertainty of used method, then we can obtain completely false result and what is even more dangerous – without knowing about it and its implications. A popular joke (a bit cruel though) illustrates such possibility very well. The researcher was investigating an insect. He tore one leg off and said to the insect "*Fly!*" The insect flew. The researcher tore another leg off and repeated the order. The insect flew again. Next, the researcher tore the wing off and repeated the order. This time the insect did not fly. Thus, the researcher noted the results of the investigation: "*Removing one wing impairs the insect's hearing.*"

Contemporary measuring devices offer to the investigator performances much better than formerly. In the past the uncertainty of a measurement of *0.1%* was regarded as excellent. Today cheap and simple digital device provide the uncertainty of measurement of *0.05%*. Such good performances may lead to misunderstandings. The lack of knowledge and experience in measurements is especially apparent, when the uncertainty of a measurement needs to be defined. It happens very often that the measurement is carried out with too accurate device and the result is presented with nonsensical number of digits. And another example – the researcher using the digital instrument of excellent quality may believe that the uncertainty given by manufacturer guarantee the same uncertainty of measurement even if the measured signal is disturbed by noises and interferences. Although the measuring methods and devices are continuously being developed and are getting better and better this should not excuse the researchers from the analysis of the measuring accuracy – this aspect is still crucial for correct measurements.

At the beginning of this chapter we tried to explain and define the term *measurement*. Measurement is also the subject of knowledge, science, engineering and the subject of lectures at the universities as well. What is the area of interest of this subject? In the past this was well defined – specialist on measurements were designing and using the measuring devices and methods: indicating instruments, bridge circuits, potentiometers etc. Today, the range of this field is more "floating". Digital signal processing, microcomputer applications, microelectronics and nanotechnology, signal analysis and transmission are common for many other disciplines, for which other factors are of prime importance. Therefore it is necessary to describe these subjects taking into consideration the "measuring" point of view.

Also, there is other aspect of "globalization" of measurement science and techniques. Today, this is not the knowledge reserved for a narrow group of engineers. Measurements are performed by almost everyone – physicists, doctors of medicine, farmers, even housewives. It is allowed for everyone to measure – with better or worse results. Therefore, the knowledge of the measurement principles is obligatory for all, not only students of electrical engineering departments.

REFERENCES

Bosch 2002 *Automotive Sensors: Bosch Technical Instruction*, Robert Bosch GmbH

Chugani C.H., Samant A.R., Cerna M. 1998 *LabVIEW Signal Processing*, Pearson Education

Cox R. 2005 *Introduction to OBDII*, Thomson Delmar Learning

David P. 2002 *OBDII Diagnostics: Secret Revealed*, Kotzig Publishing

David P. 2004 *OBDII Diagnostics*, Kotzig Publishing

Delmar A. 2005 *Introduction to OBDII*, Thomson Delmar Learning

Jiri M., Iwao Yokomori, Trah H.P. 2003 *Sensors for automotive applications*, Wiley-VCH

Macrosensors 2005 – Macro Sensors Inc., www.macrosensors.com

Pallas-Areny R., Webster J.G. 1999 *Analog Signal Processing*, John Wiley & Sons

Pallas-Areny R., Webster J.G. 1991 *Sensors and Signal Conditioning*, John Wiley & Sons

Sydenham P.H, Hancock N.H, Thorn R.T 1989 *Introduction to Measurement Science and Engineering*, John Wiley & Sons

Tlaczala W. 2005, *Virtual Instrumentation in Physics*, Chapter 106 in Handbook of Measuring System Design Ed. Sydenham P.H. John Wiley & Sons

Travis J. 2001, *LabVIEW for Everyone*, Prentice Hall

2

Fundamentals of Electrical Measurements

2.1. MAIN TERMS AND DEFINITIONS

2.1.1 Basic terms of measurement technique

Let us start again with the definition of electrical measurements. Apart from the term *"measurements"* also other terms are in use, for example *"scientific instrumentation"* (or just instrumentation) and *"metrology"*. The *scientific instrumentation* is often used as a synonym of measurements, and the *metrology* is assumed as a science about measurements. Thus, the measurements will be treated narrower, as a technique or engineering of measurements.

There are various explanations of the term "measurements". We can expand the definition presented in Chapter 1 as: *The measurement is a cognitive process of gathering the information from the physical world. In this process a value of a quantity is determined (in defined time and conditions) by comparison it (with known uncertainty) with the standard reference value.*[1]

In the definition presented above two terms should be discussed in more details. First of all, the measurement is always connected with the term measurement standard. The *standard* is the realisation of a definition of a

---

[1] The International ISO Vocabulary proposes following definitions: *Measurement is a process of experimentally obtaining information about the magnitude of a quantity. Measurement implies a measurement procedure based on a theoretical model. In practice measurement presupposes a calibrated measuring system, possibly subsequently verified. The measurement can change the phenomenon, body or substance under study such that the quantity that is actually measured differs from the value intended to be measured and called the measurand.*

given quantity, with stated value and measurement uncertainty (ISO VIM 2004). The standard value can be represented by the material standard or by the phenomenon – both can be used to the reconstruction of the standard quantity unit with defined uncertainty. Thus, as a standard of the mass can be used the material standard, prepared as the weight made from the platinum-iridium alloy, which has been kept at the Bureau International des Poids et Mesures (BIMP) at Sèvres, France for over 100 years. But as the standard of electric current we can utilize the phenomenon of the force existing between the conductors carrying the current. The unit of electric current, *ampere* is the constant current, which in two straight, parallel conductors of infinite length and negligible circular cross section placed *1 m* apart in vacuum produces between these conductors a force of $2 \cdot 10^{-7}$ *N* per each meter of length.

The second important term to discuss is the *uncertainty of measurement*. After publishing the "*ISO Guide to the expression of uncertainty in measurement*" (ISO Guide 1993) the term uncertainty practically substituted the terms *error and accuracy* [1], which have been used more often in the past. What is the difference between these terms? According to the ISO Vocabulary (ISO VIM 2004) accuracy is the ability of the measuring system to provide a quantity value close to the *true value*, while the uncertainty characterizes the dispersion of the quantity values that is being attributed to the value to be measured (*measurand*).[2]

We always determine the measured value X with an error ΔX. Thus the result of measurement should be always presented as:

$$X = X_T \pm \Delta X \qquad \text{or} \qquad X = X_T \pm \frac{\Delta X}{X} X_R \qquad (2.1)$$

The error of measurement can be expressed as the *absolute error ΔX* or as the *relative error $\Delta X/X$*. The relative error (usually given in %) is an absolute error referred to the reference value X_R. The measured value X may be used

---

[1] In describing of the accuracy of measurement there is sometimes some kind of misunderstanding. In common talking, we can often come across a statement like: "the measurement was performed with the accuracy *0.1%*". It is of course logical mistake, because it means that the measurement was performed with inaccuracy *0.1%* (or accuracy *99.9%*). To avoid such ambiguity it is better to say "the measurement was performed with the uncertainty smaller than *0.1%*".

[2] The official documents of ISO consequently use the term measurand, which means: quantity intended to be measured. It also differentiates this value and the true value (value consistent with the definition of given particular quantity). For the sake of simplicity, and because the word measurand does not exist in Dictionaries of English, further in this book these parameters (measurand or value to be measured) are called "the measured value".

as the reference value – and we call it then as the measurement error, but as the reference value it is better to use the known range of the measuring device X_{max}-X_{min} – in this case it is the error of the measuring method. It is important to know that the true value cannot be determined, because there is always some error of measurement (although this error can be infinitely small, but it always exists[1]). The terms *error* and *uncertainty* will be discussed in details in Section 2.2.1.

The logical sequence of operations used in measurements is called as the *method of measurement*. To perform a measurement we should have established a *measuring procedure* as the detailed description of measurement according to the measuring principles and to given method of measurement. The measuring procedure is used in the definition of a measurement unit, in obtaining the quantity value and measured uncertainty (ISO VIM 2004).

As the result of measurement we obtain the information about the magnitude of the quantity – usually by the assignment it to a numerical value. As the basis of this assignment the *measurement scale* is used– the ordered set of values of quantities of a given kind arranged by the magnitude. The measurement scale is build using the standard values of this quantity with a *unit of measurement* used as the elementary scale interval. Thus the determined value of quantity should be expressed by a number and reference, meaning the unit of measurement. For example because the unit of the current is *ampere* (*A*) we present the result of measurement of the current as *X A*. When we say that the current is *10 A* it means that the measured current is 10 times larger than the unit of electric current equal to *1 A*.

Most of measuring scales are additive, which means that we can add the determined values. For example we can say that the current of *10 A* is two times larger than the current of *5 A*. To build such type of the scale it is necessary to know only the reference unit equal to unit of measurement of this quantity. But also non-additive scales are used. For example the temperature can be measured as the Celsius temperature. The Celsius scale is based on two points: *0 degrees* for the freezing and the *100 degrees* for the boiling temperature of water. However, we cannot say that temperature *40 °C* is two times larger than *20 °C*. Another example of a non-additive scale is *the Mohs scale* used to express the value of hardness. This scale has been built by the assignment of the hardness of various materials to ten numbers (starting from the softest one). For example the hardness of talc is described by the number 1, the hardness of apatite by 5 and the hardness of diamond by 10 (the hardness of fingernail in this scale is described as about

---

[1] If the error is really infinitively small then we reach a barrier, as specified in the Heisenberg theory of uncertainty.

2.5, while the hardness of knife blade as about 5.5). Also in this case we cannot say that the diamond is two times harder than the apatite.

When we say that the measurement requires a comparison to the standard value of measured quantity we do not need to apply the standard of this quantity. It would be impossible and impractical taking into account the great number of various quantities. Therefore it is sufficient to define and reproduce the standard of certain number of quantities (called *base quantities*) and next to derive other quantities as the *derivative quantities*. The derivative quantities can be determined from the mathematical dependencies deduced according to the physical laws and rules (for example to know the resistance of 1Ω it is only necessary to know the voltage *1 V* and current *1 A* according to the Ohm's law $1\Omega = 1V/1A$ – although just in this case we have well defined standards of all quantities: voltage, resistance and current). Basing on this concept it was proposed to select seven base units in form of the *International System of Units - SI* (Table 2.1)[1]. From these seven basic units of SI we can derive other units – for example the discussed above unit of resistance *ohm*, Ω, can be expressed as: $m^2 \cdot kg \cdot s^{-3} \cdot A^{-2}$.

The International System of Units was adopted by the General Conference on Weights and Measures CGPM and was described in ISO Standards: ISO 1000 – *SI units and recommendations for the use of their multiple as of certain other units* and ISO 31 – *Quantities and units*[2].

---

[1] The definitions of basic units are as follows:

One *meter* is equal to the length of the path travelled by light in vacuum during a time interval of 1/299 792 458 of a second;

One *kilogram* is equal to the mass of the international prototype of kilogram;

One *second* is a time interval equal in duration to 9 192 631 770 periods of the radiation corresponding to the transition between the two hyperfine levels of the ground state of the cesium-133 atom;

One *ampere* is equal to constant current which, if maintained in straight parallel conductors of infinite length and of negligible circular cross-section, and placed 1 m apart in vacuum, would produce between these conductors a force equal to $2 \cdot 10^{-7}$ N per each meter of length;

One *kelvin* is temperature equal to a fraction 1/273.16 of the thermodynamic temperature of the triple point of water;

One *mole* is the amount of the substance in a system, which contains as many elementary entities as there are atoms in 0.012 kg of carbon 12;

One *candela* is the luminous intensity, in a given direction, of a source that emits monochromatic radiation of frequency $540 \cdot 10^{12}$ Hz and that has a radiant intensity in that direction of 1/683 watt per steradian.

[2] The problems of electrical standards are the subject of interest of following institutions: BIMP (*International Bureau of Weights and Measures*), ISO (*International Organisation of Standardization*) and IEC (*International Electrotechnical Commission*).

Table 2.1. Base quantities and base units of SI system

base quantity	base unit	
name	name	symbol
length	meter	m
mass	kilogram	kg
time	second	s
electric current	ampere	A
thermodynamic temperature	kelvin	K
amount of substance	mole	mol
luminous intensity	candela	cd

As the result of the measurement we obtain a number – thus we express the physical value by the abstract value. In this way we describe the physical word by its mathematical model. Correct construction of the **mathematical model** of analyzed object of phenomenon is a crucial, difficult and probably most important part of the measuring procedure. This task is so difficult, because similarly as we obtain the measured value with some uncertainty also our mathematical model is always only the approximation of reality. The problem of correct design of the model is well illustrated in previous chapter in the joke about insect. We can use excellent and expensive measuring devices, but if we do not include into the model the most important factors of the investigated object, we could obtain worthless and even false results.

Consider the situation when we measure the magnetic field strength using the coil sensor. In this measuring method we apply the Faraday's law stating that the voltage induced in such coil depends on the magnitude of magnetic field (exactly magnitude of the flux density B), area of the coil, frequency of the measured field and number of turns of the coil. Moreover, we should take into account that this voltage does not depend directly on the magnetic field but it depends on the derivative of flux density with respect to time, dB/dt. Thus, we know that we should integrate the output signal of the coil. Suppose that we take into consideration all these factors but we do not include in our model the fact, that this magnetic field is non-uniform. In certain circumstances the negligence of the non-uniformity of magnetic field results in the false result of investigation, although we determined other factors (induced voltage, coil area, frequency, etc.) very accurately. This example illustrates that knowledge about the investigated object and comprehensive analysis of all circumstances decides about the quality of the measuring process.

Moreover, in construction of the model of the investigated object we should take into account that most physical phenomena are varying in time. That is why in presented earlier definition of measurement the remark "*in defined time and conditions*" was included. Correct consideration of the dynamics of the investigated phenomena and dynamic properties of the measuring devices is one of the most difficult problems of metrology.

The information about measured value is often transmitted by the *measuring electric signal*. Various parameters of such signal (magnitude, frequency, phase, etc.) can be used as the measure of the investigated quantity. The use of the electric signals for carrying of the information is very convenient because the knowledge about the *processing of electrical signals* is very well developed. We can divide the electrical signals into analogue and digital ones. The *analogue electric signals* consist of the infinite sequence of values varying in the time, while the *digital electric signals* consist of finite sequence of numbers with the interval equal to one quant (usually representing one bit[1] of information).

Because analogue and digital signals are processed using various tools and methods therefore usually *digital signal processing* and *analogue signal processing* are considered separately. Similarly often the *analogue measuring technique* and *digital measuring technique* are separately analyzed. Such division of the measuring techniques seems to be slightly outdated, because today it is rather difficult to find measuring devices, which are not taking the advantage of the digital technology.

2.1.2. The main methods of measurements

One of the simplest measurement methods is the direct comparison of measured value with the standard one. An example of the *direct measurement method* is presented in Fig. 2.1. In this method the idea of weighting it is applied (this method is sometimes called the *current weight*). One coil of the electromagnet is supplied by the measured current I_x , which causes that the ferromagnetic element on one arm of the balance is attracted. (It is possible to use also other electric mechanism, for example attracting of the magnet or attracting other coil – this last mechanism is very close to the definition of the ampere.)

On the second arm of the balance similar mechanism is placed – this time the coil is supplied by the standard reference current I_s. Changing the value of the standard current we can balance the weight – the equilibrium state is when the pointer is at the zero position. We can also determine the state of equilibrium using the electrical method – for example by the measurement of the resistance R_x of the resistor with moving slider causing the change of the resistance (the potentiometer). This idea is presented in Fig. 2.1b.

Figure 2.1b presents the *indirect method of measurement* of the value of electric current. This time, the measured current causes the change of the resistance R_x. In this circuit there is a lack of the standard of current, but this

---

[1] The word *bit* was introduced by C.E. Shannon as an abbreviation of the "**b**inary dig**it**". A single bit (abbreviated *b*) can take values zero or one, true or false, and generally two mutually exclusive states.

does not mean that this standard does not exist. It exists as the scale of resistance and it could be introduced to this method by the earlier supplying the coil by the reference standard current (in such way we introduced the dependence $R_x=f(I)$ to this method).

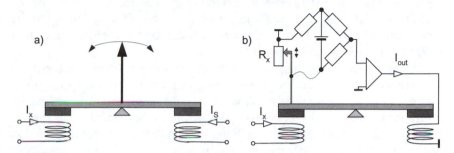

Figure 2.1. Measurement of the current I_x by the direct comparison it with the standard value I_S (a) and by the indirect comparison with the resistance value R_x (b)

Figure 2.1b presents the weight that is weighted automatically. It is because in this method we applied the idea of *feedback*. The resistor R_x (sensor of the position or displacement) is connected into the bridge circuit consisting of four resistors and supply voltage. When all resistors are the same then the output signal (connected to the amplifier) is equal to zero. When one of the resistances is changed then the equilibrium is disrupted and the signal of unbalance appears at the output of the amplifier (see Fig. 2.6).

The process of current measurement is as follows. If current $I_x = 0$ then the bridge circuit is in equilibrium state and the weight is balanced. If the current I_x changes, then the electromagnet (the coil) attracts the ferromagnetic element on the arm and the resistance R_x changes. This causes signal voltage to appear at the output of the bridge circuit. This signal after amplification supplies the second coil as the current I_{out}, which causes the movement of the second arm and balancing of the weight (similarly as it was performed manually in the example presented in Fig. 2.1a). After short period of time (the transient state) the balance returns to the equilibrium state, which is detected as zero voltage on the output of the bridge circuit. Thus by means of feedback we realize the automatic balancing of the weight and the output current I_{out} can be the measure of the tested current I_x.

The measurement of the current can be performed by the *null measurement method* – we balance the circuit and the balance state is indicated by the pointer or electrically by the zero output voltage of the bridge circuit. The same measurement can be performed by *differential measurement method*. In this case it is not necessary to balance the circuit – the deflection of the pointer or the output voltage of the bridge circuit can be used as the measure of the current value. It is also possible to use the *null-differential measurement method*. In this method we roughly balance the weight by the current I_0 and the deflection of the pointer (or change of the output voltage of the bridge circuit) is caused by the difference between equilibrium state and state after the change of the current $\Delta I = I_x - I_0$. Using the null-differential method we can obtain improvement of the sensitivity of the measurement – the movement of the pointer can be realized by the smaller current ΔI instead of current equal to I_x.

The measurement method is characterized by the *sensitivity* of the method (and related to this parameter the *resolution* of the method) and by the *range* of the measuring device. The *sensitivity* of the method informs us what value of the measured quantity is necessary to obtain output signal – the smaller this value, the larger the sensitivity. Suppose that in the example presented in Fig. 2.1 both coils are the same. Then to balance the weight it is necessary to use the same current I_s as the measured current I_x and the sensitivity $S = I_x / I_S$ is equal to *1*. We can improve the sensitivity by increasing the number of turns of the second coil. For example, if the number of turns is $n_2 = 2n_1$ then to obtain the equilibrium the reference current I_s can be 50% of the measured current I_x. Thus the sensitivity $S = I_x / 0.5\,I_s$ is two times larger.

The *resolution* informs us about the smallest value of the measured quantity, which could be possibly detected. For instance, if we can detect the deflection of the pointer equal to *5°* (represented for example by one graduation on the scale), then after improving the sensitivity by factor of two we can detect two times smaller change in the measured current. The *measurement range* of the measuring device is the maximal value of the quantity, which we can measure. In the null method described above the range is equal to I_{xmax}, while for the null-differential method it is $I_{xmax} - I_0$ (and generally $I_{xmax} - I_{xmin}$ when the beginning of the scale is not equal to zero).

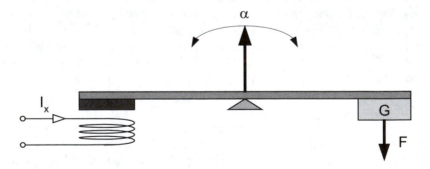

Figure 2.2. Measurement of the current I_x by indirect comparison with the gravity force F (scaled initially with the standard value of the current)

Figure 2.2 presents another idea of the indirect measurement of the current by applying the weight. In this case the movement α of the pointer (caused by attracting of the arm by the force depending on the measured current I_x) is balanced by the force of gravity F of the weight G. The standard of the current is not presented in the Fig. 2.2, but it does not mean that it does not exist. It could be introduced to this method by the earlier scaling of the measuring device. The *scaling process* could be realized by supplying the coil with the standard values of the current and determination of the pointer deflection. In this way we can determine the dependence $\alpha = f(I_x)$. For scaling of the device we can used the source of standard values called the *calibrator*.

In the example presented in Fig. 2.2 for scaling purposes standard of the weight G could be used instead of the current source. Before we perform the measurement, we could determine the dependence between the current I_x and the weight G as $G = K I_x$. This time we *de facto* measure the force (or rather the mass G) and we determine the investigated current knowing the constant K.

The indirect measurement methods are employed in almost all *indicating instruments*. In these instruments the measured value is expressed by the deflection of the pointer (in analogue instruments). This pointer indicates the measured value at the point on the scale – thus we compare the measured value and the deflection of the pointer. But also in such instrument the standard of the measured quantity exists in the background, because the points on the scale were marked by the scaling process earlier.

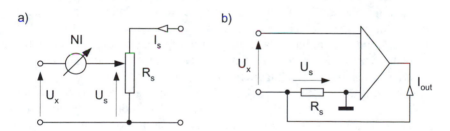

Figure 2.3. The compensation method for measurement of the voltage U_x – manual balancing of the circuit (a) and the automatic method with electronic feedback (b)

The important method of comparison of the measured value and the standard one is the *compensation measurement method* illustrated in Fig. 2.3. In the compensation method we determine the difference between two values (measured and standard) and we can precisely determine the state when this difference is equal to zero. Thus the compensation state (the equilibrium state) means that both values are the same and cancel each other. As the results, in the state of equilibrium the signal at the output of the circuit is zero.

In the example presented in Fig. 2.3 the measured voltage U_x is compensated by the voltage drop U_s on the resistor R_s. The state of equilibrium is indicated as the zero of the output signal by the *null indicator* *NI* (for this purpose a very sensitive voltmeter called *galvanometer* can be used).

The measuring procedure consists of two steps. In the first step, the standard value of the current I_s is established (for example by comparing it with the standard value). In the second step, the resistance R_s is being changed until the null indicator *NI* indicates the state of equilibrium. For known value of the current I_s the resistor R_s can be scaled directly in the voltage units.

The compensation methods enable the measurements of voltage with excellent accuracy, because we are capable of manufacturing the resistors very precisely (with very small uncertainty). Therefore the devices called *potentiometers*[1] were one of the most accurate measuring instruments – nowadays the potentiometers are substituted by accurate digital voltmeters often using the idea of compensation.

The compensation method exhibits very important advantage. In the state of equilibrium the measuring device does not take the energy from the tested source – the measurement is performed in truly non-invasive way with the infinite input resistance of the measuring instrument.

Figure 2.3b presents the realization of the compensation idea performed automatically by applying the feedback technique. The amplifier works as the null indicator amplifying the difference between the measured voltage U_x and the drop voltage U_s. This difference causes the output current I_{out}, which is increased until the input voltage of the amplifier again returns to the zero.

Apart from the compensation method there is also used the *comparative measurement method*. The magnetic *direct current comparator* (DCC) is currently used for the most precise reconstruction of the resistance standard (NIST 1458 – 2003). The term *comparator* is not always correctly interpreted, because comparison is in the definition of the measurements and practically all measuring instruments compare the measured value with the standard one. The comparative method is defined in narrower sense – as the compensation method uses the difference between two values the comparative method uses the ratio of two values[2]. An example of the comparator is presented in Fig. 2.4.

---

[1] Potentiometer is the devices that realize the compensation of two voltages.

[2] Commonly in electronics and informatics an electronic integrated circuit called comparator is used. This device amplifies the difference between two input signals – thus it uses compensation technique rather than the comparative method.

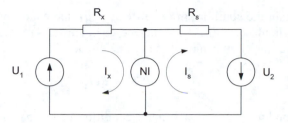

Figure 2.4. The comparator of two resistances

In the circuit presented in Fig. 2.4 we can obtain the equilibrium by the compensation of the currents I_x and I_s

$$I_x - I_s = 0 \qquad (2.2)$$

This state of equilibrium can be realized by the change of the voltage U_1 or U_2. The condition of the equilibrium is

$$\frac{R_x}{R_s} = \frac{U_1}{U_2} \qquad (2.3)$$

Thus we can use this circuit for resistance measurement. The current compensation circuit was used as the resistance comparator.

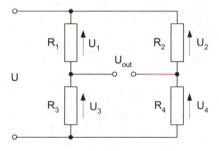

Figure 2.5. The bridge circuit

Figure 2.5 presents the *bridge circuit*, which is very often used for the measurements of the resistance (or impedance). For the circuit presented in Fig. 2.5 the voltage drops across the resistors R_1, R_2, R_3 and R_4 are respectively

$$U_1 = U\frac{R_1}{R_1 + R_3} ; \ U_2 = U\frac{R_2}{R_2 + R_4} ; \ U_3 = U\frac{R_3}{R_1 + R_3} ; \ U_4 = U\frac{R_4}{R_2 + R_4} \qquad (2.4)$$

To obtain the equilibrium (when $U_{out} = 0$) the conditions: $U_1 = U_2$ and $U_3 = U_4$ should be fulfilled. From the equations (2.4) we obtain following condition of the equilibrium of the bridge circuit

$$R_1 R_4 = R_2 R_3 \qquad (2.5)$$

The condition (2.5) is a general condition of the balance of the bridge circuit: *the products of the resistances of opposite arms of the bridge circuit should be equal.*

By applying the condition (2.5) we can determine the measured resistance $R_x = R_1$ from the following equation:

$$R_x = R_2 \frac{R_3}{R_4} \qquad (2.6)$$

In practice there are two kinds of bridge circuits: *null-type bridge circuit* (or *balanced bridge circuit*) and *deflection-type bridge circuit* (or *unbalanced bridge circuit*). In the null-type bridge a null indicator is connected to the output and the bridge is balanced – for example by the change of one of the resistors. Most often the resistor R_2 is used for balance, while change of the ratio R_3/R_4 is used for the range selection.

Currently, the null-type bridge is rather not often used as a measuring device, while the deflection-type bridge circuit is commonly used as the conditioning circuit enabling the conversion of the change of the resistance (or generally impedance) of sensor into the voltage signal. This type of the bridge circuit is first balanced and next the output voltage (voltage of unbalance) is used as the output signal of the $U = f(R)$ *(resistance – voltage transducer)*.

Figure 2.6b presents the *transfer characteristic* $U_{out} = f\ (\Delta R_x/R_x)$ of such transducer. We can see that this characteristic is *nonlinear*. There are various methods of the linearization of the conversion – they are described in Chapter 3. One of the methods of linearization is presented in Fig. 2.6c.

Figure 2.6c presents the bridge circuit with automatic balancing realized by means of feedback circuit. We can see from Fig. 2.5 that the bridge circuit is balanced by changing one of the resistances and thus changing the voltage drop across it. Therefore, the change of resistance causes the same effect as change of the voltage on one of arms. In the circuit presented in Fig. 2.6c the output signal of the bridge circuit after amplification causes that the output current I_{out} creates additional voltage drop U_w across the resistance R_w. This additional voltage can drive balancing of the bridge circuit until it returns again to the balanced state.

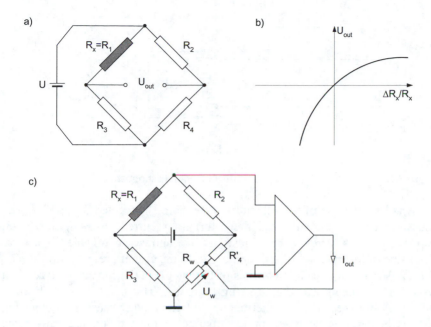

Figure 2.6. The bridge circuit as the resistance converter: a) the bridge circuit of the deflection type bridge circuit, b) its transfer characteristics, c) the bridge circuit with electronic feedback

The output current is the measure of the investigated resistance R_x. Because the bridge circuit is automatically balanced its output signal is very small – if the amplification is very large only very small signal (in range of μV) is required to generate the output current I_{out}. Therefore only small, linear part of the whole nonlinear transfer characteristic is used – thus the whole characteristic of the transducer $I_{out} = f(\Delta R_x / R_x)$ is linear.

Analyzing the equation (2.6) we see that the accuracy of the measurement of value R_x depends on the accuracy of all three other resistors. We can improve the accuracy of measurement applying the *substitution measurement method*. In that method the measuring procedure consist of two steps. In the first step the bridge circuit is balanced. Next, we substitute the measured resistor R_x by the standard resistor R_s. This time we do not balance the bridge changing the resistances of the bridge circuit but we balance it by changing the standard resistance. If the time period between these two operations is not long (to avoid potential influence of the change of the temperature or other factors) the accuracy of measurement does not depend on the accuracy of the resistors in the bridge (it depends only on the accuracy of the standard resistor). This way, the bridge circuit was used only as the device testing that after substitution of the resistors nothing changed.

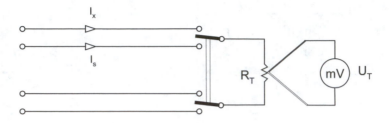

Figure 2.7. The substitution method of the current measurement

Another example of the substitution method is presented in Fig. 2.7. The measurement of the value of the alternating current is rather difficult, especially if this current is not sinusoidal. In contrast we are able to measure the value of direct current very accurately. In the circuit presented in Fig. 2.7 the measurement procedure consists of two steps. In the first step we connect to the circuit the measured alternating current I_x. This current causes heating of the resistor (heater) R_T. A thermocouple (temperature sensor) is connected to this heater – in such sensor the change of the temperature causes the change of the output voltage U_T. In the second step, we connect the standard direct current I_s to the heater. We change this current until the temperature of the heater is the same as in the first step. Because the effect of heating was the same in both cases the values of both currents are the same. Thus, we substituted the measurement of the alternating current by the measurement of the direct current.

2.2. UNCERTAINTY OF MEASUREMENTS

2.2.1 Errors, uncertainty, and reliability of signal processing

As it was discussed in Chapter 1 we are unable to determine the true value of the measured quantity, because the measurement is always performed with some uncertainty. Therefore, we can state that the measurement without the estimation of this uncertainty is worthless. For example, if we say that we determined the value of measured current as *1A* and we do not supplement this information with the estimation of the uncertainty, then it means more or less that this current could take any value (so it is not determined). Thus, the analysis of uncertainty is always accompanying the measurement and it is crucially important.

Unfortunately, the prevailing opinion is that the analysis of uncertainty is rather difficult and somewhat dull. Sometimes, people even say that the measurements would be interesting if not the theory of errors. On the other hand, if it is indispensable to use this theory better it is to grow fond of it. Moreover, in many cases the analysis of accuracy of measurement can be intellectually challenging and even can be more important and interesting than routine measurement procedure.

The International Organization of Standardization (ISO) with collaboration of many other prestigious organizations edited in 1993 a *"Guide to the expression of uncertainty in measurement"*. This document was a result of thousands discussions in metrological milieu and many years of preparation. Today, we can say that before the *Guide* there was the theory of errors and after the *Guide* there is the theory of uncertainty in measurements. Unfortunately the *Guide* did not solve the problem of understanding of measurement accuracy, because it is written with very difficult style and it is clear only for very narrow circle of specialists. For example an explanation: *"Estimate – the value of an estimator obtained as a result of an estimation"* – § C.2.26. No wonder that after the *Guide* the frustration of people active in measurements deepened and the milieu divided to the initiated peoples, who understand the *Guide,* and the rest, who don't. A lot of publications explaining the terms from the *Guide* have been published (Dieck 2002, Lira 2002, Rabinowich 1999, Taylor 1996). The *Guide* is an official document, as well as standard and law, therefore everyone is obliged to try understand it and to comply with it.

Before the *Guide* the theory of errors was divided into two parts: theory of systematic errors caused by the limited accuracy of the measuring devices and imperfection of mathematical models, and the theory of random errors utilizing the theory of probability. In *Guide* it has been assumed that such division is not justified; therefore, other coherent theory of uncertainty in measurement including these both cases of measurements has been proposed. Such idea has been accepted with satisfaction, because it organizes the theory of errors in one system. Indeed, measurements are practically always suffering from the random errors, for example caused by the variation of the measuring condition or external interferences. Thus we can only estimate the measured value with possible to determine uncertainty. Even measurements assumed as very precise are limited in accuracy by random errors. For example, when we measure voltage using digital voltmeter of excellent accuracy we never know the value between two least significant digits (if four-digit voltmeter indicates value *5005*, then all values between *5004.5* and *5005.5* are probable in the same way).

We rewrite the equation (2.1) $X = X_T \pm \Delta X$ in the form

$$X_T - \Delta X \leq X \leq X_T + \Delta X \qquad (2.7)$$

which can be read as follows: the result of measurement X is determined with the dispersion $\pm\Delta X$ around the true value X_T (bearing in mind that $\pm\Delta X$ is an absolute error of measurement).

According to the concept presented in the *Guide* the dependence (2.7) should be substituted by the dependence

$$\Pr\left(X_0 - u \leq X \leq X_0 + u\right) = 1 - \alpha \qquad (2.8)$$

which should be interpreted as follows: *the result of measurement X is determined with the uncertainty ± u around the estimated value X_0 with the level of confidence (1-α)*. Symbol *Pr* in the equation (2.8) denotes the probability.

We can see that the true value (which we never know) is now substituted by *the estimated value*. Similarly, the error is now substituted by the *uncertainty*, because we also do not know the value of that error (because we do not know the true value we can not determine the error using the equation 2.7).

A question appears: what about the errors? Indeed, the term "error" is used in everyday speech so universally that any attempt to eliminate it seems an absurd. That is why also in this book for sake of simplicity the term error will be sometimes used, but only in popularly understood cases (as the difference between the measured value and the value assumed to be true). The result of measurements should be always presented according to the definition (2.8) and the term error cannot be used as a synonym of the uncertainty.

In certain circumstances we determine the error relative to the known value. For example, we determine the *error of linearity* as the difference between the result obtained from the sensor of nonlinear transfer characteristic and the result obtained under assumption that the characteristic is linear. It would be somewhat strange to call this error as uncertainty of linearity.

The resultant uncertainty of measurement can comprise several components: *corrections* (a), *random uncertainty* (b), *uncertainty related to the imperfect accuracy of measuring devices and methods* (c), *uncertainty related to non perfect model of investigated phenomenon* (d) and *mistakes* (e).

a) The *correction* is the uncertainty ΔX_0 which we are able to determine and remove. For example the transfer characteristic of the thermoresistive sensor is often described by the dependence $R_T = R_0 \ (1+\alpha T)$ – where R_0 is the resistance in temperature $0\,°C$, R_T is the resistance in temperature T, α is the temperature coefficient. But more detailed analysis of the transfer characteristic leads to a conclusion that the thermoresitor is better described by the dependence: $R'_T = R_0 (1+\alpha T+\beta T^2)$. Thus, if we do not take into account the nonlinearity of the sensor we make error of linearity:

$$\Delta R_{T0} = R'_T - R_T = R_0 \beta T^2 \qquad (2.9)$$

Because we know the value of this error of linearity we can remove it – for example by setting it into computer memory and subtracting it every time during measurement.

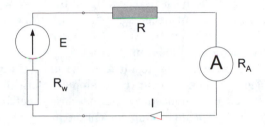

Figure 2.8. The example of the circuit used to the measurement of the current

We can also take into account the correction when our measuring method exhibits error but we are able to determine it. For example, Fig. 2.8 presents a circuit with a resistor R supplied by a source E with the internal resistance R_w. The ammeter is used to determine the current I. This current should be equal to

$$I = \frac{E}{R + R_w} \qquad (2.10)$$

But the ammeter exhibits internal resistance R_A. Therefore the current is

$$I' = \frac{E}{R + R_w + R_A} \qquad (2.11)$$

Thus imperfect ammeter introduces error $\delta I = (I'-I)/I$

$$\delta I = \frac{R_A}{R + R_w + R_A} \qquad (2.12)$$

We can remove this error and correct the result of measurement but only if we know all the resistances.

b) The *random uncertainty* exists when we can diminish it by increasing the number of measurements. If we perform several measurements and every time we obtain a slightly different result (with dispersion exceeding assumed value) we can conclude that the uncertainty is random. The uncertainty of measurement caused by the random character depends inversely on the number of measurements.

c) There are uncertainties, which we are able to estimate but we cannot remove. For example the ammeter used in current measurement exhibits limited uncertainty described usually by the manufacturer (as the accuracy of scaling). The repeating of the measurements many times would not change this error. Generally this kind of uncertainty depends on the *uncertainty of used measuring devices* and can be estimated based on the information enclosed by the manufacturers of these instruments.

d) Another kind of uncertainties can result from the imperfect model of the investigated object or phenomenon. In the example described above (Fig. 2.8) we removed the error introduced by the measuring method. But if the same circuit is supplied by the alternating current the model of the measuring circuit is much more complicated than that presented by equation (2.12). In such case we should take into account the parasitic capacitances with respect to the ground, the capacitances and inductances of resistors, influence of frequency, etc. The dependence (2.12) should be appropriately extended to include all these factors to the model.

Sometimes the model of phenomenon can be so complex that its application could be difficult, especially in industrial environment. In such case we can construct *artificial model* of the physical phenomenon as the result of the group agreement. For example, the magnetic parameter: "specific power losses" depends on a great number of factors – conditions of magnetization. It would not be reasonable to include all of them into the model. Therefore, the model of the losses has been limited to the precisely described one case – as this case the testing apparatus called the Epstein frame has been chosen. The Epstein frame has been very precise described in the international standard (EN 60404 1998) – the method of preparation of the sample, the design of the apparatus, the measuring conditions, etc. have been established in details. The standardization of the measuring procedure guarantees that all investigators perform the same measurements and obtain comparable results in every laboratory. The results of these measurements are called "the Epstein losses"; in many cases being far from the real losses in the physical sense.

e) And last but not least, the worst kind of uncertainty – the unrecognized errors. These errors are sometimes called the *mistakes*. For example, we measure the current using damaged ammeter. Or we use this ammeter in the presence of magnetic field and we do not know that this field exists and influences the measurement. We have no reason to question the manufacturer declaration of accuracy and greater number of the measurement will not help. In such case only the *validation* of the measuring procedure (by means of the standard device called calibrator) could be effective but it is not always possible.

Thus the procedure of measurement should comprise the following steps:
- analysis and determination of the mathematical model of investigated object;
- analysis and determination of the mathematical model of the measuring system;
- analysis of source of errors and determination of the resultant uncertainty of measurement.

Before the analysis of uncertainty it is reasonable to execute the cycle of the measurements. If the dispersion of the results exceeds assumed value it means that we should perform statistical analysis of these results. In such case we use the procedure called by the *Guide - type A evaluation of uncertainty.* This kind of evaluation requires certain number of measurements – this number depends on the value of dispersion and the level of confidence *1-α*.

But in certain circumstances it is not reasonable to repeat the same measurements many times. For example, we have stable supply sources, the conditions of environment are also stable (due to temperature conditioners, electromagnetic shield and grounding preventing the harmful interferences), we have very precise measuring devices. It would be just waste of time and money to repeat the measurements, especially in industrial environment. In such case we use the procedure described in the *Guide - type B evaluation of uncertainty.* In this type of evaluation of uncertainty it is not necessary to perform the statistical analysis.

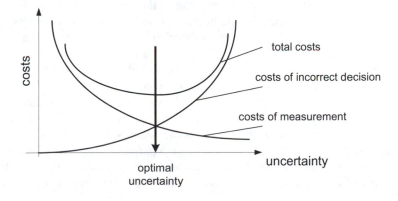

Figure 2.9. The dependence between the uncertainty and costs of measurements

It is reasonable to limit the assumed uncertainty of measurement to the certain useful level. The increase of the accuracy means higher costs – we should use more expensive measuring devices, the time necessary for measurements is longer, the qualifications of the investigators should be higher. The application of inappropriate, too precise instruments can be the uncomfortable – for example when we use five-digit instrument for the measurement of non-stable source, the last digits keep blinking and are useless.

Figure 2.9 presents the graph illustrating the relation between the costs and uncertainty. We can reduce costs of measuring procedure by decreasing the uncertainty but it is the risk that the costs of incorrect decisions can be larger, it can even cause dangerous situations. Taking this into consideration we can establish the optimal value of uncertainty.

In some cases (health service, military industry, etc.) there is a need for increased accuracy. In this case the *Guide* proposes to substitute the *standard uncertainty u* by the *expanded uncertainty ku*. The *coverage factor k* is related to the level of confidence and typically is in the range *2 – 3*.

All elements are important – construction of the mathematical model of the measuring system, analysis of the factors influencing the measure, analysis of the possible uncertainties. The authors of the *Guide* recommend: "*Although this Guide provides a framework for assessing uncertainty, it cannot substitute for critical thinking, intellectual honesty and professional skill. The evaluation of uncertainty is neither a routine task nor a purely mathematical one; it depends on detailed knowledge of the nature of the measurand and of the measurement....*" (Guide ISO – 1993).

Although uncertainty of measurements is one of the most important performances of the measuring devices we should also consider other factors influencing the *reliability of measurements*. If we use the measuring device in laboratory for exact determination of the value of quantity the reproduction of standard is most important. But if we consider the measurement wider – as the gathering of information in everyday life, industry and science other performances of the measuring equipment can be also significant.

Consider the case when we analyze the sound. Usually as the sensor of sound is used a microphone transducer processing the sound into the electrical signal. The sound (pressure of air waves) can be represented by the frequency and magnitude of alternating voltage. From Fig. 2.10 we can see that for correct processing of the sound the sensors (and whole equipment) should determine the pressure in *frequency bandwidth* about 20 - 20 000 Hz. Moreover it is required that the *magnitude dynamics* of the signal should be of even 120 dB. Indeed the dynamics of the symphonic orchestra is of about 110 dB.

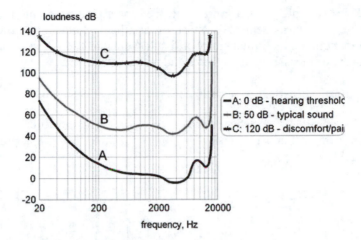

Figure 2.10. The physiological model of human hearing – the level of loudness as felt by the human ear

For correct processing of the sound these both factors: dynamics and frequency bandwidth could be more important than the accuracy of frequency determination. Thus, the uncertainty of signal processing we can widely interpret as the reliability of information and quality of reproduction of the information.

In the vinyl long-play era the sound was recorded mechanically and reproduced using electromechanical device. This kind of signal processing exhibited many drawbacks – the sound was noisy with crackles, therefore the dynamics was poor, not better than *65 dB*. Also the bandwidth was limited and the signal was distorted. No wonder that all welcomed with enthusiasm the new era of CD. This technique enabled the sound processing with excellent quality – the dynamics was even *100 dB*, and the frequency bandwidth was *5 –20 000 Hz*. To tell the truth, these parameters were much better than would be appreciated by normal human ear.

But after several years the vinyl technique fans expressed doubt if the CD technique is better – they maintained that the CD sound was reproduced less natural than the vinyl one. It was slightly irrational against all technical tests. But it is also probable that the model of human hearing represented only by dynamics and frequency(as it is presented in Fig. 2.10) is not exact. The technique of signal processing is today so effective that on the market SACD and ADVD systems with *120 dB* dynamics and up to *40 kHz* bandwidth appeared. They are almost two times better than the assumed human hearing is able to appreciate. Even such excellent performances did not stop the discussion between the admirers of various sound processors what confirm how important is model of the investigated phenomenon.

This example demonstrates that the uncertainty we should interpret widely. The measurement is most of all the technique bringing us the information about surrounding world and we cannot see it only through the accuracy of measuring instruments and methods.

2.2.2 Basic statistical terms and concepts

Consider the case when we perform a series of measurements and we obtain certain number of results in form of a table or a graph presented in Fig. 2.11.

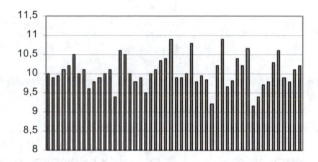

Figure 2.11. An example of the presentation of the measuring results in form a graph

We can easily analyze such set of results constructing *histogram*. The histogram can be calculated for instance with using simple *Analysis Tool* existing in popular MS Excel program. Fig. 2.12 presents the histogram of the data set presented in Fig.2.11.

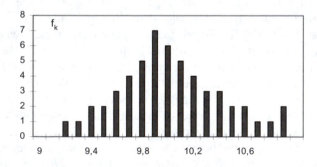

Figure 2.12. The histogram of the results of measurements presented in Fig. 2.11

On the graph of histogram, the axis x describes the value of obtained result while the y axis presents value f_k describing how often such result happened. Analyzing the histogram we can obtain the information which value revealed most often – this value is probably the closest to the true value.

Fig. 2.13 presents examples of two histograms. They represent magnetic homogeneity of the material determined by scanning the magnetic field distribution of the selected area on investigated magnetic material (Tumanski 1998). We see that one of them (Fig. 2.13a) is more uniform than the other one (Fig. 2.13b), which is expressed by the slenderness of the histogram shape.

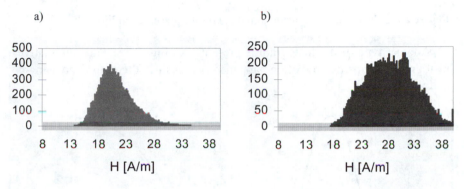

Figure 2.13. Two examples of the histograms of the test results of magnetic material homogeneity (Tumanski 1998)

It is also possible to calculate the histogram, in which the level of the bars is equal to the F_k value representing the area $f_k \Delta x$ (Fig. 2.14). On the basis of such histogram we can evaluate in which Δx range the result of measurement happened most often. Analyzing the shape of the histogram (for example its width or slenderness) we can roughly estimate the uncertainty of measurements.

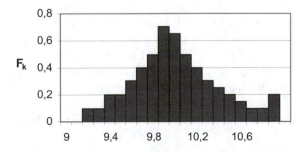

Figure 2.14. The histogram $f_k \Delta x$ of the results presented in Fig. 2.11

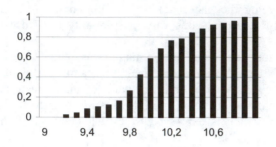

Figure 2.15. The cumulative histogram of the results presented in Fig. 2.11

Another type of the histogram is presented in Fig. 2.15. This cumulative histogram informs us how often happened the result in the range between -∞ and the *x* value.

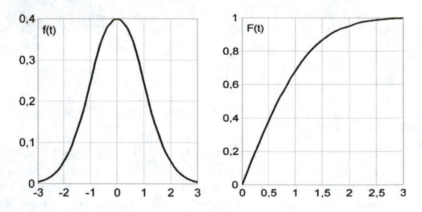

Figure 2.16. The graph of the density of probability *f(t)* and cumulative distribution for the normal distribution *F(t)* [1]

We can present the results of series of measurements as the probability distribution of the result instead of histogram. On the *y* axis we describe the probability density *f(x)* of the result of measurement (Fig. 2.16a). We can also calculate the cumulative distribution function *F(x)* as the area under the probability density function (Fig. 2.16b). The comparison of figures 2.14, 2.15 and 2.16 indicates that the probability density function is related to the histogram for infinite number of measurements (continuous function) while the cumulative distribution function is related to cumulative histogram. Very

---

[1] Figure 2.26 presents so called normalized functions of probability where $t = (x-\mu)/\sigma$ (see Eq. 2.18)

often the probability density function is represented by the *normal distribution* also called *Gaussian distribution*. The example of Gaussian distribution curves are presented in Fig. 2.16.

The *distribution function* (cumulative distribution) describes the probability that the random variable be less or equal to x

$$F(x) = Pr(X \le x) \tag{2.13}$$

The *probability density of function* is the derivative of the distribution function

$$f(x) = \frac{dF(x)}{dx} \tag{2.14}$$

thus

$$f(x) = Pr(x < X < x + dx)) \tag{2.15}$$

Knowing the probability density function we can determine the *probability* that the value X is in the range from x_1 to x_2

$$Pr(x_1 < X < x_2) = \int_{x_1}^{x_2} f(x)dx \tag{2.16}$$

and of course is

$$Pr(-\infty < X < \infty) = \int_{-\infty}^{\infty} f(x) = 1 \tag{2.17}$$

The *normal distribution* (Gaussian distribution) is described by the equation

$$f(x) = \frac{1}{\sigma\sqrt{2\pi}} \exp\left[-\frac{1}{2}\left(\frac{x-\mu}{\sigma}\right)^2\right] \tag{2.18}$$

thus it is described by two parameters: *standard deviation* σ and the *expected value* (expectation) μ.

The *expected value* is the value, around which all random variables are extended. For a continuous random variable having the probability density function $f(x)$ the expected value μ is

$$\mu = \int_{-\infty}^{\infty} x f(x) dx \qquad (2.19)$$

For the normal distribution the expected value is the symmetry axis of the function $f(x)$ and for limited number n of observations x_i is equal to the *mean value* $\bar{x}$

$$\bar{x} = \frac{1}{n} \sum_{i=1}^{n} x_i \qquad (2.20)$$

We can determine the *standard deviation* σ as the positive square root of the *variance*[1] $V(x) = \sigma^2$ given by equation

$$\sigma = \sqrt{V(x)} = \sqrt{\frac{1}{n-1} \sum (x_i - \bar{x})^2} \qquad (2.21)$$

For the normal distribution the probability that variable is equal to the expected value (in this case the mean value) with the dispersion equal to the standard deviation σ is

$$\Pr(\bar{x} - \sigma \leq x \leq \bar{x} + \sigma) = 0.6826 \qquad (2.22)$$

Thus the probability that the result of observation is in the range $\pm\sigma$ around the mean value (expected value) is *68.26%*. Similarly, we can calculate that this probability for the dispersion $\pm 2\sigma$ is *95.44%* and for $\pm 3\sigma$ is *99.73%*. We can say that the result of measurement is very close to estimated value if the uncertainty is 3σ. Therefore we sometimes say about 3σ *rule* as the rule of large probability[2].

The standard deviation of the mean value depends on the number of observations n

$$\sigma(\bar{x}) = \frac{\sigma}{\sqrt{n}} \qquad (2.23)$$

---

[1] The variance $V(x)$ describes the dispersion of the variable around the expected value and is the second-order moment about the mean $\quad V(x) = \sigma^2 = \int_{-\infty}^{\infty} (x - \mu)^2 f(x) dx$.

[2] For 4σ this probability is *99.994%*.

By increasing of the number of observations we diminish the range of uncertainty of the mean value. But the component $\sqrt{n}$ increases slowly with the increasing of n – to decrease the standard deviation by 10 it is necessary to increase the n by 100. Such effort (or waste of time) is unprofitable and therefore it is assumed that in typical cases the number of observations in the range of 20 –30 is sufficient.

When the number of observations is not large it is recommended to use the *Student's distribution* (t-distribution) instead of the normal distribution. The Students distribution is described by equation

$$p(t,v) = \frac{\Gamma\left(\dfrac{v+1}{2}\right)}{\sqrt{v\pi}\,\Gamma\left(\dfrac{v}{2}\right)}\left(1+\frac{t^2}{v}\right)^{-\frac{t+1}{2}} \tag{2.24}$$

where $v = n\text{-}1$ is the degrees of freedom and Γ is the Euler function.

The shape of the graph of the Student's distribution is similar to the normal distribution (bell shape), but it is more flat and the flatness depends on the degrees of freedom (number of operations). Practically for the $n > 30$ the students distribution is very close to the normal distribution.

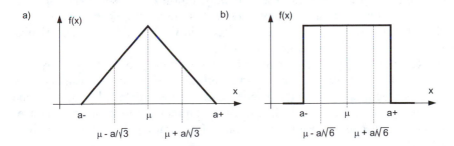

Figure 2.17. The graph of the triangular (a) and rectangular (b) density of probability function. There are indicated the range of variable corresponding with standard deviation σ of normal distribution

There are also other probability density functions, for example *rectangular (uniform) distribution* or *triangular distribution* presented in Fig. 2.17. We can assume that in the normal distribution the standard uncertainty is 1/3 of the range of variable equal to 3σ[1]. Note that in the case of triangular function the σ dispersion of normal distribution corresponds to the $1/\sqrt{3}$ of the range while in the case of rectangular function it is $1/\sqrt{6}$.

Consider the case when the resultant value is composed from various values $Y = c_1X_1 + c_2X_2 + ...$ determined with various probability distributions. In such case the *Central Limit Theorem* is helpful. This theorem states that the distribution of Y will be approximately normal with expected value equal to

$$\mu(Y) = \sum_{i=1}^{N} c_i \mu(X_i) \tag{2.25}$$

and variance

$$\sigma^2(Y) = \sum_{i=1}^{N} c_i^2 \sigma^2(X_i) \tag{2.26}$$

2.2.3. Methods of evaluation and correction of the uncertainty related to limited accuracy of measuring devices

In the case of evaluation of uncertainty of type A the calculations are relatively simple because we use well known tools of statistical analysis. The case of evaluation of uncertainty of type B is more complicated, because we should evaluate various sources of uncertainty – for this task experience, knowledge and even intuition is necessary.

Relatively easy is evaluation of the uncertainty of typical measuring devices, because we have the information about the accuracy estimated by the manufacturer. Usually the manufacturer encloses detailed documentation specifying all uncertainties. In the case of precise and expensive devices manufacturer can enclose the certificate of accuracy prepared by accredited laboratory.

The analogue indicating instruments are very well described by the standards, and they are specified by the *class of accuracy*. The digital instruments are not standardized, but there is certain universally accepted custom of describing of the accuracy of such instruments. Usually the uncertainty of digital instruments is described as:

$$\pm(\% \, rdg + \% \, FS) \qquad \text{or} \qquad \pm(\% \, rdg + \% \, range) \tag{2.27}$$

---

[1] We can assume that all measured value are in 3σ range with sufficient large probability.

which we can explain as the sum of uncertainty of indicated value (*rdg – reading*) and uncertainty of the range (*FS – Full Scale*). The absolute uncertainty is usually described using the unit of the measured value. The relative uncertainty is usually given as percentage value (%) with respect to the range value. Modern measuring devices are so accurate that the percent unit is to large for express the uncertainty. For example, presentation of the uncertainty as *0.00001%* would be inconvenient; therefore, often the description in *ppm* (*ppm – parts per million – 10^{-6}*) is used. The formula (2.27) is then presented as for example *±(ppm reading + ppm range)*.

Consider case when four-digit voltmeter with the range *10V* indicated *0.454 V* and its uncertainty is described as *± (0.05 + 0.01)%* . The uncertainty of the result is *± (0.05%·454 + 0.01·10 000)mV = ± 1.2 mV* and the relative uncertainty is *±0.26%*. This example demonstrates importance of the use of all significant digits. If for example we change the range to *1V* (if such range exists) and we obtain the result *454.5 mV* the absolute uncertainty is *± (0.04%·454.5 + 0.01%·1000 mV) = ± 0.33 mV* which is related to the uncertainty *0.07%*. Thus we improved the uncertainty more than three times only by changing the range of instrument.

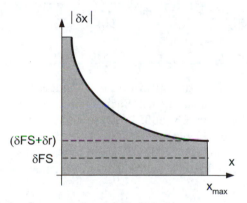

Figure 2.18. The relative δx uncertainty of the digital instrument

Table 2.2. Resolution of a digital measuring instrument as the dependence on the number of digits used

number of digits	number of counts	resolution
3-digit instrument	1000	0.1%
4-digit instrument	10 000	0.01%
4½-digit instrument	20 000	0.005%
4¾-digit instrument	50 000	0.002%

In 4½-digit instrument the first digit can be 0 or 1 and the rest 0,1,....9 while in the 4¾-digit instrument the first digit can be 0,1,2,3,4 and the rest 0,1,...9.

We see that the resultant uncertainty is the sum of the uncertainty related to the accuracy of measurement (*% FS*) and the uncertainty related to the resolution – depending on number of used digits (*% rdg*). Fig. 2.18 illustrates the dependence of the uncertainty on the measured value. The best accuracy is when we use all digits – the relative uncertainty quickly increases with the decrease of the ratio measured value to measuring range. Table 2.2 presents information about the resolution of various digital instruments.

The analysis of a data acquisition board (analogue to digital converter) is similar to the analysis of the digital instrument. Usually, the manufacturers describe the uncertainty of such devices in a form

$$(\% \ of \ rdg \pm \ LSB) \quad or \quad \% \ of \ FSR \quad or \quad \pm \ nbits \qquad (2.28)$$

which is sum of the component related to resolution and to the accuracy (*LSB –Least Significant Bit, FSR – Full Scale Range*). Table 2.3 presents information about the resolution of various analogue to digital converters.

Table 2.3. Resolution of the analogue to digital converters as the dependence on the number of bits used

number of bits	number of counts	resolution
8-bits	256	0.39%
12 bits	4 096	0.024%
16-bits	65 536	0.0015%

Another system of the specification of uncertainty is in the case of analogue indicating instruments. Such instruments are described by the standards, for example EN 60051 (EN 60051 1989). All instruments are divided in *Class of Accuracy* – for example 0.2, 0.5, 1, 2 etc. This class means that the absolute uncertainty of all enumerating graduations does not exceed the class number in % with relation to the range value[1]. Thus

$$Accuracy \ Class = \frac{|\Delta x_{max}|}{x_{max}} \qquad (2.29)$$

If we use the voltmeter of the range *100 V* and *Class 0.5* the manufacturer guarantees that all enumerated graduations have been scaled with the absolute uncertainty not larger than *±0.5%·100V = 0.5 V*. Thus the absolute uncertainty is the same for all measured values, while the relative uncertainty is smallest at the end of the range (Fig. 2.19). Similarly as in the case of digital instruments the relative uncertainty quickly increases, when the

---

[1] Sometimes the manufacturer encloses the table of corrections to all enumerated graduations.

indicated value decreases. For example if our *100 V* voltmeter indicates *25 V* (¼ of the range) the relative uncertainty increases to *2%* (in comparison with *0.5%* for *100 V*).

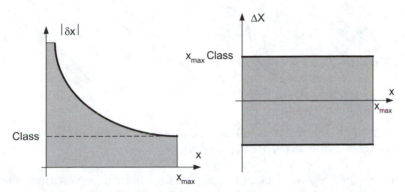

Figure 2.19. The relative δ x and absolute Δx uncertainty of the analogue instrument

The values of uncertainties presented above are determined for the *nominal conditions*. These conditions are described in standards (for example temperature *23 °C*, relative humidity *40 – 60%*, frequency *50 Hz*, etc.) Also, the *operating conditions* are determined (for example variation of the temperature *ΔT = 10 °C*, inclinations from the horizontal level *5°*, etc.), when the uncertainty can increase by the *Class* value (additionally for all factors).

Uncertainty of the measuring instruments depends not only on the accuracy of scaling procedure. The uncertainty influences also other factors, such as nonlinearity, hysteresis, zero drift, and resolution. If these uncertainties are not specified, then we can assume that they are included in the whole uncertainty declared by manufacturer. Often the manufacturers inform about these errors as the additional errors.

Fig. 2.20 presents the graphical illustration of typical errors of signal processing. The *sensitivity error* is generally related to the change of amplification, mostly due to the change of temperature, but also due to aging of the elements. When the changes are caused by the aging process it is necessary to perform re-scaling of the instrument (for example using the calibrator device). When the changes are caused by the temperature influence it is usually necessary to include the corrections to the result of measurement. The manufacturers of measuring instruments often specify the temperature influence as additional uncertainty, for example *± 50 ppm/°C*. The error of sensitivity is relatively easy to correct by calibration of the instrument. The useful method of stabilizing against the changes of sensitivity is the introduction of appropriate feedback (this method is discussed later in Chapter 4).

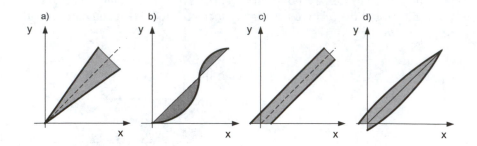

Figure 2.20. Typical errors of signal processing: a) error of sensitivity, b) error of linearity, c) error of resolution, d) error of hysteresis

The *error of linearity* (lack of linearity) results from approximation of the transfer characteristic by the straight line, when this line is nonlinear. This error is often defined as the largest difference between the straight line and the real characteristic, related to the range value. This kind of error can be eliminated by introducing the correction (if the nonlinear characteristic is known). There are various methods of linearization of the transfer characteristics. One of them is the feedback (discussed later in Chapter 4).

One of the most difficult to decrease are the *errors of resolution*. There are many sources of these errors: *noises, zero drift, limited dynamics, interferences*. The resolution errors are often defined as the smallest change of the signal that is not detected by the measuring instrument.

The *noises* are the signals containing the components of various frequencies and magnitude varying randomly. For example the so-called white noise contains signals of all frequencies. A resistor generates white noise as thermal noises, according to the formula $U_T = (4kTR\Delta f)^{\frac{1}{2}}$ (k is the Boltzman constant, Δf – the frequency bandwidth). The noises are also generated by semiconductor devices or magnetic elements (Barkhausen noise). We can reduce the influence of these noises taking into account their random character – by applying the statistical theory. There are many various methods of noises reduction – from the shielding through the applications of artificial intelligence. Modern methods of noise elimination enable *recovery of the signals* in presence of the noises even *100 dB* larger.[1] One of the most effective methods of noise reduction is the use of the *lock-in amplifiers* (discussed in Chapter 4). The noises are usually defined as *SNR* factor (*SNR – signal to noise ratio*).

---

[1] dB is *20 log (U/U₀)* – thus *100 dB* means $U/U_0 = 100\ 000$, while *120 dB* means $U/U_0 = 1\ 000\ 000$

The electromagnetic interferences can be the source of deterioration of the resolution. That is why the electromagnetic compatibility is so important. The most popular methods of reduction of the electromagnetic interferences are: shielding (magnetic or electrostatic), grounding and filtering. Mostly, interferences are the signals of industrial frequency 50 Hz. Such signals are very difficult to remove, especially when our measured signal is of the same frequency.

The zero drift is also very difficult to remove. Often, such drifts are caused by a change of temperature. This error is especially troublesome when it is accompanying a *DC* signal, because we cannot separate them. Therefore, zero drift reduction is facilitated by conversion of the *DC* signal into the *AC* signal with modulated magnitude. Recently, one of the most effective methods is the *auto-zero function*. This function is performed by periodically switching the input signal off (for example by short-circuiting the input), then determining of the offset, and then subtracting it from the signal after opening the input (The auto-zero function is more detailed discussed in Chapter 4).

The auto-zero method is possible to use only when the source of the offset is internal. More difficult is the case, when the external source influences the sensor. For example, when the resistive strain gauge sensor (sensor of strain) is influenced by temperature (which causes also the change of resistance) we cannot separate these influences (strain and temperature) because we cannot "switch off" the external temperature. In such case very useful is the differential principle.

In the *differential method* the difference of the signal of two sensors is connected to the input of the measuring device. Only one sensor is influenced by measuring quantity, but both sensors are influenced by external temperature. Thus the signals caused by the external temperature are compensated and only the signal caused by measured quantity remains.

It is possible to apply the differential method in the bridge circuit. In the circuit presented in Fig. 2.21a the output voltage depends on the changes of all four resistors

$$U_{out} \approx \frac{1}{4}\left(\frac{\Delta R_1}{R_1} - \frac{\Delta R_2}{R_2} + \frac{\Delta R_3}{R_3} - \frac{\Delta R_4}{R_4} \right) U_s \qquad (2.30)$$

Thus, if the temperature influences two identical resistors R_1 and R_2 while the measured value influences the resistor R_1, then the output signal of the bridge circuit is

$$U_{out} \approx \frac{1}{4}\left(\frac{\Delta R_1(x)}{R_1} + \frac{\Delta R_1(T)}{R_1} - \frac{\Delta R_2(T)}{R_2} \right) = \frac{1}{4}\frac{\Delta R_1(x)}{R_1} \qquad (2.31)$$

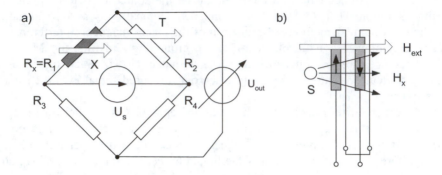

Figure 2.21. The differential method of reduction of external interferences: a) the bridge circuit, b) the gradient sensor

Fig. 2.22 illustrates several examples of the differential method. In the case presented in Fig. 2.22a two identical strain gauge sensors are glued on the surface of stressed sample. But only one of these sensors (R_1) is stressed while the other (R_2) is placed perpendicularly to the stress. The temperature influences both sensors. If we connect these two sensors in the adjacent arms of the bridge circuit we obtain the elimination of the temperature influence.

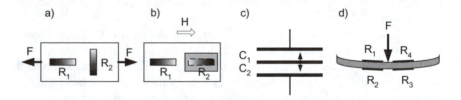

Figure 2.22. The differential method: ab) two sensors, only one is influenced by measured value, cd) two differential sensors, both influenced by the measured value

Fig. 2.22b presents the method of elimination of the temperature zero drift used in the magnetic field sensors of *Nonvolatile Electronics Inc.* (NVE 2005). Two identical magnetoresistors are influenced by external temperature, but the second sensor is isolated from the measured field H by the shield.

Instead of the pair of active/passive sensors it is better to use two or four *differential sensors*. Such sensors work differentially, which means that they fulfill the relation

$$\Delta R_1 = \Delta R(x); \; \Delta R_2 = -\Delta R(x); \; \Delta R_3 = -\Delta R(x); \; \Delta R_4 = \Delta R(x) \qquad (2.32)$$

Thus, taking into account the relation (2.30) we obtain

$$U_{out} \approx \frac{\Delta R(x)}{R} U_s \qquad (2.33)$$

We obtain elimination of the influence of temperature by the use of differential sensors, and also at the same time we improve the sensitivity, because the output signal is four times larger (in comparison with one-sensor circuit). Fig. 2.22c presents the capacitive sensor of displacement. The internal electrode is moveable. When the capacitance of the upper sensor increases (smaller distance between electrodes), then at the same time the capacitance of second sensor decreases (larger distance between electrodes).

Fig. 2.22d presents the stress measurement of the bending sample. The sensors R_1 and R_4 are compressed, while at the same time the sensors R_2 and R_3 are stretched. Therefore, all four sensors are the same but the two pairs act mutually differential.

Fig. 2.21b presents a method of elimination of the influence of external magnetic field (for example Earth's magnetic field) during the measurement of magnetic field from the source S. Such problem is common in biomedical measurements, when small magnetic field needs to be investigated, for example with magneto-cardiograph in presence of much larger Earth's magnetic field. Two sensors are connected differentially and are positioned at some distance from each other. We assume that the source of Earth's magnetic field is large and it is at long distance from the sensors; therefore, the external magnetic field H_{ext} is the same in both sensors. The investigated source of magnetic field is small and near the sensors thus sensor placed closer to this source is influenced more than the other sensor positioned at some distance from the source S. Such pair of sensors measures the gradient of magnetic field – from the distant source this gradient is very small but from the adjacent source this gradient can be significant.

The *hysteresis error* is often caused by the presence of magnetic parts. It can also be caused by the mechanical friction in indicating analogue instruments. With hysteresis we can obtain different results of measurements carried out for increasing and decreasing signal.

Additional problems appear when we process time varying signals. One of the important parameters in this case is the *frequency bandwidth* of the measuring instrument. Most of measuring devices exhibit limited bandwidth – sometimes it is limited for low frequency (the *DC* and slowly varying components are not detected), but it is always limited for high frequency. Fig. 2.23 presents the specification of the measurement uncertainty of the popular multimeter 34401 of Agilent for various ranges of frequency of measured signals.

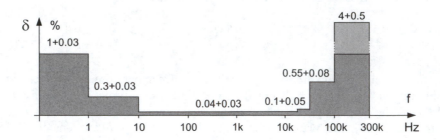

Figure 2.23. The example of specification of the measurement uncertainties depending on the frequency of the measured AC signal – the multimeter 34401 of Agilent (Agilent 2005)

The measurement of the signals at frequency other than acoustic range *20 Hz – 20 kHz* is in general less accurate. It is rather difficult to eliminate the influence of parasitic capacities at the high frequency range. Above about *1 MHz* the accuracy influences the transmission line effect and in this bandwidth special measuring instruments are used. In the case of analogue processors the bandwidth is usually defined as the frequency range in which the amplification factor K_u (or generally sensitivity coefficient K) changes no more than *3 dB* from the defined value (often for example from the value determined for frequency *1 kHz*).

In the case of digital processing the main limitations come from the *sampling frequency* which according to the Shannon theory should be at least two times larger than the greatest frequency in the signal. Recently, there are available analogue-to-digital converters with sampling frequency greater than *1 GHz*.

Usually the uncertainty is specified for a pure sinusoidal signal. The distortion from the sinusoidal signal is often defined as the presence of harmonics in the signal. Often the *THD* coefficient (*THD – Total Harmonic Distortion*) is used. This coefficient is defined as the percentage ratio of all harmonics components above the fundamental frequency to the magnitude of fundamental component:

$$THD = \frac{\sqrt{\sum_{k=2}^{N} a_k^2}}{a_1} \cdot 100\% \qquad (2.34)$$

The signal is characterized also by the *CF* factor (*CF – Crest Factor*) – the ratio of the peak value to the *rms*[1] value of the waveform. For example, in the multimeter 34401 presented above the Crest Factor of *1 – 2* causes additional error of *0.05%* of reading, while for *CF 4 –5* this error is *0.4%* of reading.

Sometimes, the *FF* factor (*FF – Form Factor*) is also used, defined as the ratio between the *rms* value and average rectified value. For pure sinusoidal waveform the *FF = 1.11*.

The term "*True rms*" is linked to distortion. Such symbol is often placed on the front panel of measuring instruments. It means that the *rms* value of the signal is measured according to definition (*rms – Root Mean Square*)

$$\hat{x} = \sqrt{\frac{1}{T} \int_{t}^{t+T} x^2(t)\,dt} \qquad (2.35)$$

The term "*True rms*" appeared as the reaction to the "non-true" measurements performed by the formerly universally used measuring instruments with rectifiers. Such instruments measure *de facto* the rectified average value, but they were scaled as *rms* devices under assumption that the dependence between these two values is the *Form Factor = 1.11*. But this condition is fulfilled only for pure sinusoidal waveforms, which is frequently not the case in typical measurements. For example, if the waveform is triangular the error resulting from the distortion is about *5.5%*, while for rectangular waveform this error is *11%*. And for pulse measurements this error is as large as around *50%* for the *crest factor = 4*. Recently the measuring devices indicated as "True *rms*" measure the distorted signal with CF up to *4* without any additional errors.

It is important to know that most laboratory multimeters do not measure the *rms* value of *AC+DC* signals. Usually the *AC* signals are separated from the input by a capacitor. Thus to obtain the *rms* value of *AC+DC* signal it is necessary to perform the measurement two times (as *DC* measurement and *AC* measurement) and then the resultant *rms* value can be calculated as

$$rms(AC + DC) = \sqrt{AC^2 + DC^2} \qquad (2.36)$$

Currently many portable instruments are indicated as *AC+DC*. It means that these instruments correctly measure the *AC* signal with *DC* component.

More complicated for analysis and corrections are the *errors of dynamics*. These errors concern the cases when the processing signal changes

---

[1] *rms* value - *Root Mean Square* value discussed later in this Chapter.

much faster than the time of measurement. For analysis of dynamic performances very useful and relatively simple is testing the response of the measuring system, when the input signal is in a form of a step of unit amplitude (there are used also other input test signals – for example unit impulse). Figure 2.24 presents two typical output signals of the circuit excited by the step unit signal.

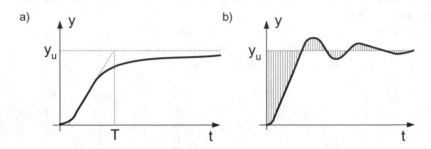

Figure 2.24. Two examples of responses to the step function: a) the device with inertia, b) the device with oscillations

The inertia type circuit (Fig. 2.24a) can be characterized by the *time constant T*. For the first order inertia this time constant can be determined as the *0.638·y_u* or by drawing a tangent line to the response curve. In the case of the circuit with oscillations (Fig. 2.24b) important parameters are the *frequency of oscillations f_o* (or resonance angular frequency ω_0 or relative frequency $\eta = \omega/\omega_0$) and *damping factor b*.

We can interpret the dynamics error as the difference between the response of investigated device and the response of the ideal device without inertia (zero order inertia) o0r oscillations. This error (difference) is indicated in the Fig. 2.24b as the area filled with lines.

The output signal of oscillating circuit can be described as

$$y(t) = y_u \left[1 - \frac{1}{\sqrt{1-b^2}} e^{-b\omega_0 t} \sin\left(\sqrt{1-b^2}\,\omega_0 t + arctg\frac{\sqrt{1-b^2}}{b} \right) \right] \quad (2.37)$$

while in the case of the first order inertia circuit this signal is

$$y(t) = y_u \left[1 - \frac{1}{\sqrt{b^2-1}} e^{-b\omega_0 t} sh\left(\sqrt{b^2-1}\,\omega_0 t + arth\frac{\sqrt{b^2-1}}{b} \right) \right] \quad (2.38)$$

The error of dynamics can be described as the difference between the output signal $y(t)$ and the steady value y_u

$$\Delta_d y = y(t) - y_u \tag{2.39}$$

or as the mean square value

$$\sigma_d = \sqrt{\int_0^\infty \left| \Delta_d y^2(t) \right| dt} \tag{2.40}$$

Sometimes it is convenient to analyze the error of dynamics not in the time domain but in the frequency domain. It is justified because in the case of linear circuit these both specifications are equivalent according to the *Parseval rule*

$$\int_0^\infty |f(t)|^2 dt = \frac{1}{2\pi} \int_{-\infty}^\infty |f(j\omega)|^2 d\omega \tag{2.41}$$

In the frequency domain the dynamic performances of oscillation type device is represented by the transform

$$G(s) = \frac{Y(s)}{X(s)} = \frac{K\omega_0^2}{s^2 + 2b\omega_0 s + \omega_0^2} \tag{2.42}$$

The amplitude and phase characteristics are described by

$$|G(j\omega)| = \frac{K}{\sqrt{\left(1 - \eta^2\right)^2 + (2b\eta)^2}} \tag{2.43a}$$

$$\varphi(\omega) = arctg \frac{2b\eta}{\eta^2 - 1} \tag{2.43b}$$

The device processes the dynamic signal without distortion if the amplitude is constant with frequency:

$$|G(j\omega)| = const \tag{2.44a}$$

Important also is the phase condition in the form

$$\varphi(\omega) = 0 \ \ or \ \pi \ or \ k\omega \tag{2.44b}$$

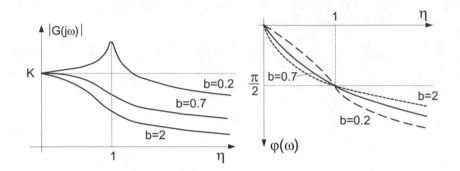

Figure 2.25. The examples of the amplitude and phase characteristics of transducer

The examples of amplitude and phase characteristics are presented in Fig. 2.25. From these characteristics we can formulate the following conclusions:
- the amplitude error is smallest in the largest bandwidth for the damping $b = 0.707$,
- for $b = 0.707$ the transducer can be used in the range $\omega < \omega_0$ (for frequency $\omega < 0.6\,\omega_0$ the amplitude error does not exceed 5%),
- the phase characteristics are close to the linear for $\omega < \omega_0$ and $b=0.7-0.8$.

The dynamics errors can be also determined as

$$\delta_d = \sqrt{\frac{1}{2\pi} \int\limits_{-\infty}^{\infty} \left| G(j\omega) - K \right|^2 \left| x(j\omega) \right|^2 d\omega} \qquad (2.45)$$

2.2.4. The estimation of uncertainty in measurements

If the random errors are dominating then we use *the method A of evaluation of uncertainty*. In this case we apply the tools of theory of statistics presented in Section 2.2.2. It is only small difference between the theory and practices of evaluation of uncertainty. The Guide differentiates between the variance σ (an abstractive term) and *estimate of variance s* (*experimental standard deviation*) related to measurements. The experimental standard deviation is determined from the same dependence as the standard deviation (see Eq. 2.21)

$$s^2(x_k) = \frac{1}{n-1} \sum_{k=1}^{n} (x_k - \bar{x})^2 \qquad (2.46)$$

The uncertainty of measurement is

$$u(\bar{x}) \equiv s(\bar{x}) \tag{2.47}$$

Figure 2.26 presents the illustration of the uncertainty evaluation for the method A.

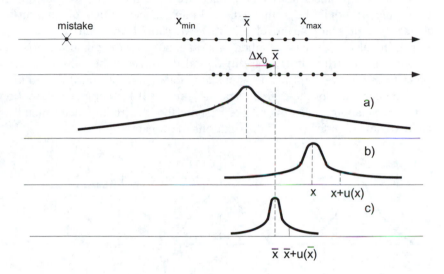

Figure 2.26. The illustration of the estimation of the uncertainty by method A: a) distribution function for the results of measurement without correction, b) distribution function for the result of measurement, c) distribution function of the mean value of series of measurements with correction

After *n* repeated measurements the corrected value is determined – the results of mistakes are removed and the correction Δx_0 is subtracted. From corrected results the estimate of measured value is determined as the mean value $\bar{x}$ (Eq. 2.20). Next, the experimental standard deviation $s(\bar{x})$ is calculated (Eq. 2.46) and later the uncertainty $u(\bar{x})$ (Eq. 2.47).
The result of measurement can be presented as

$$x = \bar{x} \pm u(\bar{x}) \tag{2.48}$$

The uncertainty of measurement can be also presented as the *expanded uncertainty u*

$$u = ku(\bar{x}) \tag{2.49}$$

Not always is it clearly specified, then for the uncertainty $u(x)$ the coverage factor k is 1 which corresponds to the confidence level 68% (for normal distribution). For expanded uncertainty the coverage factor should be specified, usually it is $K=1,\ 2$ or 3 therefore the confidence levels are 68%, 95.5% or 99.7%, respectively.

In the case of *uncertainty type B* apart from the random errors also other uncertainties should be estimated, including uncertainties of measuring devices. At present, the novelty is that we should take into account the probability also in the case of other errors, i.e. not only the random errors. As was discussed earlier, in the case of digital instruments the values between the last bits are evenly probable – thus the rectangular (uniform) distribution is recommended. For the digital frequency meters the triangular distribution is the most appropriate. Fig. 2.27 presents the uncertainty determined as standard deviation for various distributions of probability.

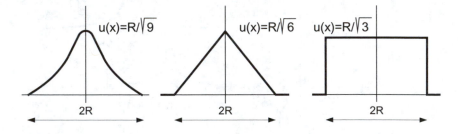

Figure 2.27. Comparison of uncertainties for various distributions of probability

After estimation of all the uncertainties we can determine the resultant uncertainty $u_B\ (x)$ as the root of squares sum of each uncertainties $u_B\ (x_i)$

$$u_B(x) = \sqrt{\sum_{i=1}^{m} u_B^2(x_i)} \qquad (2.50)$$

Similarly we can obtain the resultant uncertainty determined using both methods

$$u(x) = \sqrt{u_A^2(\bar{x}) + u_B^2(x)} \qquad (2.51)$$

When the determined value of y is function of several other quantities x_i

$$y = f(x_1, x_2 ... x_n) \qquad (2.52)$$

we can determine the *combined uncertainty*. The estimation of combined uncertainty is more complicated when component quantities x_i, x_j are *mutually correlated* (dependent). We can check this by testing the *degree of correlation*

$$r(x_i, x_j) = \frac{u(x_i, x_j)}{u(x_i)u(x_j)} \qquad (2.53)$$

which varies from *0* to *1* (*0* means that these quantities are uncorrelated, while *1* means that they are completely correlated, i.e. $x_i = k \, x_j$). In the dependence (2.53) $u(x_i) \, u(x_j)$ are the estimates of variances, while $u(x_i, x_j)$ is the estimate of *covariance* of both quantities. We can determine the combined uncertainty using the *law of propagation of uncertainty*

$$u^2(y) = \sum_{i=1}^{N} \left(\frac{\partial f}{\partial x_i} \right)^2 u^2(x_i) + 2 \sum_{i=1}^{N-1} \sum_{j=i+1}^{N} \frac{\partial f}{\partial x_i} \frac{\partial f}{\partial x_j} u(x_i, x_j) \qquad (2.54)$$

The covariance of random variables can be determined experimentally as

$$u(\bar{x}_i, \bar{x}_j) = s(\bar{x}_i, \bar{x}_j) = \frac{1}{n(n-1)} \sum_{k=1}^{n} \left(x_i^{(k)} - \bar{x}_i \right) \cdot \left(x_j^{(k)} - \bar{x}_j \right) \qquad (2.55)$$

In practice, when the component quantities are mutually weakly dependent we can neglect the second part from the dependence (2.54) and the combined uncertainty can be calculated as

$$u^2(y) = \sum_{i=1}^{N} \left(\frac{\partial f}{\partial x_i} \right)^2 u^2(x_i) \qquad (2.56)$$

Let us consider an example, when we determine the impedance Z on the basis of measurements of resistance R with uncertainty $u(R)$, inductance L with uncertainty $u(L)$ and frequency f with uncertainty $u(f)$. The relationship between these quantities is

$$Z = \sqrt{R^2 + (2\pi f L)^2} \qquad (2.57)$$

Thus from (2.56) and (2.57) we obtain

$$u(Z) = \sqrt{\frac{R^2}{R^2 + (\omega L)^2}u^2(R) + \frac{\left(\omega^2 L\right)^2}{R^2 + (\omega L)^2}u^2(L) + \frac{\left(4\pi^2 L^2 f\right)^2}{R^2 + (\omega L)^2}u^2(f)} \qquad (2.58)$$

Sometimes instead of the root of square sum we simply add the modulus of uncertainty

$$u(y) = \sum_{i=1}^{N} \left| \frac{\partial f}{\partial x_i} U(x_i) \right| \qquad (2.59)$$

but such method of calculation of uncertainty can cause the overestimation of the combined uncertainty. The sum of modulus is sometimes called the *maximal limiting uncertainty*. Instead of summing of the absolute uncertainties we can also sum the relative uncertainties as:

$$\delta^2(y) = \sum_{i=1}^{N} \left(\frac{\partial f}{\partial x_i} \right)^2 \left(\frac{x_i}{y} \right)^2 \delta^2(x_i) \qquad (2.60)$$

Figure 2.28 summarizes the main steps in evaluation of uncertainty.

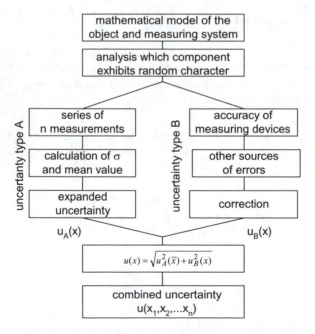

Figure 2.28. Algorithm of evaluation of uncertainty

When the result of a measurement is presented, then the form of this presentation can inform us about the uncertainty as well. For example, if we have measured the voltage as *4.565 V* with uncertainty *0.1%* (thus it is *4.565±0.005*) then it is meaningless to present the result as for example *4.565297 V* (such style of presentation is sometimes used, when the researcher used the calculator or computer to calculate the results). Of course, similarly incorrect is to present the result as *4.56 V*. Generally, the accepted rule is that: *last significant digit of result of measurement should be the same range as the last digit of the uncertainty.*

2.3. STANDARDS OF ELECTRICAL QUANTITIES

2.3.1. Standards, etalons, calibration and validation

The measurement is always related to the standard unit. The *standard* is the realization of a given quantity with stated value and measurement uncertainty, used as a reference. Using the standard we can perform calibration of the measuring instrument. The *calibration* is the operation establishing the relationship between quantity values provided by measurement standards and the corresponding indications of measuring system, carried out under specified conditions and including evaluation of measurement uncertainty (ISO VIM 2004).

The concept of preparation of the measuring standards is well known for a few centuries. The first standard of *meter* unit was prepared in 1793 in form of a brazen bar. In 1799 this bar was substituted by more accurate platinum bar, and in 1889 the standard in form of the X-shaped bar from iridium-platinum was prepared. This standard is preserved in the *Bureau International Poids at Mesures BIPM in Sèvres (Paris).*

Next, the standards of other quantities were prepared with better and better accuracy. Many countries prepared the copies of such standards and the hierarchy of standards was created: international standards → national standards → primary standards → secondary standards → working standards. The most accurate standards used only for calibration of other standards are called *etalons*. Sometimes, the group of standards are used – the resultant standard is then estimated as the average value.

Currently, the idea to substitute the material standards by the standards referred to the physical phenomena dominates. For example, the material standard of *meter* (iridium-platinum bar described above) is substituted by the standard referred to the path of the light traveled in defined time interval. Such kinds of standards are indestructible, even in the case of cataclysm. Moreover, they can be even transferred to other civilizations.

Recently, in laboratory and industrial conditions the standards are substituted by the instruments called *calibrators*. Such instruments are manufactured as accurate sources of quantities (voltage, current, power, resistance, frequency, etc.) enabling setting of these values with uncertainty better than several *ppm* or *fraction of ppm*.

The calibration procedure is used to test the accuracy of measuring instruments. Sometimes, the validation procedure is also performed. The *validation* is confirmation through examination of a given item and provision of objective evidence that it fulfils the requirement for a stated intended use (ISO VIM 2004). Thus, we can apply the validation procedure to test the reliability of the measuring method.

2.3.2. The standards of electrical quantities referred to the physical phenomena and laws

According to the definition of standard unit of *ampere* it is the current, which in two parallel conductors produce force equal *$2 \cdot 10^{-7}$ N per meter of length*. This slightly complicated definition is realized by the *Ryleigh current weight* (Fig. 2.29). Two parallel conductors are substituted by two cylindrical coils connected in series – one stationary and the other one movable, connected to the arm of the balance.

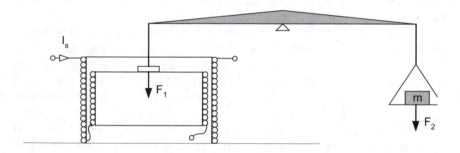

Figure 2.29. The Ryleigh weight as the standard of current

The currents in both coils cause that these coils attract each other. The force of attraction depends on the current value $F_1 = KI_s^2$ (K – constant depending on the geometry of the coils). This force is balanced by the gravity force $F_2 = mg$ (m - the mass of weight on the second arm, g – Earth's acceleration factor). The value of the current is determined as

$$I_s = \sqrt{\frac{mg}{K}}$$

$$(2.61)$$

Such standard enables the reconstruction of the current unit with uncertainty below *6 ppm*. The current unit can be reconstructed more accurately from the Ohm's law by using the standards of resistance and voltage. Uncertainty of such indirect standard is below *1 ppm*.

The *quantum Josephson effect* is utilized to reconstruct the standard unit of voltage. This effect appears at very low temperature (typically liquid helium *4.2 K*), when certain materials (for example niobium) become superconductors. The superconducting Josephson junction consists of two thin superconductors separated by very thin insulator layer.

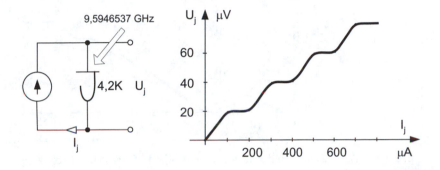

Figure 2.30. The Josephson quantum effect as the standard of voltage, after (Ibuca et al. 1983)

When Josephson junction device is irradiated by microwave energy in the frequency range *9 – 100 GHz* and it is biased by *DC* current then the voltage changes stepwise with the change of the junction current (Fig. 2.30). We can determine these steps on the volt-ampere curve very precisely. The level of the *n* step is described by the dependence

$$U(n) = nf \frac{h}{2e} = \frac{nf}{K_J} \tag{2.62}$$

Note that the voltage depends on the very well defined values: h – Planck's constant, e – electron charge and f – microwave frequency that we are able to measure very accurately. The Josephson's constant $K_J = (2e/h)$ is equal to *483 597.9 GHz/V*.

The main drawback of the quantum standard of voltage is that the output signal is relatively small and contains noise. For *100 GHz* microwave the single step of voltage is about *200 μV*. This signal can be increased by connecting many Josephson junctions in series – even up to several thousands of them. Using the quantum voltage standard it is possible to reconstruct the voltage unit with uncertainty of about 10^{-2} *ppm*.

The quantum Hall effect in the superconductors can be used to realize the standard of the resistance first described by Klaus von Klitzing in 1980. When semiconductor Hall device (Fig. 2.31) is placed in very small temperature (*1 – 2K*) and it is biased by the *DC* current the output Hall voltage depends on the magnetic field in a stepwise way (Fig. 2.31).

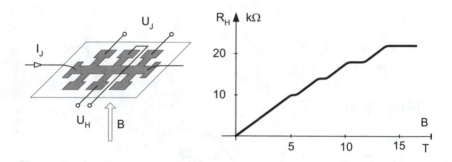

Figure 2.31. The quantum Hall effect as the standard of resistance (first steps are practically invisible)

Typical *quantum Hall device* has six or eight terminals – two for Hall output voltage U_H , two for bias current I_J and two or four for voltage U_J . The voltage U_J is used to determine the current in the device. The Hall resistance for n^{th} step is $R_H(n) = U_H /I_J$. Moreover, the voltage U_J helps in determination of the step in the $R_H = f(B)$ characteristic, because minima of U_J occur in the center of each step. The quantum Hall resistance of the nth step is described by the dependence

$$R_H(n) = \frac{h}{2e^2 n} = \frac{K_K}{n} \qquad (2.63)$$

The resistance depends only on well defined values (*h* and *e*) and it does not depend on the current I_J or the magnetic flux density *B*. The *von Klitzing constant* K_K was determined as *25 812.807 Ω*. Using the quantum resistance standard it is possible to reconstruct the resistance unit with uncertainty of about 10^{-2} *ppm*.

The main drawback of the quantum resistance standard is the very large magnetic field necessary to obtain the quantum phenomenon – for most often used forth step this field is several *T (T – tesla)*.

According to the definition of the time unit one second is the duration of *9 192 631 770* cycles of microwave light absorbed or emitted by the hyperfine transition of cesium-133 atoms in their ground state undisturbed by external fields. This idea is realized as the time/frequency standard. The example of the cesium atomic standard is presented in Fig. 2.32.

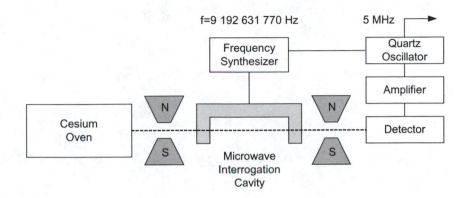

Figure 2.32. Cesium beam atomic clock – the standard of the time/frequency

The Cesium-133 atoms are heated to the gaseous state in the oven. This gas is traveling as high-velocity beam through the gate of the magnet into the microwave cavity. The magnet gate is used to select only atoms of a particular energy state. The atoms are exposed in the cavity to a microwave frequency. If the microwave frequency matches the resonance frequency of cesium the atoms change their energy state. Only atoms which changed their energy pass through the second magnet gate. The detector of these atoms tunes the quartz oscillator to the state, at which the greatest number of atoms reaches the detector. It is when the frequency of microwave cavity is exactly *9 192 631 770 Hz*. As the standard frequency the frequency of *5 MHz* of quartz oscillator is used.

Recently the most accurate atomic clock is achieved by means of fountain principle (Fig. 2.33). Such clock developed by NIST (*National Institute of Standards and Technology USA*) allows to obtain an uncertainty less than 10^{-15} (NIST TF – 2005). It is one of the most accurately determined value. Its uncertainty is better than *0.1 ns/day*, which corresponds to a change around *1s* after *30 000 000* years.

Six infrared lasers orthogonally positioned (see Fig. 2.33) in the vacuum chamber push the cesium atoms into a ball. In this process the lasers cool the atoms to the temperature a few millionths of a degree above absolute zero and reduce their thermal velocity to a few centimeters per second.

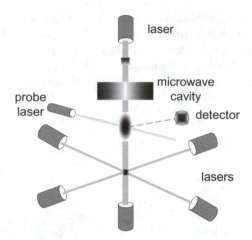

Figure 2.33. Fountain cesium atomic clock – the standard of time/frequency (NIST TF 2005)

Vertical laser tosses the ball upward and then all the lasers are turned off. Under the influence of gravity the ball falls back through the microwave cavity. During this trip the atoms interact with the microwave signal. When the ball leaves the cavity another laser beam is directed onto the ball. Those atoms, whose states were altered, emit fluorescence sensed by detector. This process is repeated many times for various microwave frequency and the frequency that causes the maximum of fluorescence is the cesium resonance.

The standard frequency is distributed by radio and satellite systems.

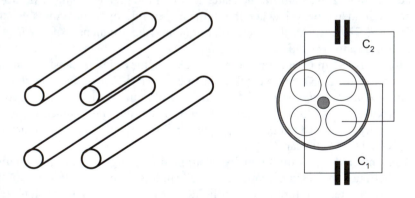

Figure 2.34. The standard of capacitance – calculable capacitor of Thompson-Lampard

Apart from the standards based on the physical phenomena other standards based on the physical laws have been developed. These standards are sometimes called as *calculable standards*. The example of a calculable standard is presented in Fig. 2.34 – it is a standard of capacitance.

The calculable standards are based on the idea that for some geometrical designs it is possible to calculate their parameters very precisely. The capacitor proposed by *Thompson and Lampard* consists of four metal cylinder bars arranged in a square and surrounded by a cylinder movable shield device. The capacitance of such capacitors can be estimated from the following dependence

$$C = \frac{\ln 2}{4\pi^2} \frac{l}{c^2} 10^7 \qquad (2.64)$$

Thus, we can calculate the capacitance with very small uncertainty because it depends only on the velocity of light c (which we know very precisely) and on the length l. And the length we are also able to measure with very small uncertainty – using the interference methods. That is why we are able to determine such capacitance with uncertainty less than 10^{-2} *ppm*. The main drawback of the air standard capacitance is its relatively small capacitance – only *1.95354904 pF/m*.

2.3.3. Material standards of electrical quantities

As the standard of voltage the *Weston cell* was used from more than 100 years. Fig. 2.35a presents the design of such cell. The Weston cell exhibits several advantages: simplicity, stability of standard voltage better than *2 ppm/year*, uncertainty of standard from *0.01%* to *0.0002%*, low level of noises. The output *EMF* is from *1.018540 V* to *1.018730 V*. Although the output voltage is varying with temperature (about *40 µV/K*) these changes are well known and can be corrected. Anyway, the standard cells are usually kept in thermostat.

The main drawbacks of the Weston cell are sensitivity to shocks and limitation of current value – even current of *1 µA* deteriorated the performance of the cell.

For that reason there is no wonder that the Weston cells are currently substituted by electronic devices. In such devices usually specially selected Zener diode is used as the voltage standard source. Electronic standards are more useful in applications, since they are resistant to shocks and can be moved without problems. They can be easily connected to a computer, because they are usually equipped with appropriate interfaces.

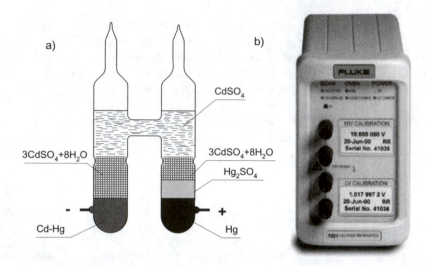

Figure 2.35. The standard of voltage – saturated Weston cell (a) and electronic standard of Fluke – model 7000 (Fluke 2005) (permission of Fluke Corporation)

Fig. 2.35b presents an example of such electronic standard – the model 7000 of Fluke. This standard enables users to obtain voltage *10 V* or *1.018V* with uncertainty less than *2 ppm*, stability better than *1 ppm/year*. The temperature influence is less than *0.03 ppm/K* in the temperature range *15 – 35 °C*. The noise level is below *0.04 ppm rms* in the bandwidth *0.01 – 10 Hz*. It is possible to consume the load the cell with output current up to about *12 mA*. Thus the electronic standard exhibits performances comparable or even better than the Weston cell.

Apart from rather expensive electronic standards of voltage, there are also available on the market relatively cheap (several to several dozen of USD) monolithic voltage references. Fig. 2.36 presents the functional block diagram of the voltage reference model AD588 of Analog Devices. This reference is designed with the accuracy suitable for 12-bit digital processing without any additional trimming elements. For better accuracy the trimming potentiometers can be used to adjust gain (Pin 5) and balance (Pin 12).

The reference consists of precise laser trimmed Zener diode source and three additional amplifiers. It is possible to obtain the output voltage *+5 V, -5 V* or *10 V* with uncertainty less than *1 mV*, temperature zero drift *1.5 ppm/K* and noises *6 µV p-p* in the bandwidth *0.1 – 10 Hz*. The stability of the output voltage is better than *15 ppm/1000 hours*. In the user notes (AD 2005) there are described among others the methods of application of this reference source to bridge circuit supply, in order to obtain precision current source or to excitation of the resistive temperature detector (sensor) *RTD*. Thus it is possible to obtain relatively cheap and useful standard reference voltage with the voltage uncertainty suitable for many measuring purposes.

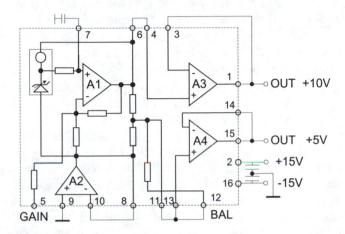

Figure 2.36. The functional block diagram of high precision voltage reference model AD588 of Analog Devices (AD 2005)

The standards of resistance are manufactured as the precise resistors prepared as the wire or strip wound on the porcelain cylinder. Such resistor is placed usually in a shielded housing. The resistance standard is usually equipped with four terminals: two (larger) terminals are used for the current excitation and second two (smaller) ones are used as the voltage (potential) terminals (Fig. 2.37).

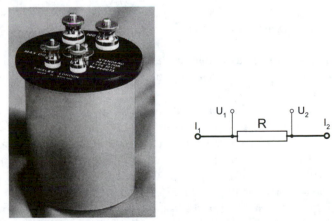

Figure 2.37. An example of the standard resistor – model 5615 of Tinsley (Tinsley 2005) (permission of Tinsley Precise Instruments)

There are certain materials suitable for precise resistors manufacturing – they should exhibit large resistivity and very small dependence of the temperature. One of the most popular is *manganin* (alloy of *84% Cu, 12% Mn, 4% Ni*) with resistivity $\rho = 0.42 \ \mu\Omega \cdot m$ and temperature coefficient $\alpha_T = (0.5 - 2) \ 10^{-5}/K$ (for comparison pure copper exhibits the temperature coefficient $\alpha_T = 4 \cdot 10^{-3}/K$). Another material used for precise resistors preparation is the *Evanohm* (*75% Ni, 20% Cr, 2.5% Al, 2.5% Cu*) with resistivity $\rho = 1.2 \ \mu\Omega \cdot m$ and excellent temperature properties (α_T less than $10^{-6}/K$). An important requirement for resistive materials is negligible thermoelectric voltage in relation to copper.

The resistors are wound bifilarly which enables users to obtain a resistor practically without the inductance. The bifilar winding (Fig. 2.38) is performed in such a way that in adjacent wires the currents are in the opposite directions and therefore the magnetic fields compensate each other. The capacitance of standard resistor is very small due to special design. Although standard resistors are with negligible inductance and capacitance usually they are used for *DC* circuits. For *AC* applications special kind of resistors (indicated *DC/AC* resistor) can be used but in the bandwidth limited to about *1 kHz*.

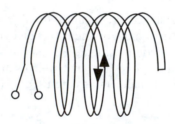

Figure 2.38. The principle of bifilar winding

Standards resistors available on the market exhibit typical uncertainty of *0.5 – 10 ppm*, long term stability of about *2 ppm/year* and temperature stability of about *2 ppm/°C*.

For everyday applications sometimes more usable can be adjustable *resistor decade box*. In such resistor is it possible to set desired value of resistance in the decade sequence: $x \times 1000\Omega + \ x \times 100\Omega + x \times 10\Omega + \ldots$ Currently, there are available resistors with adjustable resistance in range *1 mΩ – 100 MΩ*. The smallest step is *1 mΩ*. The uncertainty of the resistor decade boxes is typically of *0.01%*. It should be noted that often such uncertainty is attainable for the total resistance of the resistor and with decreasing of the decade step it gradually increases (for example: *R/step 0.001 Ω - 4%, 0.01 Ω - 2%, 0.1 Ω - 0.4%, 1 Ω - 0.1%, 10 Ω - 0.04%*, the rest of resistors *0.01%* (IET Labs 2005). Fig. 2.39 presents the typical design of resistor decade box.

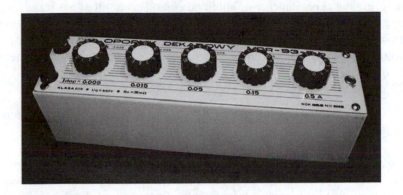

Figure 2.39. An example of the resistor decade box

The Thompson-Lampard standard described in Section 2.3.2 enables users to obtain rather small values of capacitance. For practical purposes the simplest capacitor can be used with capacitance up to *100 nF* and uncertainty about *10⁻⁵*. The standard capacitors are usually placed in the shielding and the capacitance in respect to this shielding should be taken into account. Usually, the shielding is connected to the third terminal (Fig.2.40a). In such configuration the parasitic capacitances *C', C''* can be corrected. When the terminals 0 and 1 are short-circuited than the capacitance between terminals 1 and 2 is $C_{12} = C_s + C''$ and for short-circuiting the terminals 0 and 2 this capacitance is $C_{12} = C_s + C'$.

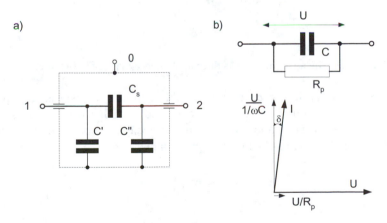

Figure 2.40. The three terminal capacitor (a) and the phasor diagram of currents in the capacitor (b)

Analyzing the capacitance we should take into account the loss resistance resulting from the evidence of finite value of the insulation between the electrodes. The loss resistance R_p (Fig. 2.40b) is usually represented by *loss-tangent* $tg\delta$ (the angle δ is the angle between total current I and displacement current U ωC). Thus

$$tg\delta = \frac{1}{\omega C R_p}$$ (2.65)

Also air capacitor is not without losses because for the air the loss-tangent value is about 10^{-5} for *1 kHz*. A dielectric material is often inserted between the electrodes in order to obtain larger value of capacitance. Such capacitors are smaller than the air capacitors but in this case the losses are not negligible. For example the mica dielectric exhibits losses $tg\delta \approx 5 \cdot 10^{-4}$ for the frequency 1 kHz. Also capacitors are available as the decade boxes.

Figure 2.41. The air-cored coil as the calculable standard of inductance

Similarly as in the case of calculable capacitors (discussed in previous section) also the inductance can be designed in such a way that it is possible to calculate its inductance with satisfying accuracy. For example inductance of air-cored coil (Fig. 2.41) can be determined as

$$L = \frac{4\pi^2 n^2 r^2}{l} 10^{-7}$$

where n is number of turns, r is the radius of the coil and l is the length of the coil.

The mutual inductance M can be represented by two coupled coils – for example coil presented in Fig. 2.41, but wound with two wires. Due to finite value of the resistance it is practically not possible to obtain the ideal inductance standard. Also, it is not possible to eliminate the self-capacitance of the coil. Therefore, the inductance of the standard is usually described for defined value of frequency, most often for *1 kHz*.

It is relative easy to obtain *standard value of frequency* (moreover, the frequency is the quantity that we are able to measure with very small uncertainty). As the standard of frequency we can use the *quartz oscillator*. Appropriately prepared quartz crystal can exhibit the stability of resonance frequency better than $10^{-8}/year$. Unfortunately this frequency depends on the temperature; therefore it is necessary to use the thermostat or special temperature corrections. It is very convenient that the standard frequency can be transmitted by the radio. The frequency standard, model 910R of Fluke controlled by the satellite signal (cesium atomic standard in the GPS system) enables users to obtain the frequency with the stability better than $10^{-12}/24$ *hours*.

2.3.4. The reference multimeters and calibrators

There are available digital measuring instruments with extremely small uncertainty. Thus, such instruments can be used as the working standards. For example *reference multimeter* model 2002 of Fluke enables to measure the DC voltage with uncertainty *(0.5+0.2)ppm*, direct current with uncertainty *(6+2)ppm*, AC voltage with uncertainty *(30+1)ppm*, AC current with uncertainty *(200+100)ppm* and the resistance with uncertainty *(1+0.25)ppm*.

As the real working standards the measuring instruments called *calibrators* can be used. Such instruments can deliver the standard signals or values enabling to scale other measuring devices. As an example we can consider the *High Performance Multi-Product Calibrator* model 5520 of Fluke presented in Fig. 2.42.

The performances of Fluke calibrator are presented in Table 2.3. This calibrator consists of two independent standard sources of *DC* and *AC* voltages or currents with controlled frequency and phase between them (for calibration of power meters and energy meters). Additionally calibrator can work as the standard resistance, inductance, capacitance and temperature (for modelling of thermoresistors or thermocouples). Thus this calibrator can deliver the standard values formerly available only by the high quality standards.

Table 2.3. The performances of calibrator 5520 model of Fluke

Functions	Ranges	Uncertainty (95% 1 year)
DC voltage	$0 \div \pm 1020$ V	12 ppm
AC voltage, 10 Hz ÷ 500 kHz	1mV ÷ 1020 V	120 ppm
DC current	$0 \div \pm 20,5$ A	100 ppm
AC current, 10 Hz ÷ 30 kHz	29 mA ÷ 20,5 A	600 ppm
Resistance	$0 \div 1100$ MΩ	28 ppm
Capacitance	0,19 nF ÷ 110 mF	0.25%
Phase between two AC signals	$0 \div \pm 179,99°$	±0.07°
Frequency	0,01 Hz ÷ 2 MHz	25 ppm
DC power	10,9µW ÷ 20,5 kW	0.023%
AC power	10,9µW÷ 20,5 kW	0.08%
Temperature (thermocouple)	-250°C ÷ 2316°C	0.14°C
Temperature (thermoresistor)	-200°C ÷ 630°C	0.03°C

Figure 2.42. Calibrator model 5520 of Fluke (Fluke 2005) (permission of Fluke Corporation)

REFERENCES

AD 2005 Analog Devices Inc., www.analog.com

Agilent 2005 Agilent Technologies, www.agilent.com

Dieck R.H. 2002 *Measurement Uncertainty: Methods and Applications*, ISA

EN 60051 1989 *Direct acting indicating analogue electrical measuring instruments and their accessories. Specification for special requirements for ammeters and voltmeters,* European Standard EN 60051

EN 60404 1998 *Magnetic Materials – Part.2 Method of Measuring of Magnetic Properties of Electrical Sheet and Strip by Means of an Epstein Frame*, European Standard EN 60404

Fluke 2005 Fluke Corporation, www.fluke.com

Ibuka M., Naito S., Furuya T. 1983 *Point-contact Josephson Voltage Standard*, IEEE Trans. Instr. Meas., 32, p. 276

IET 2005 IET Labs. Inc., www.ietlabs.com

ISO VIM 2004 *International Vocabulary of Basic and General Terms in Metrology*, International Organization for Standardization ISO (2003, revised edition 2004)

ISO Guide 1993 *Guide to the expression of uncertainty in measurement*, ISO

Lira I. 2002 *Evaluation of the Measurement Uncertainty: Fundamentals and Practical Guidance*, IOP Publ.

NIST 1458 2003 *NIST Measurement Service for DC Standard Resistors*, NIST Technical note 1458

NIST TF 2005 NIST – Time Frequency Division, www.tf.nist.gov

NVE 2005 Nonvolatile Electronics Inc. www.nve.com

Rabinowich S.G. 1999 *Measurement Errors and Uncertainty: Theory and Practice*, Springer Verlag

Taylor J.R. 1996 *An Introduction to Error Analysis: The study of Uncertainty in Physical Measurements*, University Science Books

Tinsley 2005 Tinsley Precise Instruments, www.tinsley.co.uk

Tumanski S. Stabrowski M. 1998 T*he Magnetovision Method as a Tool to Investigate the Quality of Electrical Steel*, Meas. Sci. Technol., 9 , pp. 488-495

3

Classic Electrical Measurements

3.1. INDICATING MEASURING INSTRUMENTS

3.1.1 Electromechanical instruments versus digital measuring systems

There are many commonly used measuring instruments, which we could call as traditional, classic or even old fashioned. The electromechanical indicating instruments, cheap manually balanced bridge instruments or induction type watt-hour meters are still present everywhere.

There are several advantages of traditional electromechanical instruments: simplicity, reliability, low price. The most important advantage is that the majority of such instruments can work without any additional power supply. Since people's eyes are sensitive to movement also this psycho-physiological aspect of analogue indicating instruments (with moving pointer) is appreciated.

On the other hand, there are several drawbacks associated with electromechanical analogue indicating instruments. First of all, they do not provide electrical output signal, thus there is a need for operator's activity during the measurement (at least for the reading of an indicated value). Another drawback is that such instruments generally use moving mechanical parts, which are sensitive to shocks, aging or wearing out. Relatively low price of moving pointer instruments today is not as advantageous as earlier, because on the market there are available also very cheap digital measuring devices with virtual pointer.

Regrettably, it can be stated that most of the electromechanical analogue instruments are rather of poor quality. In most cases these instruments are not able to measure with uncertainty better than 0.5%. The accuracy is also affected by so-called parallax error, in which the reading result depends on the position of the user's eye. The measurement is often invasive, because such mechanisms may need relatively large power consumption to cause the movement. Thus, electromechanical voltmeters exhibit insufficiently large resistance, while the resistance of electromechanical ammeters is not sufficiently small.

There is no doubt that the future is for automatic, computer supported measuring systems. But electromechanical instruments are still present in our lives (for example the attempts to substitute such instruments in cars finished with not a success). Moreover without understanding of the principles of old analogue measuring methods it can be difficult to understand usually more complicated digital instruments which often use traditional principles of operation. This chapter is called "*Classic Electrical Measurements*" as analogy to the classical music – old classical music is still appreciated and present but it is rather closed era of the history of music. Similarly, classic measurements are still present in our life, and the progress is observed mostly in more modern measuring techniques.

3.1.2 The moving coil meters

The *moving coil* instrument is the most popular indicating electromechanical device. An example of such an instrument is presented in Fig. 3.1.

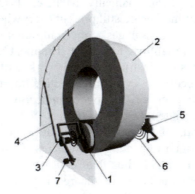

Figure 3.1. The example of moving coil indicating instrument (1- moving coil, 2 – permanent magnet, 3 – axle, 4 – pointer, 5 – bearings, 6 – spring, 7 – correction of zero)

A rectangular coil with the pointer fixed to its axle is used as the moving part in such instruments. The conic ends of axles are pressed against the bearings. The current is delivered to the coil by two springs – these springs are also used as the mechanisms generating returning torque for the pointer.

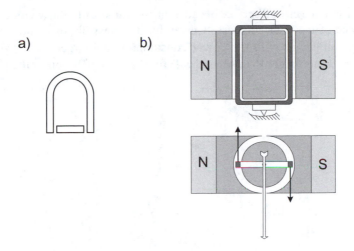

Figure 3.2. The moving coil mechanism: a) the symbol of such instrument, b) the principle of operation

The moving coil is placed into the gap between the magnet poles and soft iron core, shaped in such a way as to produce uniform magnetic field. The movement of the coil is caused by the interaction between the magnetic field of the magnet and the magnetic field generated by the coil. The rotation of the coil (and the pointer attached to it) is due to the torque M, which depends on the flux density B of the magnet, on dimensions d and l of the coil, on number of turns z of the coil and of course on the measured current I:

$$M = Bzdl \cdot I \qquad (3.1)$$

The position of the moving element (the angle of rotation, α) results from the balance between the torque and the returning torque of the springs $M_z = k\,\alpha$ (k is the constant of the elasticity of the spring). Thus from the condition $M = M_z$ we find that the amount of rotation is

$$\alpha = \frac{Bzdl}{k} \cdot I = c \cdot I \qquad (3.2)$$

The angle of rotation is proportional to the measured current I, which is advantageous, because it means that the scale is linear. The larger is the constant c in Equation (3.2) the more sensitive (thus better) is the measuring device, because less current is required to cause the movement of the coil. The best way to improve the sensitivity of the device is to use large magnetic flux density B. The increasing of the number of turns or the dimensions of the coil is not very effective, because at the same time the weight and the resistance of the coils increase. Currently, it is possible to manufacture the moving coil device with the power consumption not larger than several μW (and current not larger than several μA) for the full deflection of the pointer.

Figure 3.3. The movement of the pointer after connection of the device to the measured current

The elasticity of the spring plays important role because it influences the character of the pointer movement. It is convenient if this movement is with small oscillation (see Fig. 2.24b). If it would be purely inertial without an overshoot then the observer would not be sure when the pointer reaches final position. It is important to obtain the oscillatory movement with a short period and with reasonable damping of oscillation. Ideally, only one oscillation period should be visible – the next one should be damped.

The parameters of the movement depends on the mass m of the moving part and on the elasticity coefficient, k

$$T = \frac{T_0}{\sqrt{1-b^2}}, \qquad T_0 = 2\pi\sqrt{\frac{m}{k}}, \qquad b = \frac{P}{2\sqrt{mk}} \qquad (3.3)$$

where T is the time constant, T_0 is the period of oscillations of the moving element, b is the degree of damping and P is the damping coefficient.

Thus the character of the movement depends on the ratio between the mass, the elasticity of the springs and the damping. Special air damper can be used in order to obtain correct damping of the movement. In the case of a moving coil device the aluminum frame of the coil can work as the damper – the eddy currents induced in this frame interact with the magnetic field of the magnet slowing down the velocity of the movements.

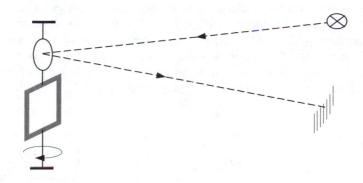

Figure 3.4. The principle of the light indicator used in very sensitive measuring devices

A microammeter is a typical moving coil device. There is also a special type of moving coil device called a galvanometer. The *galvanometers* are designed in such a way, as to obtain extremely large sensitivity. The moving element is bearingless – the coil is suspended on ribbons. This ribbons act as the current supplying wires and also as the springing parts. The coil is wound without a frame (to reduce the moving mass), which causes that the element can move with significant oscillations. The damping of the movement depends on the current induced in the coil, hence on the resistance of the external circuit. For that reason the manufacturers often give information about recommended resistance of the external circuit – the *critical resistance*, with which the movement of the pointer is optimal. To obtain large sensitivity of the galvanometer the pointer is enlarged – it is often substituted by the light indicator, the luminous beam reflecting from the mirror fixed to the coil (Fig. 3.4).

Another special kind of galvanometer is a device with a large moment of inertia – called *ballistic galvanometer*. Due to this feature the ballistic galvanometer measures quantity of electrical charge rather than current and it can be used for the measurement of the magnetic flux density. There is also another special type of galvanometer – *fluxmeter*, in which the moving coil is suspended without the returning torque. Due to such feature the galvanometer acts as integrating device and was used for measurements of magnetic flux[1]. Currently, both these types of galvanometers are substituted by electronic integrating instruments.

The moving coil device can be used as the microammeter without any additional elements (Fig. 3.5a). If in series with the moving coil device an additional resistor R_d (*series resistor*) is connected, then we obtain the

---

[1] Integration is necessary because induced voltage is $U \approx dB/dt$

millivoltmeter or voltmeter (Fig.3.5b) (because the current I in the device is directly proportional to the voltage U). When the millivoltmeter is connected in parallel with another resistor R_b, called a *shunt resistor* we obtain the ammeter (Fig.3.5c), because voltage U_b is proportional to the measured current I_x (the resistance of millivoltmeter is much larger than resistance of the shunt resistor R_b thus we can assume that $U_b \cong I_x R_b$).

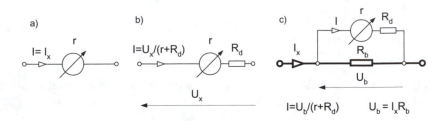

Figure 3.5. The concept of microammeter (a), voltmeter (b) and ammeter (c)

The temperature influences the flux density B of the permanent magnet and the elasticity of the springs k. Fortunately, both of these influences result in opposite changes of the α (see Eq. 3.2). Therefore, their influences are negligible when the device is used as the microammeter (Fig. 3.5a).

The case of the millivoltmeter (Fig. 3.5b), and also indirectly of ammeter (Fig. 3.5c), is more complicated. The change of temperature causes change or the resistance r of the coil (the changes of resistance of the other resistors R_d and R_b are negligible, because they are prepared from *manganin* – special temperature independent alloy). Thus the current I, in the device changes with the temperature for fixed value of the measured voltage U, according to relation $I = U / (r + R_d)$. This change is significant, because copper wire of the coil exhibits change of the resistance of about 4%/10 °C. The temperature error of the millivoltmeter circuit presented in Fig. 3.5b we can describe as follows:

$$\delta_T = \frac{\dfrac{U}{r+R_d} - \dfrac{U}{r+\Delta r + R_d}}{\dfrac{U}{r+R_d}} \cong \frac{1}{1+\dfrac{R_d}{r}} \frac{\Delta r}{r} = \frac{4\%}{1+\dfrac{R_d}{r}} \tag{3.4}$$

Thus the error caused by the change in temperature depends on the ratio R_d /r. It is easy to calculate that if the millivoltmeter is designed for measurements with uncertainty better than 0.5% then it is necessary to use the resistors with values $R_d = 7 \cdot r$. This means deterioration of the sensitivity of the millivoltmeter. Let us consider a case of a moving coil device with resistance *10 Ω* and nominal current *1 mA*. Theoretically, such device could be used to design a millivoltmeter with a minimal range $U_{nom}=Ir = 10\ mV$. But if we are planning to design a millivoltmeter of the class of accuracy 0.5% it is necessary to use additional resistance $R_d = 70\ \Omega$, which limits the minimal range of such millivoltmeter to *80 mV*. For the voltmeters, the problem of temperature errors correction is usually easy to solve, because it is necessary to use the series resistor. For example, in order to design a *10 V* range voltmeter with a device described above it is necessary to connect a resistor of about *10 kΩ* much larger than is required for the temperature error correction.

The ammeter instrument can be designed similarly to the millivoltmeter – by measuring voltage drop on the shunt resistor R_b (Fig. 3.5c). For example, if we use the moving coil device with the parameters described above and we would like to design an ammeter with a range *1 A* and the accuracy class *0.5%* then it is necessary to use a shunt resistor which would result in voltage drop larger than *80 mV* (thus $R_b = 80\ m\Omega$). It is possible to design measuring instruments of better performance with more sensitive devices. For example, for non-invasive current measurements it is necessary to design the ammeter with the resistance as small as possible. If the device exhibits the nominal current ten times smaller (in our case *100 μA* instead of *1 mA*) then the shunt resistor (and approximately the resistance of the whole ammeter) can be also ten times smaller – in our case *8 mΩ*).

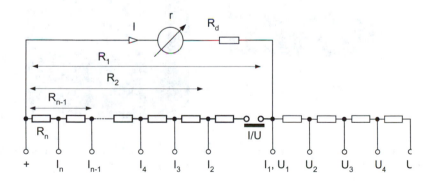

Figure 3.6. The design of universal multi-range voltammeter

Fig. 3.6 presents the design of universal ammeter and voltmeter (voltammeter) with selectable ranges. To obtain the multi-range ammeter the special design of *universal shunt resistor* is very useful. The universal shunt resistor is designed to obtain the same current I for various input currents. Thus it should be

$$\begin{cases} (I_n - I)R_n = I[(r + R_d) + (R_1 - R_n)] \\ (I_{n-1} - I)R_{n-1} = I[(r + R_d) + (R_1 - R_{n-1})] \end{cases} \tag{3.5}$$

After simple calculations we obtain the condition of universal shunt resistor in form

$$\frac{I_n}{I_1} = \frac{R_1}{R_n} \tag{3.6}$$

The moving coils measuring instruments are usually manufactured as the panel meters (with class of uncertainty typically 1, 1.5 or 2.5%) and as the laboratory meters with class of uncertainty typically 0.5%. Fig. 3.7 presents examples of analogue indicating meters.

Figure 3.7. The examples of analogue meters in panel and laboratory design (meters of EraGost) (Eragost 2005) (permission of EraGost)

The main disadvantage of the moving coil meters is that they indicate only *DC* values of the signals. In the past, these devices were also used for measurements of *AC* values with the aid of rectifiers. Although such devices measure the average value it is possible to scale it in *rms* values, knowing that $X_{rms}/X_{AV} = 1.11$. But this dependence is valid only for pure sinusoidal signals. Thus the rectifying AC measuring devices can be used only for the measurements of poor accuracy.

3.1.3 The moving iron meters

The main advantage of moving iron meter is that such instrument measures the *rms* value of the signal. Therefore, it can be used for *AC* measurements. The design of moving iron meter is presented in Fig. 3.8.

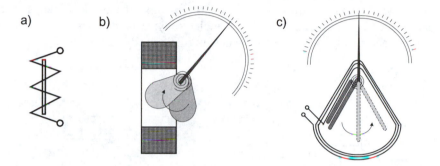

Figure 3.8. The moving iron meter: a) the symbol of such instrument, bc) the principle of operation

The measured current is connected to the stationary coil and the magnetic field generated by this coil interacts with the moving iron element. The iron vane changes its location as to increase the magnetic flux flowing through it. In the device presented in Fig. 3.8b the iron element is attracted into the coil. In the device presented in Fig. 3.8c the movable iron element is repelled from the additional stationary iron element. The angular deflection α depends on the measured current I and the change of the inductance dL caused by this deflection:

$$\alpha = \frac{1}{2k}\frac{dL}{d\alpha}I^2 \qquad (3.7)$$

Although the deflection is a nonlinear function of measured current it is possible to design the device (the component $dL/d\alpha$) in such a way that the expression $(dL/d\alpha)\ I^2$ is close to linear. Because the response of the device depends on the squared value of the current it is possible to obtain the meter of *rms* value. Due to the error caused by magnetic hysteresis (when *DC* current is measured) these devices are used almost exclusively for *AC* measurements.

The moving iron meter exhibits several advantages: simplicity of the design – no need to supply the moving element, easy change of the range by selecting the number of the turns in the coil. The drawbacks of moving iron devices are relatively large power consumption *(0.1 – 1VA)* and small sensitivity (in comparison with moving coil device). The smallest obtainable range of moving iron milliammeter is several *mA*. Also, the frequency bandwidth is limited to about *150 Hz*.

3.1.4. Electrodynamic meters – wattmeters

The electrodynamic meters were formerly the most accurate indicating instrument. Today for accurate measurements these instruments are substituted by the digital devices. The electrodynamic devices are still used as wattmeters. Fig. 3.9 presents the design of electrodynamic device.

a)   b)

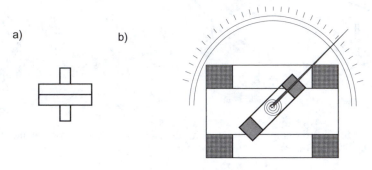

Figure 3.9. The electrodynamic meter: a) the symbol of such instrument, b) the principle of operation

The electrodynamic device design is based on two coils: a stationary and a moving one. The currents flowing through these coils induce a force, which causes rotation of the movable coil. Thus the concept of an electrodynamic device is very close to the definition of ampere – it acts due to the force between two wires (see Fig. 2.29). The torque M resulting from the interaction between two coils depends on currents: I_1 in stationary coil, I_2 in movable one and the phase shift φ between these currents:

$$M = cI_1I_2\cos\varphi \qquad (3.8)$$

Thus if one coil is connected to the current and the second to the voltage we can directly measure the power, because $P = UI\cos\varphi$. Fig. 3.10 presents typical connection of electrodynamic meter as the wattmeter. The wattmeter has two pairs of terminals – the current and the voltage terminals. In the voltage circuit there is usually introduced a series resistor R_d. Thus the torque can be calculated from the following equation:

$$M = c\frac{1}{R_d}IU\cos\varphi = k \cdot P \qquad (3.9)$$

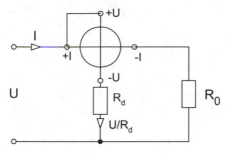

Figure 3.10. The connection of the wattmeter for the measurement of electric power

The wattmeters are often used for measurements in three-phase systems. Two examples of such connections for the system with and without neutral wire are presented in Fig. 3.11. The three-phase power meters can be used as three separate meters or one meter with three electrodynamic devices with a common axle.

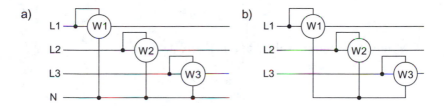

Figure 3.11. The connection of wattmeters in three-phase system with neutral wire (a) and without neutral wire (b)

Since each wattmeter measures the power of one of three phases the sum of indications of three wattmeters is the total power of a three-phase system:

$$P = P_1 + P_2 + P_3 \tag{3.10}$$

If the load is symmetrical the value of three phase currents is the same. In such case it sufficient to connect one wattmeter and the total power is $P=3P_1$. In a three-phase system without the neutral wire (three-wire system) it is possible to use the measuring system consisting of two wattmeters (so-called the *Aron method*). Fig. 3.12 presents the connection of wattmeters in the Aron method.

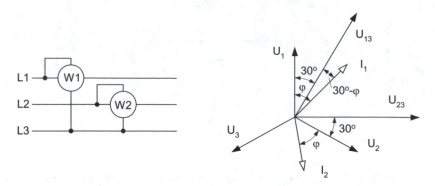

Figure 3.12. Power measurement in three-wire, three-phase system: the circuit and corresponding phasor diagram for symmetrical load

It can be simply proved that total power is measured correctly by means of the Aron method. The total instantaneous value of power in three-phase system is

$$p = u_1 i_1 + u_2 i_2 + u_3 i_3 \tag{3.11}$$

The sum of the current values in three-wire network is

$$i_1 + i_2 + i_3 = 0 \tag{3.12}$$

thus

$$i_3 = -(i_1 + i_2) \tag{3.13}$$

By substituting i_3 value (Eq. 3.13) in the equation (3.11) we obtain

$$p = (u_1 - u_3)i_i + (u_2 - u_3)i_2 = u_{13}i_1 + u_{23}i_2 \tag{3.14}$$

Thus the power indicated by the wattmeters is

$$P = \frac{1}{T}\int_0^T pdt = U_{13}I_1 \cos(U_{13}I_1) + U_{23}I_2 \cos(U_{23}I_2) = P_1 + P_2 \tag{3.15}$$

The sum of powers indicated by two wattmeters is equal to the total power in the system. This dependence is universal and holds for symmetrical or non-symmetrical circuits. The only limitation is that the circuit should be three-wire, i.e. without the neutral wire (condition 3.12).

For the symmetrical load (balanced circuit) according to the phasor diagram presented in Fig. 3.12 the wattmeters indicated values are

$$P_1 = UI\cos\left(30° - \varphi\right)$$

$$P_2 = UI\cos\left(30° + \varphi\right)$$

(3.16)

where U is phase-to-phase voltage and I is the phase current.

It is easy to prove that using the Aron circuit for balanced system we can determine the reactive power Q and the phase angle φ because

$$tg\varphi = \sqrt{3}\frac{P_1 - P_2}{P_1 + P_2}; \qquad Q = \sqrt{3}\left(P_1 - P_2\right)$$

(3.17)

We can also measure the reactive power by appropriate connection of the wattmeters into the three-phase circuit, as shown in Fig. 3.13.

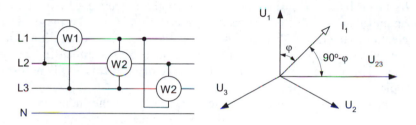

Figure 3.13. Reactive power measurement in three-phase system: the circuit and corresponding phasor diagram

For balanced systems the indication of the first wattmeter can be calculated from the following dependence

$$P_1 = I_1 U_{23}\cos\left(90° - \varphi\right) = \sqrt{3}I_1 U_1 \sin\varphi = \sqrt{3}Q_1$$

(3.18)

For three-phase three-wattmeter system the total reactive power can be determined as

$$Q = \frac{P_1 + P_2 + P_3}{\sqrt{3}}$$

(3.19)

For the reactive power measurement we can also use the two wattmeter method (Aron method for reactive power). The Aron circuit has several drawbacks, the primary one being that it requires the fulfillment the condition (3.12). Every short circuiting to ground or leakage current causes the incorrect measurement. But in the power electrical systems two wattmeters measuring systems are often used for economical reasons. The wattmeters are relatively cheap but the measuring transformers necessary to isolate and reduce the currents and voltages are often relatively expensive.

The electrodynamic meters can be used also for current and voltage measurement (in such cases both coils are connected in series). But the main drawback of electrodynamic devices is large power consumption (several *VA*). Therefore nowadays they are practically used only as wattmeters, especially for three phase power systems where several *VA* power consumption is negligible.

3.1.5. Induction type watt-hour meters

The induction *watt-hour meters* (energy meters) are still present in our houses, although they exhibit serious drawbacks. First of all the reading must be taken by a person in order to account the energy used (i.e. there is no output signal which could be read automatically). Moreover, these meters are electromechanical with quite complex system of error correction. Thus, in the future the mechanical energy meters will be substituted by electronic ones. This process is slow due to the range of problems – it is necessary to replace millions of devices.

Fig. 3.14 presents the principle of operation of the induction watt-hour meter (*Ferrari's system*). Two independent cores are supplied by the currents proportional to the current and the voltage. These two cores generate magnetic fluxes ϕ_i and ϕ_u, which flow through a rotating aluminum disc, in which eddy currents are induced.

The rotating torque M_r is due to the interaction between the eddy currents and the fluxes. The torque depends on the values of the currents in the cores and the phase angles between them

$$M_r = c\omega I_1 I_2 \sin(I_1, I_2) \tag{3.20}$$

The first current is proportional to the measured current $I_1 = I$, while the second current is proportional to the voltage $I_2 = kU$. Due to large inductivity of the voltage core the current I_2 is shifted in phase by almost 90° with respect to the supplied voltage and $sin(I_1, I_2) \cong cos(U, I) = cos\varphi$. Thus the torque is dependent on the measured power

$$M_r \cong ck\omega \cdot IU \cos\varphi \tag{3.21}$$

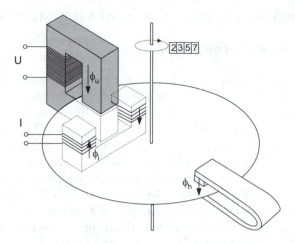

Figure 3.14. The principle of operation of the induction watt-hour meter

Additionally, the induction meter is equipped with the braking magnet. Interaction between the magnetic field of the permanent magnet and the eddy currents induced by this field causes a braking torque proportional to the angular speed of the disk. Under the influence of both torques the watt-hour meter acts as the asynchronous motor with the speed of the disk proportional to the power supplied to the load. As a result, the number of revolutions n in the time period t (angular speed) is the measure of power

$$\frac{n}{t} \cong KUI \cos \varphi \qquad (3.22)$$

The mechanical register counts the number of revolutions and hence indicates the energy consumed by the load.

The principle of operation described above is significantly simplified. In the real instruments the phase shift in voltage coil is not exactly 90° thus additional phase correction winding is necessary. The braking torque is caused not only by the magnet, but also by the two cores and additional magnetic shunt is necessary for correction of this effect. Also additional correction is necessary to compensate for the effect of friction in the aluminum disc bearings. The total error of the induction meter is various for various measured power and it is described by the error characteristic. All corrections should be precisely set to ensure that the characteristic of errors does not exceed required limits. The main weakness of the induction watt-hour meters is that these corrections, hence generally the performance of the meter, changes with the aging process resulting in the risk that consumer or energy distributor are deceived.

3.2. RECORDING AND DISPLAYING MEASURING INSTRUMENTS

3.2.1. Fundamentals of oscilloscopes

The first oscilloscopes were introduced more than hundred years ago by Ferdinand Braun[1] in 1897. And until now, these instruments are one of the most important tools in engineering and scientific research. Modern digital oscilloscopes allow not only displaying the signals, but also recording, analyzing them (for example spectral analysis) and even performing measurements of the signals (Kularatna 2003). On the market, there are still available analogue *cathode ray tube CRT* oscilloscopes (also called *cathode ray oscilloscopes CRO*) (Sherman 2004) as well as *LCD analogue or digital devices*. The prices of analogue and digital types of oscilloscopes are at present at the same level, thus most likely in the near future only more versatile digital instruments will remain on the market.

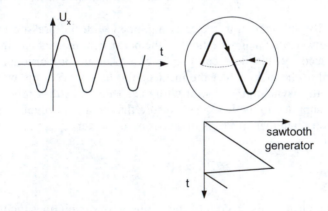

Figure 3.15. The principle of displaying of the signal in the oscilloscope

Fig. 3.15 illustrates the principle of displaying the signal in the oscilloscope. The picture on the screen is obtained in such a way that during the horizontal movement of luminous point it is deflected vertically proportionally to the value of the detected signal. The horizontal movement is obtained by the sweep oscillator generating a sawtooth signal (Fig.3.15). If the frequency of the signal is larger than several Hz due to the inertia of our eyesight it is not possible to see such a picture of the signal. Therefore the main function of the oscilloscope is to somehow stop the picture on the screen. If the period of oscillation of the sawtooth signal is the same (or

---

[1] Braun was awarded the Nobel Prize in Physics in 1009 with Marconi for wireless telegraphy.

multiple) as the period of investigated signal the successive pictures appear to be the same. This creates the illusion that the picture is standing still. We say that the investigated signal is *synchronized* with the sweep signal.

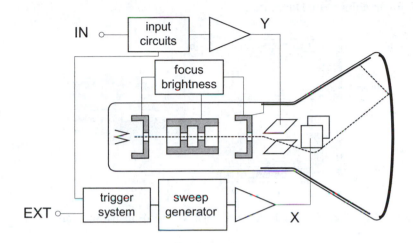

Figure 3.16. The block diagram of a classical oscilloscope

The block diagram of a typical CRT oscilloscope is presented in Fig. 3.16. The electron beam from the cathode is passing through the acceleration anode and focusing area. This beam strikes the phosphor coating on the screen causing the light. Depending on the composition of the phosphor coating it is possible to obtain various colors of the light (green, blue or yellowish green) and time of the lighting. Typically the light decay occurs after about *40µs*, but there are materials with prolonged time of lighting even to several ms. Usually the system of deflection of the electron beam consists of two pairs of plates: horizontal and vertical ones. The electrostatic force depending on the voltage causes the deflection of the electron beam and the amount of deflection is directly proportional to the voltage applied to the plates.

Recently, the cathode ray tube is often substituted by the *liquid crystal display LCD* (Sarma 2004), especially in hand held oscilloscopes. Liquid crystal material consists of rod shaped molecules that can be ordered by electrostatic field. The display is divided in small areas called pixels, each of which is driven by an individual electrode.

Fig. 3.17 presents the design of typical LCD, in which another method of movement of lighted pixels is used than in the case of CRT. Pixels are organized in the matrix with scanning/row electrodes for horizontal movement control and data/column electrodes for vertical movement control.

In the simplest case these electrodes are driven in passive way, but to obtain better parameters the *Active Matrix Addressing AMLCD* is used. In the active system each pixel is driven by thin film transistor *TFT* connected to two capacitors for voltage storage. If a system of filter is added, then it is possible to obtain colored LCD monitor.

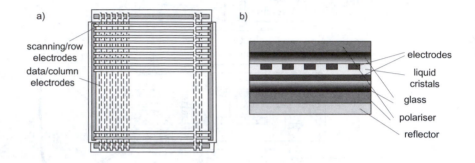

Figure 3.17. Planar (a) and cross-sectional view (b) of matrix addressed in LCD (Sarma 2004)

Fig. 3.18 presents the principle of synchronization of the investigated signal. The sawtooth voltage is initiated by the pulse from triggering system. In the simplest case of the automatic trigger mode this start can occur for the zero value of signal – it is important to start every time in precisely defined point of signal, because only in such case we can obtain still picture on the screen.

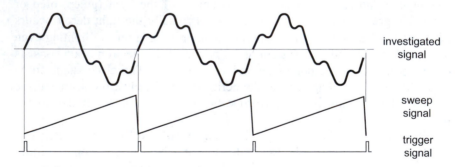

Figure 3.18. The automatic mode of synchronization of investigated signal

Of course it is possible to set manually the moment of triggering – for example for defined value or for rising or falling slope. This moment of start can be observed directly on the screen as the very initial point of the signal.

During the falling slope of the sawtooth signal the luminous spot is being blanked and comes back to the initial point on the screen. Of course the user can manually change the frequency of time base in order to observe one or more periods of investigated signal.

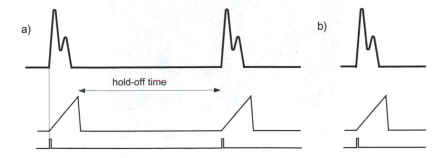

Figure 3.19. The triggering mode of synchronization of investigated signal

More complicated is the case when investigated signal is periodical but with short pulse time in comparison with the period time of the whole signal (Fig. 3.19a) or when it is a single pulse (Fig. 3.19b). In such case it is better to use the trigger mode of synchronization. In this mode the trigger circuit is connected to the source of trigger signal – this can be either internal or external source. The sweep generator controlled by trigger system starts exactly in defined point, often in initial point of the investigated signal. It is possible to set additionally the hold-off time when the trigger system does not start again. The triggering mode can be periodical (for example in the case presented in Fig. 3.19a) or a single sweep mode (for example in the case presented in Fig. 3.19b).

In the single sweep model the sawtooth signal is composed from only one tooth – this means that the luminous spot is traveling through the screen horizontally only once. This mode is very convenient to observe non-periodical signals, for example single pulses.

Modern oscilloscopes, also analogue ones, are not just tools designed for observation the signals. The switching devices for setting *time base* and *vertical deflection* are scaled in *sec/div* and *volts/div*, respectively. Because on the screen these divisions are clearly indicated as the vertical and horizontal scales (Fig. 3.20) it is relatively easy to determine the value of the signal (for example the value of magnitude) or value of the time (for example to determine the phase shift between two signals or frequency of the signal). Thus the oscilloscope is a complete measuring device, especially valuable in the high frequency range.

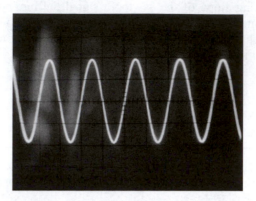

Figure 3.20. The example of the screen of analogue oscilloscope

In typical oscilloscopes, it is possible to use the dual channel mode, although the display uses only a single beam of electrons. The displaying of two signals is realized by using the *alternate* or *chopper* mode as it is illustrated in Fig. 3.21. In the alternate mode both signals are displayed on the screen alternately – due to the inertia of human eyesight and due to the persistence of lighting the illusion that both signals are shown at the same time arises.

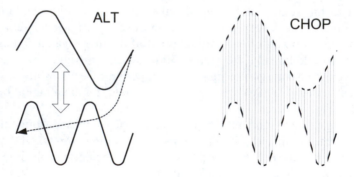

Figure 3.21. The dual channel operations of single beam oscilloscope.

The alternate mode cannot be used for the low frequency signals because it can result in blinking of the picture. For low frequency signals the chopper mode should be used. In such case both signals are chopped – we see alternately small part of each signal. If the chopping rate is much higher than the frequency of the signals, then the observer sees both signals as continuous lines on the screen.

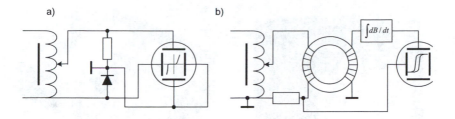

Figure 3.22. The X-Y model of operation of oscilloscope: a) for determination of diode characteristic, b) for determination of hysteresis loop of soft magnetic material

Usually the oscilloscope displays the time varying signals, thus the horizontal axis is the time axis. But it is also possible to connect another signal to the horizontal deflection system. In such mode of operation we can display on the screen the function $Y=f(X)$. Fig. 3.22 presents two examples of X-Y operation. In the first example, the signal proportional to the current (voltage drop on the resistor) is applied to the vertical plates while the signal proportional to the voltage on the diode is connected to the horizontal plates. Thus we obtain the characteristic $I=f(U)$ of the diode on the screen of oscilloscope.

The second example (Fig.3.22b) illustrates the possibility of investigating the hysteresis loop (*B-H* loop). To the horizontal plates the signal proportional to the magnetic field strength (that is proportional to the magnetizing current) is connected while to the vertical pair of the plates the signal proportional to the flux density (the voltage E induced in the winding after integration, because $E = f(dB/dt)$) is used. Thus we can observe the $B=f(H)$ loop on the screen.

3.2.2. Recorders and data storage devices

Today practically all digital oscilloscopes enable storage and recall the investigated signal. Also most measuring instruments are equipped with the communication interface, through which is possible to save the signals and data in computer systems as well. For that reason electromechanical recording devices very popular in the past are currently going out of use. They are substituted by the devices saving data to transportable memory – data loggers, digital recorders and also computer measuring systems. The set of data saved for example on the hard disc of a computer can be easily converted later on to the graphical form of signal and print using high quality laser printer.

Figure 3.23. Two examples of the recorders: x(t) recorder of Lumel company (Lumel 2005) and X/Y recorder of Graphtec Company (Graphtec 2005) (permissions of Lumel and Graphtec Company)

Fig.3.23 presents two examples of recorders – for time varying signals and for $Y=f(X)$ signals. The recorder of time varying signals is recording them in a form of an ink trace on the movable paper. But also in this kind of operation several novelties appeared. For example, in newer model of the recorder presented in Fig. 3.23 the paper ribbon is substituted by the LCD screen and magnetic memory.

One of the features of mechanical chart recorders is very important. This kind of monitoring and recording creates the "hard copy" in the form of a paper document. In some circumstances (ISO 9000 quality assurance procedures, ISO 14000 environmental protection systems, military and health systems, inspection of car drivers) it is required keep the results of register for a long period of time (even up to several years). For such purposes ordinary paper data is the most non-corruptible.

3.3. BRIDGE CIRCUITS

3.3.1. Balanced and unbalanced bridge circuits

The bridge circuits were used as the most accurate devices for the measurements of resistance (and generally impedance). Nowadays, the bridge circuits are not as important as they used to be, because new, more effective methods of impedance measurement are developed (Agilent 2003). The bridge circuits are commonly used as the resistance (impedance) to voltage converters and the bridge principle is utilized in digital RLC meters.

Two main bridge circuits: supplied by the voltage source or the current source are presented in Fig. 3.24.

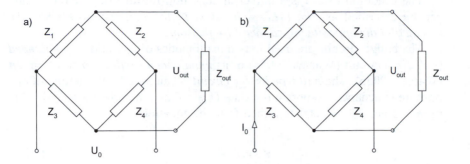

Figure 3.24. The bridge circuit: a) supplied by the voltage source, b) supplied by the current source

For the bridge circuits presented in Fig. 3.24 the dependence of the output voltage U_{out} on the circuit parameters are as follows:

$$U_{out} = \frac{Z_1 Z_4 - Z_2 Z_3}{(Z_1 + Z_2)(Z_3 + Z_4) + \dfrac{Z_1 Z_2}{Z_{out}}(Z_3 + Z_4) + \dfrac{Z_3 Z_4}{Z_{out}}(Z_1 + Z_2)} U_0 \qquad (3.23)$$

$$U_{out} = \frac{Z_1 Z_4 - Z_2 Z_3}{(Z_1 + Z_2 + Z_3 + Z_4) + \dfrac{(Z_1 + Z_3)(Z_2 + Z_4)}{Z_{out}}} I_0 \qquad (3.24)$$

When the load impedance is infinitely large (no-load state) we can write the dependences (3.23) and (3.24) as

$$U_{wy} = \frac{Z_1 Z_4 - Z_2 Z_3}{(Z_1 + Z_2)(Z_3 + Z_4)} U_0 \qquad (3.25)$$

$$U_{wy} = \frac{Z_1 Z_4 - Z_2 Z_3}{Z_1 + Z_2 + Z_3 + Z_4} I_0 \qquad (3.26)$$

Thus the condition of the balance $U_{out} = 0$ of the bridge circuit is

$$Z_1 Z_4 = Z_2 Z_3 \quad \text{or} \quad Z_1 Z_4 - Z_2 Z_3 = 0 \qquad (3.27)$$

The condition (3.27) is a universal condition for all bridge circuits, and can be described as: *the bridge circuit is in the balance state when the products of the opposite impedances are the same.*

The bridge circuits are used in two main modes of operation: as *balanced* (*null type*) circuit (Warsza 2005a) or as *unbalanced* (*deflection type*) circuit (Warsza 2005b). The *null type bridge circuit* is balanced by the setting of one or more impedances to obtain the state $U_{out} = 0$ and then the measured value of impedance $Z_x = Z_1$ is determined from the equation

$$Z_x = Z_2 \frac{Z_3}{Z_4} \tag{3.28}$$

In the deflection type of bridge circuit we first balance the bridge circuit and then we can determine the change of resistance from the output signal as

$$U_{out} \cong S \frac{\Delta Z_x}{Z_x} U_0 \tag{3.29}$$

Thus the unbalanced bridge circuit operates as the transducer of the change of impedance to the voltage (S is the sensitivity coefficient of the unbalanced bridge circuit).

3.3.2. Null type DC bridge circuits

The Wheatstone bridge circuit presented in Fig. 3.25 is usually used as the *DC* bridge circuit. The condition of the balance for such circuit is

$$R_x = R_2 \frac{R_3}{R_4} \tag{3.30}$$

Usually, the ratio of the resistances R_3/R_4 is set in sequence: *0.001– 0.01 0.1 – 1– 10* etc. and in this way the range of the bridge circuit is changed. The adjusting resistor R_2 is used for balancing the bridge circuit and the value of measured resistance R_x can be determined directly from equation (3.30).

The uncertainty of the determination of the measured value of R_x depends on the accuracy of the component resistors. According to Eq. (2.60) from (3.30) we obtain

$$\delta R_x = \sqrt{(\delta R_2)^2 + (\delta R_3)^2 + (\delta R_4)^2} \tag{3.31}$$

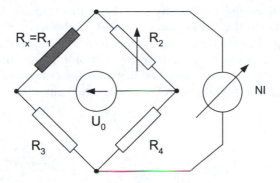

Figure 3.25. The Wheatstone bridge circuit

The quality of determination of the balance state (the same the uncertainty of measurement) depends on the resolution Δr. The resolution Δr is the smallest change of the adjusting resistor that causes noticeable change of indication $\Delta\alpha$ of null indicator (it is assumed that human eye can detect the change of the pointer equal to about 1/5 of the smallest graduation). Of course the resistor R_2 should have a sufficient number of adjustable decades in order to obtain sufficient resolution.

The resolution of bridge circuit depends on: sensitivity of the null indicator S_g, sensitivity of the bridge circuit S ($U_{out} \approx S \, \Delta R_2$), resistance of the null indicator R_o, resistance of the bridge circuit R_b (resultant resistance at the output terminals) and the supply voltage U_o

$$\Delta r = \frac{R_o + R_b}{U S_g S} \qquad (3.32)$$

An obvious conclusion can be drawn from the equation (3.32) – the resolution and sensitivity are better for more sensitive null-indicators ($S_g = \Delta\alpha/I_{out}$). Similarly, the supply voltage should be as large as possible – the limitation is the dissipation power of the used resistors. This limitation depends on the bridge configuration – the condition of the balance is the same for $R3/R4 = 10\Omega/100\Omega$ and $R3/R4 = 10k\Omega/100k\Omega$ but the permissible supply voltage is for both configurations quite different[1]. Also the sensitivity of the bridge circuit depends on the bridge configuration – this subject is discussed in Section 3.3.5.

In the case of the measurements of very small (less than 1Ω) resistances other sources of uncertainty have to be taken into account. For small

---

[1] Sometimes simple change of the diagonals (supply and output) can results in significant change of the sensitivity.

measured resistances the uncertainty can be influenced by the contact resistances, thermoelectric voltages and most of all the resistances r of the wires connecting the resistance to be measured. The influence of the thermoelectric voltages can be reduced by performing the measurement procedure in two steps – for positive and negative polarization of the supply voltage, and then by calculation of the average value from these two measurements.

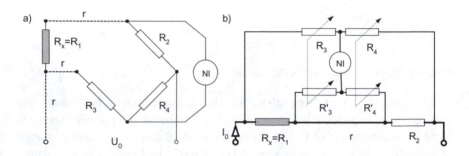

Figure 3.26. The methods of reduction of the influence of the connecting wires in the case of the measurements of small resistances: a) three-wire connection, b) four-wire connection in the Kelvin bridge circuit

It is possible to reduce the influence of the resistance r of the connecting wires by using the three-wire connection presented in Fig. 3.26a. If all three wires exhibit the same resistance (the same length) we can write that

$$(R_x + r)R_4 = R_2(R_3 + r) \tag{3.33}$$

and

$$R_x R_4 = R_2 R_3 + r(R_2 - R_4) \tag{3.34}$$

If additionally the condition $R_2 = R_4$ is fulfilled then the influence of the resistance r of the connecting wire is negligible.

For very small resistance, very useful is the modification of the Wheatstone bridge in the form presented in Fig. 3.26b (*the Kelvin bridge*). The condition of the balance for this bridge is as follows

$$R_x = R_2 \frac{R_3}{R_4} + r \frac{R_3 R'_4 - R'_3 R_4}{R_4(R'_3 + R'_4 + R_p)} \tag{3.35}$$

First of all, the resistance of connection wire r should be small – therefore such wire is prepared as a short and large diameter wire. The second term in the equation (3.35) as negligible if the following condition is fulfilled

$$R_3 R_4' = R_3' R_4 \qquad \text{or} \qquad \frac{R_3}{R_3'} = \frac{R_4}{R_4'} \qquad\qquad (3.36)$$

The condition (3.36) is relatively easy to achieve by mechanical coupling of the resistors R_3/R_3' and R_4/R_4'. In such case, the condition for balance of the Kelvin bridge (3.35) is the same as for the Wheatstone bridge (3.30).

The Kelvin bridge enables measurement of the resistances in the range $0.0001\ \Omega - 10\ \Omega$.

3.3.3. The AC bridge circuits

Taking into account that the impedance Z is $Z = |Z| e^{j\varphi}$, the condition of balance of AC bridge circuit (3.28) can be written in a form

$$\begin{cases} |Z_1||Z_4| = |Z_2||Z_3| \\ \varphi_1 + \varphi_4 = \varphi_2 + \varphi_3 \end{cases} \qquad\qquad (3.37)$$

Thus to obtain the balance of AC bridge circuit two conditions should be fulfilled: magnitude and phase (3.37). This means that in order to balance such bridge circuit two independent adjusting elements are necessary. The process of balancing is therefore more complicated than in the DC bridge circuit.

In the case of the AC bridge circuits another problem appears – it is difficult to eliminate influence of the stray and to earth capacitances (Fig. 3.27). For that reason, it is necessary to shield all the elements in the AC bridge circuits (Fig. 3.27a). Shielding does not eliminate the capacitive coupling but enables investigators to establish their level during the balancing. More effective is to use the *Wagner earth* (*Wagner ground*) with additional elements Z_5, Z_6 connected as presented in Fig. 3.27b.

The bridge circuit with the Wagner elements consists *de facto* of two bridges. First, the bridge $Z_1 Z_3 Z_5 Z_6$ is balanced (the switch of null indicator in position 1), and then the bridge $Z_1 Z_2 Z_3 Z_4$ (the switch of null indicator in position 2). (Sometimes it can be necessary to balance both bridges many times – approaching the equilibrium state in a stepwise manner.) In the state of balance of both bridges the potentials of points a,b and e are the same and equal to the potential of earth. Therefore the capacitances C_{ae} and C_{be} do not influence the distribution of currents. The capacitances C_{de} and C_{ce} are connected to the Wagner elements and also do not influence the balance condition of the main bridge circuit.

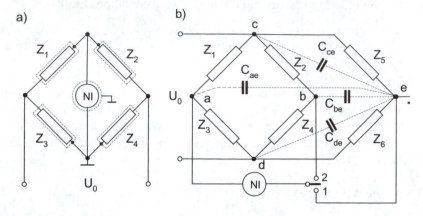

Figure 3.27. The methods of reduction of the influence of parasitic capacitances: a) shielding of the elements b) the Wagner earth additional elements

A huge number of various AC bridge circuits were designed and developed: *Maxwell, Wien, Schering, Hay, Owen, Anderson, de Sauty,* etc. Moreover, all these bridges exist in various mutations and modifications. *(Hague 1971, Warsza 2005a)*. Historically the oldest and most known are the Wien bridge (Fig. 3.28a) and Maxwell bridge (Fig. 3.28b) circuits.

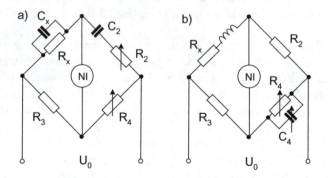

Figure 3.28. The Wien bridge circuit (a) and Maxwell-Wien bridge circuit (b)

The conditions of the balance state of *Wien bridge* circuit are as follows:

$$C_x = \frac{C_2 R_4}{R_3(1+\omega^2 C_2^2 R_2^2)}, \qquad R_x = \frac{R_3(1+\omega^2 C_2^2 R_2^2)}{\omega^2 R_2 R_4 C_2^2} \qquad (3.38)$$

The condition of the balance are frequency dependent. Therefore the Wien bridge is rather seldom used for capacitance measurement, but it is frequently used as the frequency-dependent part of the oscillator, according to the dependence:

$$\omega^2 = \frac{1}{R_x C_x R_2 C_2} \qquad (3.39)$$

The conditions of the balance state of the *Maxwel-Wien* bridge (called also often as the *Maxwell bridge*) are as follows:

$$L_x = R_2 R_3 C_4, \qquad R_x = R_2 \frac{R_3}{R_4} \qquad (3.40)$$

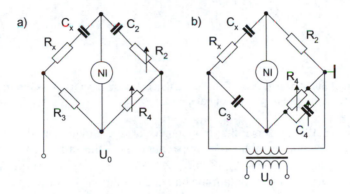

Figure 3.29. Two examples of the bridge circuits for capacitance measurements: de Sauty-Wien bridge (a) and Schering bridge (b)

Fig. 3.29a presents the AC bridge circuit for capacitance measurements (the *de Sauty-Wien bridge*). From the Eq. (3.28) the conditions of the balance state can be described as

$$C_x = C_2 \frac{R_4}{R_3}, \qquad R_x = R_2 \frac{R_3}{R_4} \qquad (3.41)$$

The $tg\delta_x = \omega C_x R_x$ can be calculated as

$$tg\delta_x = \omega C_2 R_2 \qquad (3.42)$$

Fig. 3.29b presents special kind of the bridge circuit – *the Schering bridge* designed for high voltage and cable testing. The main part of supply high voltage is on the capacitances C_x and C_3, and adjustable elements R_4, C_4 are additionally grounded. The measured parameters can be determined from the equations

$$C_x = C_3 \frac{R_4}{R_2}; \quad R_x = R_2 \frac{C_4}{C_3}, \quad tg\delta_x = \omega C_4 R_4 \qquad (3.43)$$

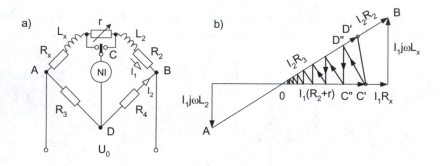

Figure 3.30. An example of the inductance bridge circuit and simplified diagram illustrating the circuit process of balancing

Fig. 3.30a presents the inductance bridge circuit (modified Maxwell bridge). Assuming that the impedances of the arms are as follows: $Z_x=R_x+j\omega L_x$, $Z_2=R_2+j\omega L_2$, $Z_3=R_3$, $Z_4=R_4$ and using the Eq. (3.28) after simple calculations we obtain the balance conditions in form

$$L_x = L_2 \frac{R_3}{R_4} \qquad (3.44)$$

$$R_x = (R_2 + r)\frac{R_3}{R_4} \quad \text{or} \quad R_x = R_2 \frac{R_3}{R_4} - r \qquad (3.45)$$

To obtain the balance of the bridge circuit presented in Fig. 3.30a the Q factor ($Q=\omega L/R$) of the inductances measured L_x and standard one L_2 should be the same. For that reason, an additional resistor r is used. This resistor is connected to L_x or L_2 element (this connection is chosen experimentally – only in one position of the switch it is possible to balance the bridge). Depending on the position of the switch one of the equations (3.45) can be used to determine the value R_x.

The inductance bridge circuit presented in Fig. 3.30 is useful for measurements of inductance $L_x R_x$ as well the Q factor $\omega L_x/R_x$. It is also possible to measure the mutual inductance M_x. For determination of the M_x

value the measurements are performed two times – with the coils connected in the same directions of the flux L' and with the coils connected in opposite directions L''

$$L' = L_1 + L_2 + 2M \qquad \text{and} \qquad L'' = L_1 + L_2 - 2M \qquad (3.46)$$

Then, the mutual inductance can be calculated as

$$M = \frac{L' - L''}{4} \qquad (3.47)$$

As it was mentioned above, one of the drawbacks of AC bridge circuits is the complicated process of balancing. Let us consider the vector diagram presented in Fig. 3.30b. The bridge is balanced by the so called successive approximation. Assume that for balancing we use the elements r and R_3 and that the unbalanced voltage is represented by the line $C'D'$. By changing the R_3 we move from the point D' to D''. Note that it is not possible to obtain the zero value of the $C - D$ distance because in this step of balancing the $C' - D''$ distance is the local minimum of the output voltage. If we now change the r value we will move from the point C' to C'' (this time the distance $C'' - D''$ is the local minimum of the output voltage). We can see that to obtain the balance it is necessary to perform many steps of approximation.

By appropriate design of bridge circuit it is possible to improve the balancing process – even down to two steps. It is also possible to use two null indicating devices with a $90°$ phase shift between them, which enable practically mutually independent balancing of both components. But generally the time of measurement using the AC bridge circuit is limited due to the complex problem of searching for the balanced state conditions.

The uncertainty of the AC bridge circuits can be analyzed similarly as in the case of DC bridge circuits (see Section 3.3.2). For example the uncertainties of determination of C_x according to the conditions (3.43) is

$$\delta C_x = \sqrt{(\delta R_3)^2 + (\delta R_4)^2 + (\delta C_2)^2} \qquad (3.48)$$

Of course this dependence is valid if other conditions are fulfilled, for example the influence of parasitic capacitances is eliminated and the error of resolution is significantly smaller than the uncertainty resulting from Eq. (3.48).

The necessity of application of various bridge circuits for measurements if RLC parameters is rather inconvenient. Fig. 3.31 presents the examples of the universal RLC bridge circuits that are composed from the same elements $(R_2 R_3 R_4 C_w)$.

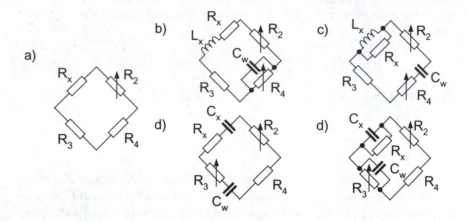

Figure 3.31. The examples of the universal RLC bridge circuits composed from the same elements

Depending on the measured parameter one of the five versions of circuit can be chosen. The parameters can be calculated according to the following conditions:

circuit a)
$$R_x = R_2 \frac{R_3}{R_4}$$
(3.49)

circuits bc)
$$L_x = R_2 R_3 C_w ; \quad R_x = R_2 \frac{R_3}{R_4} ;$$
(3.50)

circuits de)
$$C_x = C_w \frac{R_4}{R_2} , \quad R_x = R_2 \frac{R_3}{R_4}$$
(3.51)

3.3.4. The transformer bridge circuits

The change of impedance during the balancing process results in the change of the voltage drop across this impedance. Thus the bridge can be balanced by inserting additional source of voltage instead of changing the value of the impedance. For example in the circuit presented in Fig. 3.32a the condition of the balance is

$$R_x = R_w \frac{E_1}{E_w}$$
(3.52)

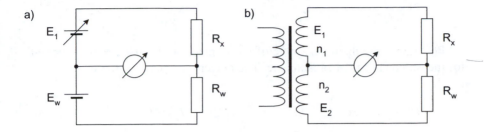

Figure 3.32. The bridge circuits balanced by the change of the voltage

Similarly for the bridge circuit presented in Fig. 3.32b the condition of the balance can be written in the form

$$\frac{R_x}{R_w} = \frac{E_1}{E_2} = \frac{n_1}{n_2} \tag{3.53}$$

The bridge can be balanced not only by the change of impedance or voltage but also by the change of the number of turns n. This is very convenient, because the number of turns can be precisely adjusted. Especially in the case of the digital bridge circuits it is much easier to connect the windings than to change the resistors or capacitors.

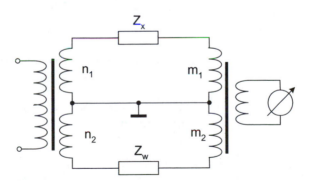

Figure 3.33. An example of the transformer bridge circuit

Fig. 3.33 presents the example of the transformer bridge circuit with two transformers. The output transformer acts in this circuit as the current comparator – the null indicator points to zero, when the resultant flux in the transformer is also equal to zero. The condition of the balance of this circuit is

$$\frac{Z_x}{Z_w} = \frac{n_1}{n_2} \frac{m_1}{m_2} \tag{3.54}$$

Fig. 3.35 presents the example of the transformer bridge circuit designed for measurement of the capacitance. The currents in this circuit are

$$I_1 = U_1 \left(\frac{1}{R_x} + j\omega C_x \right) ; \quad I_{2C} = U_1 \frac{n_2}{n_1} j\omega C_w ; \quad I_{2R} = U_1 \frac{n_2}{n_1} \frac{1}{R_w} \tag{3.55}$$

The null indicator points to zero if the following condition is fulfilled

$$n_3 I_1 = n_4 I_{2C} + n_5 I_{2R} \tag{3.56}$$

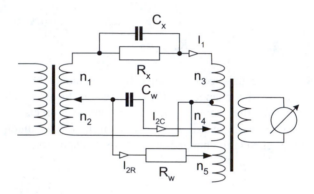

Figure 3.34. An example of the transformer bridge circuit designed for capacitance measurement

From the equations (3.55) and (3.56) the conditions of the balance of the circuit presented in Fig. 3.34 are as follows

$$C_x = C_w \frac{n_2 n_4}{n_1 n_3} ; \quad R_x = R_w \frac{n_1 n_3}{n_2 n_5} ; \quad tg\delta_x = \frac{n_5}{n_4 \omega R_w C_w} \tag{3.57}$$

The transformer bridge circuits exhibit several important advantages in comparison with impedance bridge circuits. As was mentioned earlier, the balancing is possible by the change of the number of turns. In transformer bridges the parasitic capacitances shunt the transformer turns and practically do not influence the conditions of the balance. Also the sensitivity of the transformer bridges is significantly better than in the case of impedance bridges. In order to make use of these advantages it is necessary to construct the transformers very precisely, with minimal stray fields. Therefore the transformer bridges are usually more expensive than classic circuits without transformer coupling.

3.3.5. The unbalanced bridge circuits

The unbalanced bridge circuits are used as the transducers converting the change of the resistance (and generally impedance) into the output voltage

$$U_{out} \cong SU_o \frac{\Delta R_x}{R_{xo}} = SU_o \varepsilon \tag{3.58}$$

where S is the sensitivity of the transducer and ε is the relative change of the resistance

$$R_x = R_{xo} \pm \Delta R_x = R_{xo}(1 \pm \varepsilon) \tag{3.59}$$

where R_{xo} is usually the resistance in the balance state.

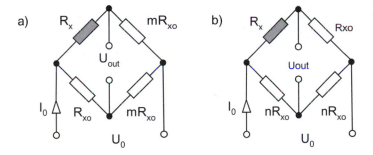

Figure 3.35. Two kinds of symmetry of the unbalanced bridge circuit

Usually, various kinds of resistance sensors are used as the R_{xo} element, for example the temperature sensor $R_T = R_{T0}(1 + \alpha \Delta T)$. It is very convenient to use two *differential sensors* instead of just one sensor. In the differential sensors the changes of the resistances are in the opposite direction:

$$\begin{cases} R_{x1} = R_{xo}(1 + \varepsilon) \\ R_{x2} = R_{xo}(1 - \varepsilon) \end{cases} \tag{3.60}$$

The unbalanced bridge circuits are usually designed with symmetry in respect to the output diagonal (Fig. 3.35a) or to the supply diagonal (Fig. 3.35b).

Substituting the relation (3.59) into the equations (3.25) or (3.26) after simple calculations we can derive the dependencies of the transfer characteristics of unbalanced bridge circuits:

for the circuit a)

$$U_{out}/U_0 = \frac{m}{(1+m)^2 + (1+m)\varepsilon}\varepsilon \; ; \qquad U_{out}/I_0 = R_{xo}\frac{m}{2(1+m)+\varepsilon}\varepsilon \quad (3.61)$$

for the circuit b)

$$U_{out}/U_0 = \frac{n}{2n(2+\varepsilon)}\varepsilon \; ; \qquad U_{out}/I_0 = R_{xo}\frac{n}{2(1+n)+\varepsilon}\varepsilon \qquad (3.62)$$

We can see that all analyzed circuits are nonlinear. The nonlinearity depends on the value of the resistances (coefficients m or n). Fig. 3.36 presents the example of the dependences $U_{out}=f(\Delta R_x/R_{xo})$.

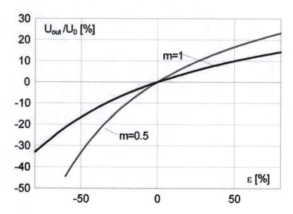

Figure 3.36. The example of the transfer characteristics of unbalanced bridge circuit

The nonlinearity of the bridge transducer is not always a drawback – in some circumstances this bridge nonlinearity can be used to correct the nonlinearity of the sensor. Let us consider the example presented in Fig. 3.37. We use a thermistor sensor with very nonlinear characteristic $R=f(T)$ (Fig.3.37a) in order to measure the temperature. If the bridge characteristic would be linear, then the resultant characteristic of the transducer $U_{out} = f(T)$ would also be nonlinear – curve 1 in Fig. 3.37b. By appropriate choice of the bridge configuration (bridge nonlinearity), in our case by applying *m=0.3* we obtain almost linear processing of the temperature into the output voltage – curve 3 in Fig. 3.37b.

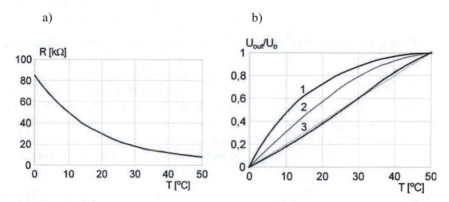

Figure 3.37. The transfer characteristic of the typical thermistor sensor (a) and the resultant characteristics of the bridge circuit with thermistor sensor connected into one of the arms (b) (1 – calculated under assumption that the bridge circuit is linear; 2 – calculated for *m = 1*; 3 – calculated for *m = 0.3*).

There are various methods of linearization of the unbalanced bridge circuit. One of them is applying of the multiplier device (Fig.3.38) (Tran 1987).

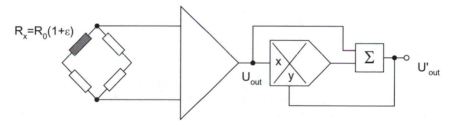

Figure 3.38. The linearization of the bridge circuit by applying the multiplier device

Assuming *n=1*, $U_o = 2E$ and $\varepsilon = 2X$ the equation (3.62) can be written in the normalized form

$$U_{out} = \frac{X}{1+X}E \qquad\qquad (3.63)$$

After connection of the multiplier circuit as it is presented in Fig. 3.38 (assuming that *K = E*) the output voltage is

$$U'_{out} = U_{out} + \frac{U_{out}U'_{out}}{K} = \frac{U_{out}}{1-U_{out}} = \frac{\dfrac{X}{1+X}}{1-\dfrac{X}{1+X}} = X = \frac{1}{2}\varepsilon \qquad (3.64)$$

Thus the transfer characteristic of the bridge circuit with multiplier is linear.

Other method of linearization presents Fig. 3.39 (Kester 1999). This method is similar to described in Section 2.1.2. method of auto-balancing bridge circuit (Fig. 2.6). If the bridge circuit is automatically balanced it uses only a small linear part of the transfer characteristic. Because output signal of the bridge is very small an additional amplifier on the output is necessary.

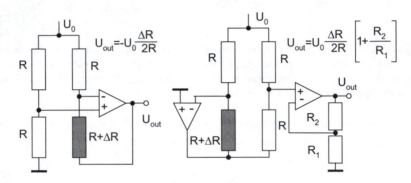

Figure 3.39. Two examples of the auto-balancing method of linearization (Kester 1999)

Fig. 3.40 presents the methods of connection of the differential sensors into the unbalanced bridge circuit. For these circuits the transfer characteristics are as follows:

for the circuit a)

$$U_{out}/U_0 = \frac{2m}{(1+m)^2 - \varepsilon^2}\varepsilon\ ; \qquad U_{out}/I_0 = R_{xo}\frac{m}{1+m}\varepsilon \qquad (3.65)$$

for the circuit b)

$$U_{out}/U_0 = \frac{1}{2}\varepsilon\ ; \qquad U_{out}/I_0 = R_{xo}\frac{n}{1+n}\varepsilon \qquad (3.66)$$

We can see that the unbalanced bridge circuit with differential sensors are linear (with exception of the circuit (a) supplied from the voltage source).

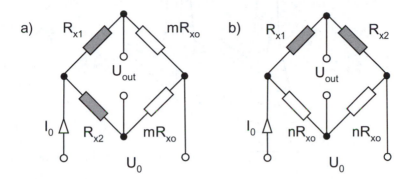

Figure 3.40. Two kinds of symmetry of the unbalanced bridge circuit with differential sensor

Let us consider the sensitivity of the unbalanced bridge circuit. Neglecting the nonlinearity (calculating the S factor in the balance state) and assuming the load resistance is infinitively large from Eq. (3.61) or (3.65) we obtain for the single sensor:

$$S = \frac{m}{(1+m)^2} \tag{3.67}$$

and for two differential sensors:

$$S = \frac{2m}{(1+m)^2} \tag{3.68}$$

The bridge circuit with differential sensors is two times more sensitive than the bridge with one sensor. If the bridge circuit is an AC bridge then the sensitivity is a complex value, because

$$m = \frac{Z_2}{Z_1} = \frac{|Z_2|e^{j\varphi_2}}{|Z_1|e^{j\varphi_1}} = \frac{|Z_2|}{|Z_1|}e^{j(\varphi_2-\varphi_1)} = |m|e^{j\theta} \tag{3.69}$$

Thus the sensitivity S for the differential sensors is

$$S = \frac{2|m|}{1+2|m|\cos\theta + m^2} \tag{3.70}$$

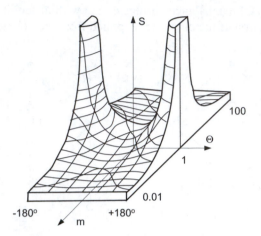

Figure 3.41. The dependence of the sensitivity S of the bridge circuit on the circuit configuration (Karandiejew 1975)

Fig. 3.41 presents a graphical representation of the dependence (3.70). From this figure we can conclude that:
- the sensitivity is largest when the ratio m is equal to 1,
- the sensitivity can be larger, when the phase difference between impedances Z_1 and Z_2 is larger.

The general dependence $U_{out}=f(\varepsilon)$ of the unbalanced bridge circuit with four sensors is

$$U_{out} = \frac{1}{4}(\varepsilon_1 - \varepsilon_2 - \varepsilon_3 + \varepsilon_4)U_o \qquad (3.71)$$

Thus for one sensor the sensitivity is $S=1/4$, for two sensors it is $S=1/2$, while for four sensors we obtain four times larger sensitivity in comparison with the one-sensor case

$$U_{out} = \varepsilon\, U_o \qquad (3.72)$$

3.3.6. The alternatives for bridge circuits – Anderson Loop

Wheatstone bridge is more than 150 years old (Wheatstone 1843). In some cases it can be successfully substituted by the differential amplifier. The circuits with differential amplifiers are presented in Fig. 3.42.

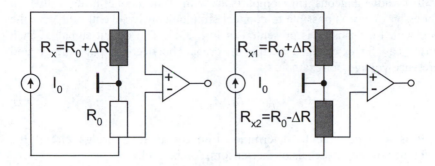

Figure 3.42. The resistance/voltage converter with differential amplifier

The circuits presented in Fig. 3.42 exhibit similar performances as the bridge circuit – compensation of the offset voltage (zero output signal for $\Delta R = 0$), linear conversion, compensation of interferences, as for example changes of external temperature. A similar circuit was patented in 1994 by NASA (Anderson 1994, Anderson 1998) and today is sometimes called the *Anderson Loop*. Two examples of the Anderson loop are presented in Fig. 3.43.

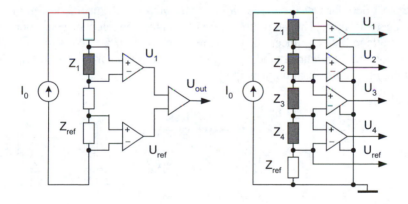

Figure 3.43. The Anderson loop with one sensor and with four sensors

In Anderson loop presented in Fig. 3.43 for the case of one sensor the output signal is

$$U_{out} = I_0 \, \Delta Z \tag{3.73}$$

It is advantageous (in comparison with Wheatstone bridge) that in Anderson loop it is possible to connect simultaneously several sensors – the loop with four sensors is presented in Fig. 3.43. The output signal of each sensor can be determined as the difference between output voltage and reference voltage, for example

$$U_1 - U_{ref} = I_0 \, \Delta Z_1 \qquad (3.74)$$

It is also possible to determine similarly as in a bridge circuit the difference of output signals of the sensors, for example

$$U_1 - U_2 = I_0 \left(\Delta Z_1 - \Delta Z_2 \right) \qquad (3.75)$$

Especially in the case when we use several sensors the Anderson loop can be alternative for bridge circuit.

3.4. POTENTIOMETERS AND COMPARATORS

The devices utilizing the idea of compensation of two voltages, called *potentiometers*, for many years were the best and most accurate instruments for measurements of voltage (directly) and current or resistance (indirectly). The most popular was the *Feussener potentiometer* with special design of the decade resistance. The laboratory potentiometers enabled measurement of the voltage with uncertainty better than *0.01%*. Nowadays, the potentiometers are substituted by high quality digital voltmeters (often also realizing the compensation principle).

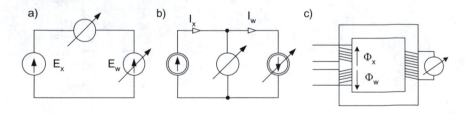

Figure 3.44. The compensation circuits: a) compensation of the voltages, b) compensator of the currents, c) compensation of the magnetic fluxes

The compensation can be realized by means of mutually neutralization of two voltages (Fig. 3.44a), but also other values, for example two currents (Fig.3.44b) or two magnetic fluxes (Fig.3.44c) can be compensated. Fig. 3.45 presents the typical design of the compensating transducer $I_{out} = f(U_x)$.

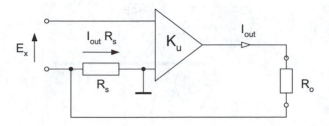

Figure 3.45. The transducer $I_{out} = f(U_x)$ utilizing the compensation of two voltages

The transducer presented in Fig. 3.45 exhibits many important advantages of the compensation principle – very large input resistance (practically infinitively large) and excellent accuracy. For correct design of the transducer the uncertainty of the signal processing depends only on the accuracy of the standard resistor R_s (it does not depend on the amplification value of the amplifier – more details in Section 4.6).

In the comparator devices a comparison of two values is performed. As a tool for comparison the compensation principle can be utilized. For example in the circuit presented in Fig. 3.46a we compare two resistances by the compensation of two currents while in the circuit presented in Fig. 3.46b we utilize the compensation of two voltages. The current comparator presented in Fig. 3.46c utilizes the compensation of two magnetic fluxes.

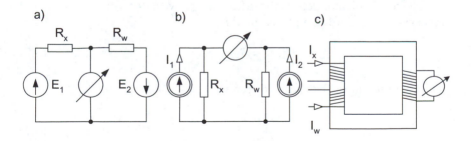

Figure 3.46. The comparators of the resistances (a,b) or currents (c)

The *direct current comparator DCC* is currently used by the NIST as the most precise method of the reconstruction of the resistance standard from the Hall quantum resistance device (NIST1458 2003). An example of such a comparator is presented in Fig. 3.47.

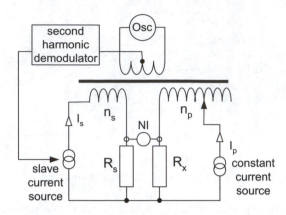

Figure 3.47. The direct current comparator used by NIST as the resistance standard (NIST1458 2003)

In the current comparator presented in Fig. 3.47 two simultaneous balances are required – ampere-turn balance and voltage balance. The ampere-turn balance is performed automatically using the feedback circuit to control the slave current source (as the state of balance the second harmonic induced in the transformer is used). Thus automatically the condition is fulfilled

$$n_s I_s = n_p I_p \qquad (3.76)$$

The null-indicator detects the difference between the voltage drops on the resistances: standard R_s and measured R_x

$$R_x I_p = R_s I_s \qquad (3.77)$$

The condition of the balance is therefore:

$$R_x = \frac{n_p}{n_s} R_s \qquad (3.78)$$

The state of balance can be achieved by adjustment of the number of turns n_p. It enables the investigator to determine the measured resistance with excellent accuracy because we can count the number of turns practically without error. The presented circuit looks as not very complicated but to correct its operation the transformer should exhibit excellent quality – it is prepared from high permeability ring cores, with perfect symmetry, shielding, etc.

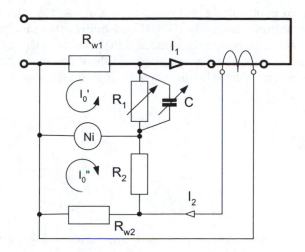

Figure 3.48. The AC current comparator used as the testing device for the measuring transformer

Fig. 3.48 presents the example of the AC current comparator used for the testing of the current transformer. The primary current I_1 and the secondary current I_2 are compared as the voltage drops on the standard resistors R_{w1} and R_{w2}. In the case of the AC comparator it is necessary to balance the circuit in two steps – by changing the amplitude (element R_1) and by changing the phase (element C). In the state of balance both current I_o' and I_o'' are the same and null indicator is pointing to zero. The transformation ratio K and the angle error $tg\gamma$ of transformer can be determined as

$$K_I = \frac{I_1}{I_2} = \frac{R_{w2}R_1}{R_{w1}R_2} \; ; \qquad tg\gamma \cong \omega R_1 C \qquad (3.79)$$

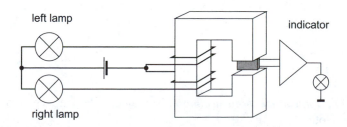

Figure 3.49. The current comparator as the device for testing the bulbs in the car lighting

Another example of the current comparator used as the car lighting system tester is presented in Fig. 3.49. If both bulbs in the car lighting system work, then both currents are the same and the resultant magnetic flux in the yoke is equal to zero. The damage of one of the bulbs causes the unbalance of fluxes, which is detected by the magnetic field sensor placed in the gap.

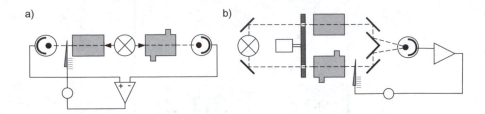

Figure 3.50. The light comparator in the devices to the measure the light absorption

Fig. 3.50 presents two examples of the light comparators. The light passing through two absorption cells is measured by two photodetectors (Fig. 3.50a) or by one photodetector alternately. In one cell is the standard liquid while in the second is the measured liquid. The difference between two lights after amplification controls the movement of the wedge. The position of the wedge can be used as a measure of the absorption properties of investigated liquid, for example for the testing of the concentration of such liquid.

Presented examples of old classic measuring principles such as bridge circuit, compensation, and comparator circuit in modern measuring instruments proved that these methods are still important for measurement technique. The special area of application of such methods is the measurement equipment of extremely high accuracy (as described above direct current comparator). As an example we can point the old principle of so called *Warshawski bridge*. This bridge was first time described in 1946 (Warshawski 1946). In 1990 this bridge was again discovered by NIST and today it is one of the most accurate methods of resistance measurement (NIST1458 2003).

REFERENCES

Agilent 2003 *Impedance Measurement Handbook*, Agilent Technologies Application Note

Anderson K.F 1994 *Constant Current Loop Impedance Measuring System that is Immune to the Effects of Parasitic Impedances*, US patent No. 5 371 469

Anderson K.F. 1998 *Your Successor to the Wheatstone Bridge? NASAs Anderson Loop*, IEEE Instrumentation and Measurement Magazine, 1, p.5

Eragost 2005 EraGost, www.eragost.pl

Graphtec 2005 Graphtec Corporation, www.graphteccorp.com

Hague B. 1971 *Alternating Current Bridge Circuits*, Pitman

Kester 1999 *Practical Design Techniques for Sensor Signal Conditioning*, Analog. Devices Corp.

Kularatna N. 2003 *Digital and Analogue Instrumentation Testing and Measurement*, Chapter 5 (Fundamentals of Oscilloscopes), IEE

Lumel 2005 Lumel S.A., www.lumel.com.pl

NIST 1458 2003 *NIST Measurement Service for DC Standard Resistors*, NIST Technical Note 1458

Sarma K.R. 2004 *Liquid Crystal Displays*, Chapter 32 in Electrical Measurements, signal Processing and Displays, CRC Press

Sherman C.J. 2004 *Cathode Ray Tube Displays*, Chapter 31 in Electrical Measurements, signal Processing and Displays, CRC Press

Tran Tien Lang 1987 *Electronics of Measuring Systems*, Wiley

Warsza Z. 2005a *Electrical Bridge Circuits*, Chapter 126 in Handbook of Measuring System Design, Wiley

Warsza Z. 2005b *Unbalanced DC Bridges*, Chapter 127 in Handbook of Measuring System Design, Wiley

Warshawski I. 1946 *A multiple bridge for elimination of contact-resistance errors in resistance strain-gauge measurements*, National Advisory Committee for Aeronautics NACA – Technical note. 1031

Wheatstone Sir Charles 1843 *An Account of Several New Instruments and Processes for Determining the Constants of a Voltaic Circuit*, Philosophical Transactions of the Royal Society of London, 133, pp. 303-329

4

Processing of the Analogue Measurement Signals

4.1. SIGNAL CONDITIONING

4.1. 1. Analogue measurement signals

The information obtained as the result of measurement is usually processed as a measurement signal. As the *measurement electric signal* we mean the time varying electric signal representing measured value. Various signal parameters can be used as the representation of the measured value: magnitude, frequency, phase, etc. Usually electric current (or voltage) with sufficiently large magnitude is preferred. Often to obtain good quality of electric signal for further applications (control systems, digital processing, computing etc.) it is necessary to perform additional processing of this signal.

The signal processing is performed for various purposes:

a) *Signal conditioning* (Pallas Areny 1991) – the change of the signal from a primary form to a convenient form (for example conversion of the change of resistance of thermoresistive sensor into the change of voltage). Usually signal conditioning circuits enable additional correction of linearity. Sometimes term signal conditioning is understood more widely as the conversion of the sensor signal to the final form (including amplification, conversion to the standard interface, etc.).

b) *Signal acquisition* (Park 2003, James 2000) – usually typical data acquisition board realize the following function: mixing input signal, converting into digital form and transmission to the computer acceptable interface (for example PCI or USB).

c) *Signal conversion* – change of one form of the signal to another, for example analogue to digital conversion or alternating current to direct current conversion.

d) *Signal recovery* – recovery or improvement of the signal quality, for example elimination of the influence of noises.

e) *Signal amplification* – increase of signal value with preservation other parameters, as frequency bandwidth, wave shape, etc.

f) *Signal separation* – physical disconnection of the input and output circuits.

g) *Signal filtering* – transmit from the input to output only selected part of signal harmonics.

h) S*ignal harmonization* – conversion of the signal from the arbitrary form to commonly or standard used form, for example conversion of the measured signal into $\pm 5V$ signal, acceptable by most digital circuits.

i) *Signal modulation* – for example conversion of the measured signal into high frequency signal with varying magnitude carrying the information (magnitude modulation).

Some of these processes can be performed digitally, but it is better to use good quality analogue signal for further digital processing. The output signals of most of the sensors are analogue, i.e. the information is represented by the magnitude of the electric signal.[1] *The analogue signal* theoretically can take an infinite number of values. In contrary *the digital signal* is represented by limited number of values – *quanta*. Due to different technique of processing of these signals we usually consider these both techniques separately as *analogue measurement technique* (analogue signal processing) (Pallas Areny 1999) and *digital measurement technique* (digital signal processing – DSP) (Lai 2004, Smith 2003). Additionally, we can analyze the analogue signals as *DC signals* (*DC – direct current*) and *AC signals* (*AC – alternating current*).

The comparison of both digital and analogue techniques indicates many advantages of the former one. These advantages are as follows:

- better immunity to the interferences and disturbances (the digital signal is usually sufficient large, the "0" and "1" states can be determined with large dispersion of the signal value);
- the signals are easy to transmit and register;
- it is easy to perform necessary calculations;
- signals are naturally accepted by the computers.

---

[1] When we measure the rotational speed by counting the pulses we also receive analogue signal – sequence of pulses. Only if we use special kind of sensor (encoder) we would receive directly the digital output signal.

It does not mean that we can consider both techniques as excellent digital and less excellent analogue ones. Both techniques are complementary and special importance should be attributed to the analogue techniques in the initial state of signal processing. The amplification techniques are still analogues and play an important role in the signal processing.

The signals can be deterministic or stochastic. The *deterministic signals* can be predicted with absolute certainty and are reproducible. In the case of the *stochastic signals* we can only predict (estimate) them with some level of probability. We use tools of theory of probability to describe and analyze the stochastic signals.

The DC signal is described by one parameter – its value. The AC signal can be described by various parameters: the magnitude U_m or peak value U_p, mean value U_0, average (rectified) value U_{AV}, effective (*rms*) value U_{rms}, peak-to-peak value U_{pp}, instantaneous value $u(t)$. Moreover, we should know the frequency f (or $\omega = 2\pi f$ or period $T = 1/f$) and the phase φ.

If the voltage signal is described by the equation

$$u(t) = U_m \sin(\omega t + \varphi) \tag{4.1}$$

then its main parameters are as follows

$$U_0 = \frac{1}{T} \int_{t_0}^{t_0 + T} u(t)dt \tag{4.2a}$$

$$U_{AV} = \frac{1}{T} \int_{t_0}^{t_0 + T} |u(t)| dt \tag{4.2b}$$

$$U_{rms} = \sqrt{\frac{1}{T} \int_{t_0}^{t_0 + T} u^2(t)dt} \tag{4.2c}$$

It is easy to calculate that for sinusoidal signal these parameters are

$$U_0 = 0; \qquad U_{AV} = 0{,}637\ U_m; \qquad U_{rms} = 0{,}707\ U_m; \qquad U_{pp} = 2\ U_m$$

We can analyze the AC signals in the time domain or in the frequency domain (both methods are complementary). The conversion between signal described in time domain and frequency domain is possible using Fourier transform

$$x(t) = \frac{1}{2\pi} \int_{-\infty}^{+\infty} X(\omega)e^{j\omega t}d\omega \qquad (4.3a)$$

$$X(\omega) = \int_{-\infty}^{+\infty} x(t)e^{-j\omega t}dt \qquad (4.3b)$$

The periodical non-sinusoidal signal can be described as the sum of its harmonics (by Fourier series)

$$x(t) = a_0 + 2\sum_{k=1}^{\infty} a_k \cos(k\omega_0 t) + b_k \sin(k\omega_0 t) \qquad (4.4)$$

where:

$$a_0 = \frac{1}{T}\int_{t_1}^{t_1+T} x(t)dt, \qquad a_k = \frac{1}{T}\int_{t_1}^{t_1+T} x(t)\cos(k\omega_0 t)dt, \qquad b_k = \frac{1}{T}\int_{t_1}^{t_1+T} x(t)\sin(k\omega_0 t)dt$$

or in exponential form

$$x(t) = \sum_{k=-\infty}^{+\infty} c_n e^{jk\omega_0 t} \qquad (4.5)$$

where: $c_n = \dfrac{1}{T}\displaystyle\int_{t_0}^{t_0+T} x(t)e^{-jk\omega_0 t}\,dt$.

When the function $x(t)$ is even (*in mathematical sense*) then coefficients $b_k = 0$ and when the function $x(t)$ is odd then $a_k = 0$. Table 4.1 presents the Fourier representation of some typical signals.

As a measure of the distortion of non-sinusoidal signals the coefficient h can be used – it describes the content of harmonics with respect to the fundamental harmonic, and can be calculated as

$$h = \frac{\sqrt{U_2^2 + U_3^2 + U_4^2 \ldots}}{U_1} \qquad (4.6)$$

In the case of non-periodic signal instead of Fourier series (4.5) we can use Fourier integral in a form

$$X(f) = \int_{-\infty}^{\infty} x(t) e^{-j\omega t} dt \qquad (4.7)$$

Table 4.1. Fourier representation of typical signals

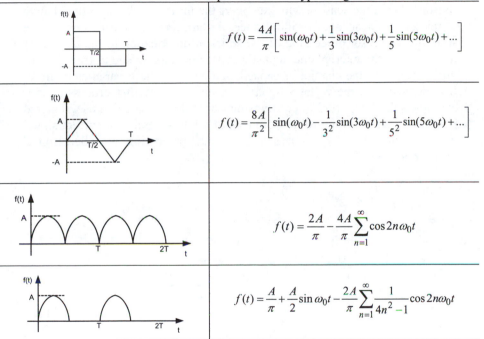

$$f(t) = \frac{4A}{\pi}\left[\sin(\omega_0 t) + \frac{1}{3}\sin(3\omega_0 t) + \frac{1}{5}\sin(5\omega_0 t) + ... \right]$$

$$f(t) = \frac{8A}{\pi^2}\left[\sin(\omega_0 t) - \frac{1}{3^2}\sin(3\omega_0 t) + \frac{1}{5^2}\sin(5\omega_0 t) + ... \right]$$

$$f(t) = \frac{2A}{\pi} - \frac{4A}{\pi}\sum_{n=1}^{\infty} \cos 2n\omega_0 t$$

$$f(t) = \frac{A}{\pi} + \frac{A}{2}\sin \omega_0 t - \frac{2A}{\pi}\sum_{n=1}^{\infty} \frac{1}{4n^2 - 1}\cos 2n\omega_0 t$$

The distorted signal can be presented as a Fourier series also in a graphical form. Usually the signals are presented in form a line spectrum where the individual harmonics are represented by vertical lines (Fig. 4.1).

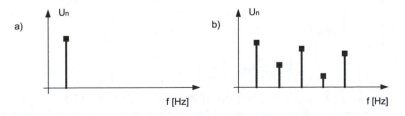

Figure 4.1. An example of the spectral analysis of the sinusoidal signal (a) and distorted signal (b)

The position of the line on the frequency axis describes the frequency of this harmonic while the magnitude of the line describes the magnitude of given harmonic. Of course the undistorted signal is represented just by one line (Fig. 4.1a).

4.1.2. Conditioning of resistance, capacitance and inductance

It is possible to measure many non-electric quantities using parametric sensors, that is the sensors which convert the measured value to the change of resistance, capacitance or inductance. As an example we can consider the strain-gauge sensor processing of the stress or deformation to the change of resistance or the microphone acoustic sensor processing sound (pressure of the air wave) to the change of capacitance. Usually these parameters are at first converted to the voltage signal, convenient for further processing, for example amplifying, filtering or transmission. It is assumed that useful signal is a voltage signal of the value from tens of mV to several V or current signal $4 - 20$ mA. This very first step of signal processing is called *signal conditioning*.

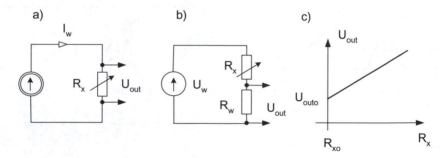

Figure 4.2. The typical converters of the resistance into the voltage: voltage drop (a), voltage divider (b) and their transfer characteristic (c)

Figure 4.2 presents two methods of conversion of the resistance to the voltage. The first one seems to be the most obvious – it utilizes the *Ohm's law* – the resistance is supplied by the stabilized current I_w and the voltage drop U_{out} is proportional to the sensed resistance

$$U_{out} = I_w R_x \tag{4.8}$$

Thus we have linear conversion of the resistance into voltage signal. But the dependence (4.8) is valid only if the resistance connected to the output of transducer is infinitively large. Consider finite resistance of the output load

R_0 . In such case we obtain a nonlinear characteristic $U_{out} = f(R_x)$ and this nonlinearity depends on the ratio R_x/R_0. The *Ohm's law* is also used in the converter presented in Fig. 4.2b. The measured resistance is connected in the circuit of the voltage divider supplied by the voltage source U_w. The output signal U_{out} is described by the equation

$$U_{out} = U_w \frac{R_w}{R_w + R_x} = U_w \frac{1}{1 + \dfrac{R_x}{R_w}}$$

(4.9)

The conversion is nonlinear, which is not always a drawback, because in certain cases it can be used for linearization of a non-linear sensor. The main disadvantage of the circuits presented in Fig. 4.2 is that the dependence $U_{out} = f(R_x)$ does not start from zero because the resistance of the sensor usually also does not start from the zero value but from the certain R_{x0} value

$$R_x = R_{xo} \pm \Delta R_x = R_{xo}\left(1 \pm \frac{\Delta R_x}{R_{xo}}\right) = R_{xo}\left(1 \pm \varepsilon\right)$$

(4.10)

Similarly the output signal of the converters presented on the Fig. 4.2 includes the constant component U_{outo} (Fig. 4.2c), because

$$U_{out} = U_{outo}\left(1 + \varepsilon\right)$$

(4.11)

This offset component is disadvantageous, because more convenient is the case when the output signal of the transducer is zero for starting point of the range of the sensor. If this condition is fulfilled then we can connect the typical voltmeter as the measuring instrument. Moreover, large offset component can cause the saturation of the amplifier (if any amplifier is used).

Taking into consideration the offset problem, the circuits presented in Fig. 4.3 are more convenient as the resistance to voltage transducers. These circuits enable the compensation of the offset component (Fig. 4.3c). The first one is the unbalanced bridge circuit described in Chapter 3 (deflection type bridge). If this bridge circuit is in the balance state for the starting point of the sensor resistance then the output signal is offset-free, because

$$U_{out} \cong S\frac{\Delta R_x}{R_{xo}} = S \cdot \varepsilon$$

(4.12)

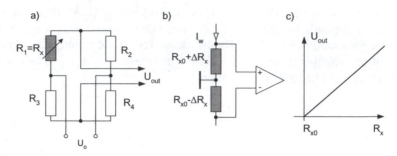

Figure 4.3. Converters of the resistance into the voltage with elimination of the offset component: unbalanced bridge (a), with differential amplifier (b) and their transfer characteristic (c)

The deflection type bridge circuit exhibits several other important advantages. First of all such a circuit is immune to the variation of the external interferences, for example changes of the ambient temperature. If all resistors in the bridge circuit are the same and only one of them is the sensor ($\varepsilon_1 = \delta x$), then with the change of the ambient temperature all resistors change their resistances: $\varepsilon_2 = \varepsilon_3 = \varepsilon_4 = \delta T$ while $\varepsilon_1 = \delta x + \delta T$. Thus according to Eq. (3.71) δT components are eliminated and only useful δx component remains at the output of the bridge

$$U_{out} = \frac{1}{4}\left(\delta x + \delta T - \delta T - \delta T + \delta T\right) = \frac{1}{4}\delta x \qquad (4.13)$$

It is much better to use the differential type of the sensor ($\varepsilon_1 = +\delta x$, $\varepsilon_2 = -\delta x$ and $R_1 = R_{xo}(1+\varepsilon)$, $R_2 = R_{xo}(1-\varepsilon)$) connected to the adjacent arms of the bridge circuit. In this case we obtain elimination of interference effects with two times larger output signal. And of course the best case is to use four differentially connected sensors, because in such circuits we obtain four times larger output signal.

The unbalanced bridge circuit usually exhibits non-linear transfer characteristic (for more details see Chapter 3). In some cases this feature can be profitable, because this way we can perform linearization of the sensor performances (see example presented in Fig. 3.37).

Similar performances as the unbalanced bridge circuits (Fig. 4.3a) exhibits the circuit with differential amplifier (Fig. 4.3b). The differential amplifier converts the difference of the input signals ($U_{out} = K_u \, (U_1 - U_2)$). Thus if one of the input resistors is active (a measuring sensor) and the second one is the same passive resistor we obtain elimination of the offset voltage and also elimination of the interferences.

It is advantageous to connect four sensors into bridge circuit because we obtain the maximal value of the sensitivity. But how to balance the bridge circuit with four active arms? Theoretically, we can connect parallel balancing resistor to one of the arms. But in this case one of the resistors exhibits different performances than the other three resistors, which can cause incomplete elimination of the external influences.

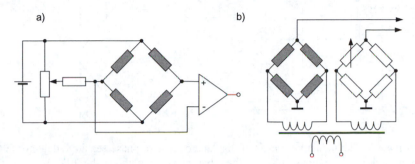

Figure 4.4. The balancing methods of the four-sensors bridge circuits: supplied by DC signal (a) and by AC signal (b)

Fig. 4.4 presents two examples of balancing methods of the four-sensor bridge circuit. The circuit presented in Fig. 4.4a enables balancing of the bridge circuit supplied by the DC, and Fig. 4.4b by AC voltage. Circuit from Fig. 4.4b is often used in the strain-gauge bridge systems. The output signal of the bridge circuit is compensated by the output signal of the second bridge circuit supplied by the same voltage.

The methods of conversion of the resistance to the voltage signal described above are also suitable for the conversion of the capacitance, inductance or generally impedance. In such case the bridge circuit should be supplied by the AC voltage. The Wheatstone bridge can be substituted by the special AC bridge circuit (Maxwell, Wien or other – see Chapter 3). But in this case we have to eliminate parasitic capacitances and inductances. It is also necessary to consider the influence of the cable's capacitance. The AC circuits require balancing of two components – magnitude and phase.

Even if we balance both components of the unbalanced AC bridge circuit we do not solve all problems. The output signal contains two components – one in phase with the supply voltage, and the other one shifted by 90 degrees. The shifted component can be effectively eliminated by phase-sensitive rectifier (more details will be given in Section 4.2.4). The phase-sensitive rectifier realizes the following dependence

$$U_{out} = kU_x \cos\varphi \qquad (4.14)$$

where φ is the phase shift of the output signal with respect to the reference voltage U_{ref}. As this reference we can use the supply voltage and then we can eliminate the 90 degrees shifted component of the output signal (because if $\varphi = 90^o$ then $cos\varphi = 0$).

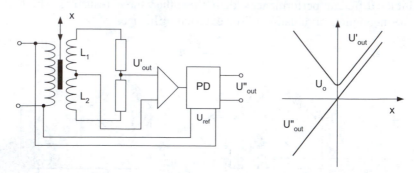

Figure 4.5. The inductive sensor of displacement with phase-sensitive detector (PD) connected to the output (U_{out}' – the characteristic without the PD, U_{out}'' – the characteristic after application of the PD)

Figure 4.5 presents the example of application of the phase sensitive detector for the correction of the transfer characteristic of a differential inductive displacement transducer. If the moving iron element is in the middle position, then inductances L_1 and L_2 are the same and the bridge circuit is balanced. After displacement of the iron element one inductance increases and the other one decreases. The output signal of the circuit is proportional to the displacement. The transfer characteristic of the sensor is nonlinear and contains some offset (line U_{out}' – Fig. 4.5). With the phase sensitive detector we can obtain a linear output characteristic without the offset (line U_{out}'' – Fig. 4.5). Moreover, with the detector we obtain another positive effect – taking into account the polarization of the signal we can detect the direction of displacement (which was not possible in the case of the AC output signal).

We can convert the variation of the resistance, inductance and especially the capacitance to the frequency using a typical oscillator circuit. Fig. 4.6 presents two such circuits – *Hartley* and *Colpitts* oscillators. In these circuits, the nonlinear dependence of the frequency on the measured parameter is inconvenient, because usually $f \approx \sqrt{1/X}$, where X is C or L.

It also is possible to adapt the oscillator utilizing the operational amplifier presented in Fig. 4.7a. The frequency depends on the capacitance C (or resistance R) as follows

$$f = \frac{1}{2RC\ln(1 + 2R_1/R_2)} \qquad (4.15)$$

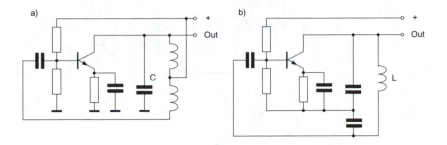

Figure 4.6. The conversion of the capacitance C or inductance L to the AC signal with frequency dependent on the measured parameter: a) Hartley oscillator, b) Colpitts oscillator

Fig. 4.7b presents the hybrid oscillator circuit developed by Analog Devices. The *OSC1758* circuit converts the capacitance C ($C=C_1=C_2$) to the frequency of the range *0 – 10 kHz* according to the relation $f = (1/C) \cdot 10^{-5}$ (where C in Farads, f in Hz).

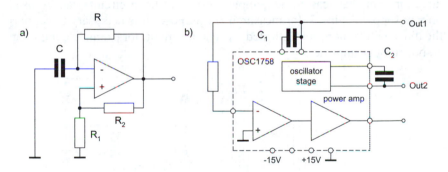

Figure 4.7. Converters of capacitance to frequency: a) with operational amplifier, b) hybrid power oscillator OSC1758 of Analog Devices[1]

4.1.3. AC/DC conversion

It is relatively easy to convert the value of AC signal to the DC signal using diode rectifiers. A typical rectifying diode conducts the current for only one direction of applied voltage. An example of the transfer characteristic of a silicon rectifying diode is presented in Fig. 4.8.

---

[1] OSC1758 is not currently manufacture by AD.

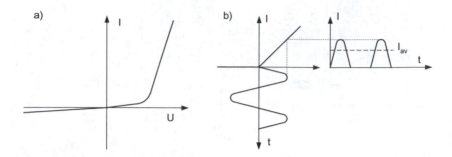

Figure 4.8. The transfer characteristic of typical rectifying diode (a) and the principle of the AC/CD conversion by means of such diode (b)

A typical transfer characteristic of the rectifying diode is presented in Fig. 4.8a. In the conducting direction the characteristic is nonlinear and the conductance starts from a certain non-zero threshold voltage ($U_p \cong 0.5\ V$). Moreover, a small inverse current will be conducted in the "non-conducting" direction. For that reason, the simple rectifying diode circuits can be used only in supply devices – for measurement purposes it is necessary to support the diode with a more sophisticated electronic circuit, for example containing an operational amplifier.

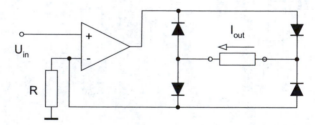

Figure. 4.9. AC/DC converter with rectifying diodes and the operational amplifier

Figure 4.9 presents the typical AC/DC converter with rectifying diodes and operational amplifier. The amplifier operates as a current transducer – rectified current flows in the feedback circuit. For strong feedback the threshold voltage $U_p{}'$ decreases to negligible small value $U_p{}' \approx U_p/K_u$. (K_u – gain of the amplifier). In such case the converter behaves as a practically linear transducer of input voltage u_{in}

$$I_{wy} = \frac{|u_{in}|}{R} \tag{4.16}$$

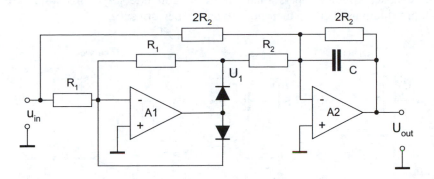

Figure 4.10. Converter of average value of AC voltage to the DC voltage (Tran Tien Lang 1978)

Figure 4.10 presents another converter of average value of AC voltage to DC voltage. The amplifier *A1* works a half-wave rectifier according to the following conditions

$$U_1 = \begin{cases} -u & for \quad u \ge 0 \\ 0 & for \quad u \le 0 \end{cases} \tag{4.17}$$

The second amplifier (*A2*) adds two voltages

$$U_{out} = -(u_{in} + 2U_1) \tag{4.18}$$

thus, the full-wave rectification is achieved:

$$U_{out} = \begin{cases} u & for \quad u \ge 0 \\ -u & for \quad u \le 0 \end{cases} \tag{4.19}$$

The converters presented in Fig. 4.9 and 4.10 allows us to determine averaged rectified value of AC signal. The conversion of the *rms value* of the AC signal to the *DC* voltage is more difficult (Kitchin 1986). According to the Eq. (4.2c) we should perform the following operations: square, mean value calculation and root operation (*root-mean-square*). Theoretically these mathematical operations are possible to perform using two multipliers, as is schematically shown in Fig. 4.11.

In the circuit presented in Fig. 4.11 the first multiplier is squaring the input voltage, then this voltage is averaged in the integrator circuit and next rooting is performed. The root operation is achieved by second multiplier in the feedback circuit of operational amplifier. Converters with this design are rather seldom used due to their poor dynamic performance.

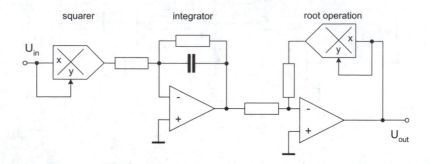

Figure 4.11. The true *rms* converter using two multipliers

Many manufacturers offer ready-to-use monolithic converters of true *rms* value. As the examples consider two types of *rms* converters developed by the Analog Devices: AD636 and AD637.

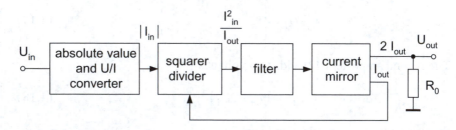

Figure 4.12. The principle of operation of the AD636 *rms* converter of Analog Devices (Kitchin 1986)

In the AD636 *rms* converter (Fig. 4.12) the input voltage is rectified and converted to the current. In the squarer/divider block the function I_{in}^2/I_{out} is realized. This block acts as the squarer and root calculator due to the feedback (the squarer/divider circuit utilizes the log/antilog device described below). Next, the active low-pass filter allows us to obtain the mean value of the signal. The rectified I_{out} signal is proportional to the *rms* value of the input voltage. This transducer converts the input voltage with the uncertainty not larger than *1mV ± 0.5% rdg*. The more detailed circuit illustrating the principle of operation of the AD636 converter is presented in Fig. 4.13.

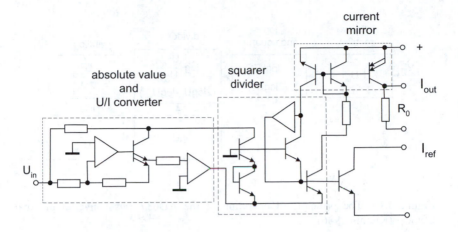

Figure 4.13. The simplified circuit of the AD636 *rms* converter of Analog Devices (Kitchin 1986)

Fig. 4.14 presents an example of the application of the *rms* converter – as the *true rms* voltmeter (proposed by the Analog Devices).

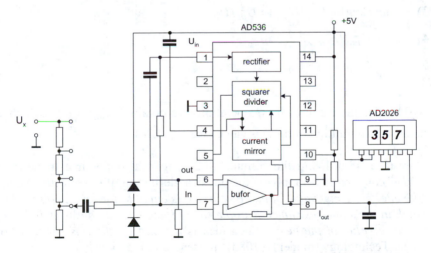

Figure 4.14. The design of the *true rms* voltmeter design – the proposal of Analog Devices (Kitchin 1986)

Figure 4.15 presents the principle of operation of another true *rms* converter of Analog Devices – model AD737. This device utilizes the log-antilog circuit (logarithmic and exponential amplifier – described in Section 4.4.6).

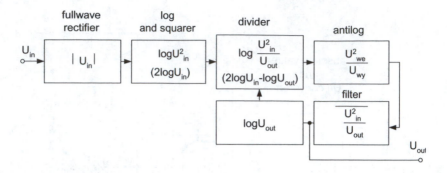

Figure 4.15. The principle of operation of the AD637 *rms* converter of Analog Devices (Kitchin 1986)

The AD737 *rms* converter (Fig. 4.15) performs the following operations:

1) $2\log U_{in}$ which corresponds with: $\log U_{in}^2$

2) $2\log U_{in} - \log U_{out}$ which corresponds with: $\log\left(U_{in}^2 / U_{out}\right)$

3) $anti\log(\log U_{in}^2 / U_{out}) = U_{in}^2 / U_{out}$

4) $\overline{U_{in}^2 / U_{out}}$

5) $U_{out} = \overline{U_{in}^2} / U_{out}$

Thus, as the result of these operations we obtain: $U_{out} = \sqrt{\overline{U_{in}^2}}$ which is exactly the definition of the *rms* value (see Eq. 4.2c).

Instead of utilizing the formula (4.2c), it is also possible to perform *AC/DC* conversion by applying the physical definition of the effective value: *"the effective (rms) value of the AC current is equal to the DC current value, which in the same resistance R in the time of one period causes emission of the same value of the heat"*. Such idea is realized by *the rms* converter of Linear Technology (model LT1088)[1] presented in Fig. 4.16.

The voltage generated by the thermocouple is a measure of the heat emission on the resistor and is compared with the voltage generated by the second thermocouple sensor heated by the AC current. Thus, we compare the heat effects of both currents. Such a transducer enabled *AC/DC* conversion in very large bandwidth (up to *300 MHz*) with a large crest factor.

---

[1] LT1088 is not currently manufactured by LT.

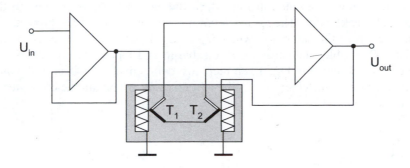

Figure 4.16. Thermal-effect converter of the *rms* voltage (Linear Technology 2005)

Table 4.2 presents the comparison of the performances of various *rms* converters.

Table 4.2. The parameters of typical true *rms* converters

Model	Manufacturer	max error of conversion	bandwidth	crest factor
AD636	Analog Devices	5mV±0,5%rdg	450kHz	7
AD637	Analog Devices	1mV±0,5%rdg	1MHz	10
AD737	Analog Devices	0,2mV±0,5%rdg	190kHz	5
LH0091	National Semic.	0,5mV±0,05%rdg	800kHz	10
LT1088	Linear Techn.	0,01FS	300MHz	50

Figure 4.17 presents the principle of the conversion of the peak value of the AC signal. The rectified voltage is connected to the capacitor *C*. If the time constants of the *RC* circuits (circuits of charge and discharge respectively) are designed in such a way that process of charging is fast and discharging is slow then the voltage across the capacitor is equal to the peak value of the supplying voltage. Figure 4.18 presents two examples of peak value converters.

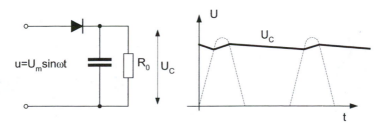

Figure 4.17. The principle of operation of the peak-value converter

In the circuit presented in Fig. 4.18a the capacitor C_1 is charged to the maximal (peak) value. The feedback circuit (elements R_1 and D_1) protects the amplifier from the saturation when $U_{in} < U_c$. Periodical connection of the resistor R_p enables the capacitor to discharge and repeat the conversion. In the circuit presented in Fig. 4.18b there are two converters of the peak value connected to the differential amplifier. In this way we obtain the converter of the peak-to-peak U_{pp} value.

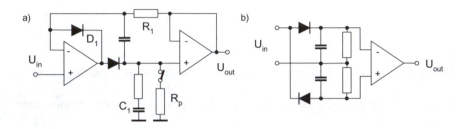

Figure 4.18. The examples of peak value (a) and peak-to-peak value (b) converters (Pallas-Areny 1999)

The *phase-sensitive rectifier (phase-sensitive demodulator* or *phase-sensitive detector)* plays a very important role in the measurements and instrumentation. It helps in the separation of two signal components – in phase and 90 degrees out of phase, as it was presented in the example earlier (see Fig. 4.5). Moreover, this converter is the main element of the lock-in amplifier – described in Section 4.5.4).

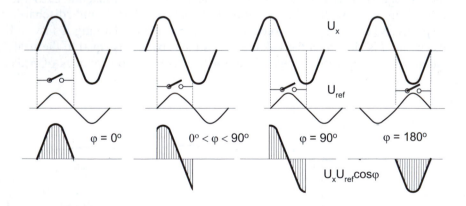

Figure 4.19. The principle of operation of the phase-sensitive rectifier

The phase-sensitive rectifier converts the measured voltage U_x with respect to additional reference voltage U_{ref}. Thus, it is equipped at least with two inputs and one output. The reference voltage switches the rectifying elements in such a way that the mean value of the output signal depends on the phase shift between both input signals. This principle of operation is illustrated in Fig. 4.19.

The phase-sensitive rectifier works according to the following equation

$$U_{out} = kU_x \cos\varphi \qquad (4.20)$$

where $\cos\varphi$ is the cosine of the phase shift between U_x and U_{ref}.

A typical diode polarized as conducting and non-conducting element can be used as the switch in phase sensitive rectifier. An example of such a rectifier is presented in Fig. 4.20.

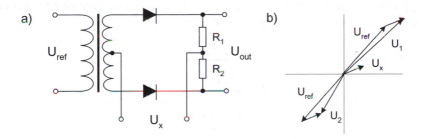

Figure 4.20. The simplest diode phase-sensitive rectifier

In the phase-sensitive rectifier circuit presented in Fig. 4.20 the voltage signal U_x is added to the reference voltage U_{ref} in the first half-period of this signal and is subtracted from this voltage in the second half of the period. According to the phasor diagram presented in Fig. 4.20b the output signal U_{out} is

$$U_1 = \sqrt{U_{ref}^2 + 2U_{ref}U_x\cos\varphi + U_x^2} \; ; U_2 = \sqrt{U_{ref}^2 - 2U_{ref}U_x\cos\varphi + U_x^2} \qquad (4.21)$$

and

$$U_{out} = U_1 - U_2 = 2U_x \cos\varphi \left(1 - \frac{U_x^2}{U_{ref}^2}\sin^2\varphi\right) \cong 2U_x \cos\varphi \qquad (4.22)$$

For correct operation of this rectifier it is required that $U_{ref} > U_x$. The full-wave version of diode phase-sensitive rectifiers is presented in Fig. 4.21a.

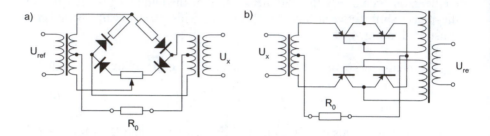

Figure 4.21. Two examples of the full-wave phase-sensitive rectifiers

Also, a transistor can be used as the phase sensitive switch (Fig. 4.21b), because to obtain the conduction both voltages U_{BE} and U_{BC} should be appropriately polarized. The mean value of the collector current depends on the phase shift between U_{BE} and U_{BC} voltages.

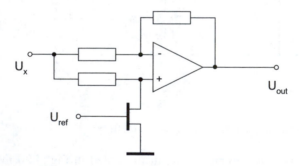

Figure 4.22. The phase-sensitive rectifier with the operational amplifier

The advantage of diode and transistor phase sensitive rectifiers is that it is not necessary to use any additional supply voltage. However, a phase-sensitive rectifier with the operational amplifier (Fig. 4.22) exhibits better performance. In such a case one of the inputs of the circuit is connected to the V_{ref} reference voltage.

There are also available ready-to-use monolithic phase-sensitive rectifiers. As an example a functional block of the AD630 circuit developed by the Analog Devices is presented in Fig. 4.23.

The monolithic phase-sensitive detector AD630 includes two precise input amplifiers and a third amplifier switched by the reference voltage connected to the input 9 (see Fig. 4.23).

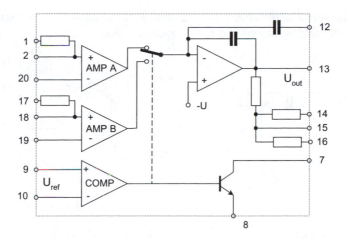

Figure 4.23. The functional block diagram of phase-sensitive detector model AD630 of Analog Devices (Analog Devices 2005)

Figure 4.24 presents an example of the application of the phase sensitive detector. This device can be used as the output circuit of the unbalanced type AC bridge circuit. This application enables us to separate only the in-phase component of the bridge output signal.

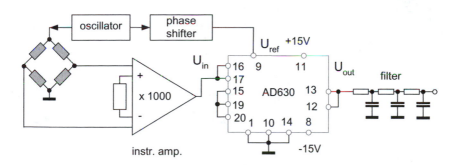

Figure 4.24. The application of the phase-sensitive rectifier as the output circuit of the unbalanced type AC bridge (Analog Devices 2005)

4.1.4. Voltage to frequency conversion

The time-varying signals, as frequency, period, phase shift, width of the pulse prove to be advantageous in comparison with amplitude-varying signals. Similarly as digital signals they are more immune to interferences. Frequency is a value, which we can measure easily and with high precision.

Thus in some applications it is convenient to convert information from amplitude-varying signal into the frequency value.

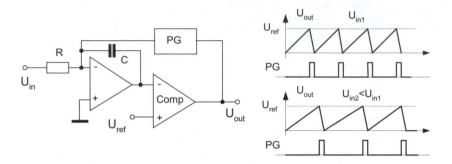

Figure 4.25. The voltage to frequency converter (Tran Tien Lang 1987)

Figure 4.25 presents a typical design of a *V/f* converter with an integrator circuit charging a capacitor *C* at a rate proportional to the amplitude of input voltage U_{in}. Each time when the output voltage of integrator circuit reaches certain value equal to the reference voltage U_{ref} the comparator switches into a reset mode and discharges the capacitor. The pulse generator *PG* can be used for this goal. The frequency of the output signal depends on the amplitude of input voltage

$$f = \frac{U_{in}}{RCU_{ref}}$$ (4.23)

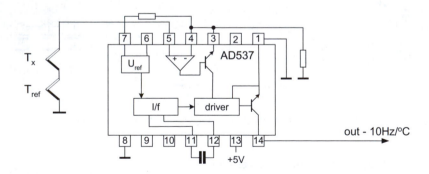

Figure 4.26. An example of the temperature transducer utilizing AD537 voltage-to-frequency converter of Analog Devices

Figure 4.26 presents the hybrid voltage to frequency converter of Analog Devices – model AD-537.

The voltage-to-frequency converter AD537 of Analog Devices enables us the conversion of the input voltage to the frequency up to *100 kHz* with nonlinearity error less than *0.05%*. The conversion factor $K=U_{in}/f$ can be set by connecting appropriate external *R* and *C* elements to the device. Figure 4.26 presents the example of application of this transducer to the temperature conversion with the conversion factor *10 Hz/°C* (for chromel-constantan thermocouple sensor).

4.2. AMPLIFICATION OF THE SIGNALS

4.2.1. The differential, operational and instrumentation amplifiers

The amplifiers are generally used for the amplification of the voltage signal. However, the amplification process enables also improvement of the signal quality – mostly the signal-to-noise ratio. A good amplifier should exhibit sufficiently large and steady amplification factor $K_u=U_{out}/U_{in}$, large input resistance and small output resistance.

It is also required to perform the amplification without the distortion. The amplifier can process the signal with *frequency distortion* (*linear distortion*) caused by the unequal gain of all frequencies comprising the signal. To obtain small linear distortion the bandwidth of frequency should be sufficient. Also the gain K_u should be the same for all frequencies in the bandwidth. There is also *amplitude distortion* (*non-linear distortion*) caused by non-linear transfer characteristics and introduction to the signal additional harmonics. The non-linear distortion appears when the input signal is large. Therefore it is recommended to limit input signal of the amplifier by applying the negative feedback.

Usually, it is assumed that the *frequency bandwidth* is the frequency range for which the K_u factor does not roll off more than *3dB*. And another important factor – the *dynamics of the amplifier* (ratio between the smallest detectable signal and the largest – limited by the supply voltage signal). Dynamics (the smallest signal) depends mostly on the level of noises and zero drifts.

The real revolution in the amplification technique was the development of the integrated operational amplifiers, and later the instrumentation amplifiers (Coughlin 2000, Franco 2001, Jung 2004, Stanley 2001). The main part of these devices is the *differential amplifier* (Fig. 4.27). The important advantage of such an amplifier is the possibility of suppression of the parasitic signals. The input signal is processed as the difference of two inputs signals

$$U_{out} = K_u(U_1 - U_2)$$ (4.24)

The parasitic interference signals ΔU are the same on both inputs. Therefore the output signal is

$$U_{out} = K_u[(U_1 + \Delta U) - (U_2 + \Delta U)] = K_u(U_1 - U_2) \tag{4.25}$$

Thus it is possible to amplify the voltage difference with the large common signal ΔU in the background.

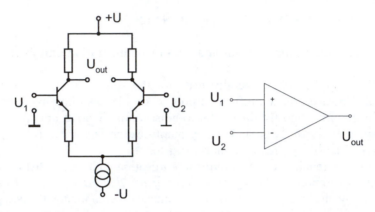

Figure 4.27. The differential amplifier and its symbolic representation

The possibility of the rejection of the common parasitic component is described by the coefficient *CMRR – Common Mode Rejection Ratio* defined as

$$CMRR = \frac{K^-}{K^+} \tag{4.26}$$

where K^- is the amplification of the voltage difference and K^+ is the amplification of the common signal. In use is also the parameter *CMR – Common Mode Rejection* defined as

$$CMR = 20 \log CMRR \tag{4.27}$$

Taking into account this parameter the output voltage is

$$U_{out} = (U_1 - U_2)K^-\left[1 + \frac{1}{CMRR}\frac{\Delta U}{(U_1 - U_2)}\right] \tag{4.28}$$

The second component in the square brackets of the equation (4.28) describes the error caused by the presence of the common component.

The operational amplifier (*OpAmp*) is a special kind of the integrated amplifier with the following performances: very large amplification factor – larger than 10^5, very large input resistance and small output resistance. Additionally, small temperature errors coefficient, small level of noises, small zero drifts and large *CMRR* factor are required.

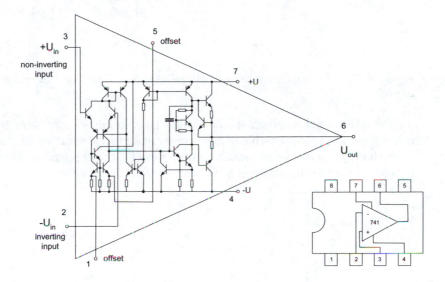

Figure 4.28. The typical circuit of the operational amplifier

Fig. 4.28 presents the typical design of the operational amplifier, which is in fact a developed version of the simple circuit presented in Fig. 4.27. The operational amplifier practically does not work intrinsically – with such large amplification it would be immediately saturated. Therefore the operational amplifier works always with suitable feedback. We can say that the operational amplifier is some kind of semi-finished product designed for the construction of various electronic circuits. By connecting an appropriate feedback circuit we can obtain the required device: multiplier, adder, integrator, generator, filter and of course voltage amplifier.

Figure 4.29 presents two typical circuits of the voltage amplifiers based on the operational amplifier – with asymmetrical input and symmetrical differential one.

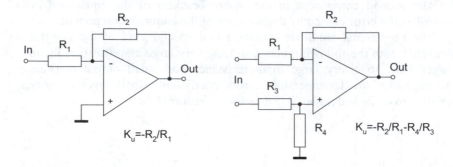

Figure 4.29. Two main examples of voltage amplifiers based on the operational amplifier

The operational amplifier should exhibit large amplification factor but it is not absolutely indispensable to have steady and precise value of this factor – which is obligatory for measurement purposes. This requirement is fulfilled for special kind of the amplifier – *the instrumentation amplifier*.

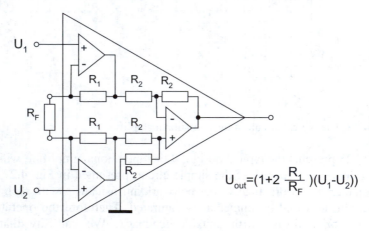

Figure 4.30. Typical circuit of the instrumentation amplifier

The instrumentation amplifier is also the integrated circuit with differential amplifier and its amplification factor should be stable, precise in setting and immune to interferences. Also other parameters should be of high quality. The *CMMR* factor should be larger than *110 dB*, temperature zero drift smaller than *1 μV/°C*, input impedance larger than 10^9 *Ω*, input bias current smaller than *10 nA* (for bipolar transistors) or *10 pA* (for FET transistors), noises below *10 nV/√Hz*, nonlinear error below *0.01%*

(attainable *0.0001%*), frequency bandwidth not smaller than *100 kHz* (attainable larger than *100 MHz*).

We can see that the instrumentation amplifier is a special kind of amplifier suitable for measurement purposes. It consists of several operational amplifiers – the typical circuit of the instrumentation amplifier is shown in Fig. 4.30. The amplification factor is set by appropriate external resistor R_F.

4.2.2. Isolation amplifiers

There are circumstances, when it is necessary to ensure the galvanic separation of the input and output circuit. For example, when we measure very small differential signal superimposed on large common signal. Such a case is presented in Fig. 4.31 where the relatively low voltage (with the level of tens *mV*) across the shunt resistor R_b is measured in the presence of relatively high voltage (several hundred *V*), which is used for supplying the load R_o. Between the ground of the indicating instrument connected to the amplifier and the ground of the supply source there could be a voltage difference dangerous for the servicing personnel (and also the amplifier could be damaged due to this large potential difference).

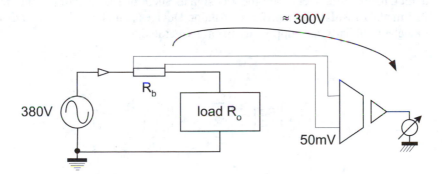

Figure 4.31. The application of the isolation amplifier for separation supply source and receiver when the potential difference of grounding can be large

As another example we can consider the case when during detection of fetal heartbeat signal V_B with level of about *50 μV* we cannot separate accompanying much larger (about *1 mV*) signal of maternal heart beat V_M and additionally large signal of ground coupling, power line *50 Hz* interference, etc. Especially dangerous is the situation when large difference between ground potential of input circuit and output circuit exists. Therefore in medical equipment it is indispensable to use the isolation amplifiers.

Also dangerous is the situation when the input circuit is distant from the rest of the circuit and even more when these both parts are supplied from different sources. The difference between the ground potential of both circuits can be as large as hundreds of volts. Without galvanic separation such difference can cause destruction of the amplifier (and of course could be dangerous to the servicing personnel).

In the isolation amplifier the connection between input terminals and output terminals does not exist (and the same applies for input ground and output ground). Sometimes, also the supply sources are physically separated. The rejection of the isolation voltage U_{IM} (voltage between common parts of input and output circuits) is described by *IMRR – Isolation Mode Rejection Ratio*

$$U_{out} = K_u \left(U_{in} + \frac{\Delta U}{CMRR} \right) + \frac{U_{IM}}{IMRR} \tag{4.29}$$

It is convenient to use a small transformer as the isolation device. However, the transformer does not transform *DC* signals, which is sometimes necessary. For that reason it is needed to use the modulator (signal is transformed as the magnitude varying carrier *AC* signal) and the demodulator to recover again the *DC* signal. Such idea is realized in the *AD 215* model of isolation amplifier of Analog Devices (Fig.4.32). In the *AD210* model additionally the supply circuits are separated.

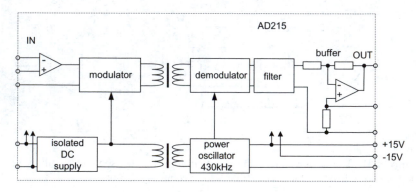

Figure. 4.32. The isolation amplifier with transformer galvanic separation – model AD215 of Analog Devices

Another principle of galvanic separation is applied in the isolation amplifier of *Texas Instruments* – model *ISO124* (Fig. 4.33). In this case a capacitor with capacitance *1 pF* is used as the isolation device. The input signal is converted to *AC* signal using integrating analogue to digital circuit

A1. Next, this signal transmitted by the capacitance is again converted to the *DC* signal using integrating amplifier A2. Additional sample-and-hold circuit is used to obtain the output signal without the carrier components.

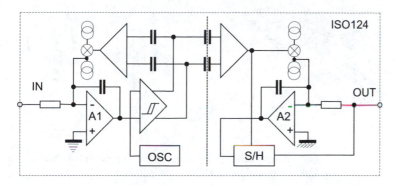

Figure 4.33. The isolation amplifier with capacitive galvanic separation - model ISO124 of Texas Instruments (Texas Instruments 2005)

It is also possible to use the light transmission as the separation device. Such a principle is utilized in the isolation amplifier of Burr-Brown – model 3652 (Fig. 4.34). The LED diode D3 illuminates two identical photodiodes D2 in the output circuit (converter current/voltage) and D1 in the input circuit (converter voltage/current). The diode D1 is in feedback circuit, which enables us to minimize the nonlinearity errors. Since both diodes are illuminated identically therefore $I_1 = I_2 = I_{in}$.

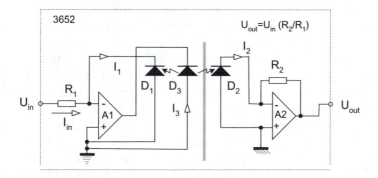

Figure 4.34. The isolation amplifier with light galvanic separation – model 3652 of Burr-Brown (Burr Brown 2005)

The isolation amplifiers described above exhibit *IMRR* factor as large as *130 – 140 dB* for permissible voltage difference between input and output as large as *1500 – 3000 V*.

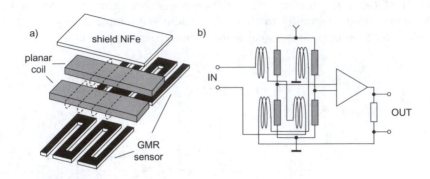

Figure 4.35. The isolation amplifier with magnetic galvanic separation – model developed by Nonvolatile Electronics Inc (Nonvolatile 2005)

Figure 4.35 presents the galvanic separation device developed by Nonvolatile Electronics Inc. (Nonvolatile 2005). The current in the planar coil generates the magnetic field detected by the *GMR* (giant magnetoresistance) sensor. The *GMR* sensor is in the feedback circuit – it only detects the presence of magnetic field and thus its nonlinearity does not influence the linearity of the device. This device can work in analogue (Fig. 4.36b) or in digital mode (with transmission speed even above *1 GB/s*).

4.2.3. Amplifiers of very small DC signals

The main problem in the amplification of small DC signals is the temperature zero drift of the amplifier, because it is very difficult to separate this zero drift signal and the useful *DC* signal. The temperature zero drift is caused by non-ideal technology – even small differences in various parts of the circuit can be the source of *DC* offset signal. This effect is minimized by laser trimming to the level of *1 – 50 μV/°C*, but even such small offset is still limiting the amplification of small *DC* signals.

Recently, there are two main techniques of decreasing the zero drift influence. The first one is based on the conversion of the *DC* signal to the *AC* signal. Then it is possible to separate *DC* zero drift and *AC* useful signal. The *AC* signal can be amplified with ease and converted back to the *DC* one. This technique is called a *chopper amplifier*, because the conversion to the *AC* signal is achieved by very fast connecting and disconnecting the signal to the input.

The second technique employs the *auto-zero* principle. The amplifier is periodically disconnected from the input, then the input is short-circuited to ground and the zero component is compensated. The auto-zero technique enables us to eliminate the zero drift down to the level of *5nV/°C*.

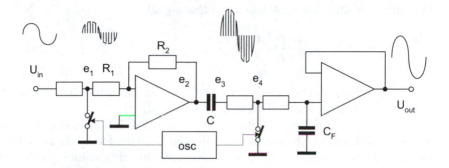

Figure 4.36. The principle of the chopper amplifier

The principle of operation of the chopper amplifier is illustrated in Fig. 4.36. The signal is switched with frequency of up to tens of kHz. To the amplifier are delivered pulses with amplitude modulated by the input signal. After amplification this signal is demodulated by the synchronous switching (or using the phase-sensitive detector).

If the input signal is $u_{in}(t)$ and the commutation frequency is ω_o, then the signal at the input of amplifier is

$$e_1 = u_{in}(t)\left(\frac{1}{2}+\frac{2}{\pi}\cos\omega_o t - \frac{2}{3\pi}\cos 3\omega_o t + \frac{2}{5\pi}\cos 5\omega_o t...\right) \qquad (4.30)$$

If we describe the zero component as U_o and the zero drift as ΔU_o and the analysis is limited to the first harmonics (we can eliminate higher harmonics using the filter at the output) the output voltage of the amplifier is

$$e_2(t)=-\frac{R_2}{R_1}u_{in}(t)\left(\frac{1}{2}+\frac{2}{\pi}\cos\omega_o t\right)+\left(1+\frac{R_2}{R_1}\right)U_o+\left(1+\frac{R_2}{R_1}\right)\Delta U_o \qquad (4.31)$$

The DC components are blocked by the capacitor *C*, thus the e_3 signal is

$$e_3(t)=-u_{in}(t)\frac{R_2}{R_1}\left(\frac{2}{\pi}\cos\omega_o t\right) \qquad (4.32)$$

After the second commutation the signal is

$$e_4(t) = e_3(t)\left(\frac{1}{2} + \frac{2}{\pi}\cos\omega_0 t - \frac{2}{3\pi}\cos 3\omega_0 t \ldots\right) \qquad (4.33)$$

Thus after the filtration we obtain the output signal

$$u_{out}(t) = -u_{in}(t)\frac{R_2}{R_1}\frac{4}{\pi^2} + U_{o2} \qquad (4.34)$$

The U_{o2} component is the zero drift of the last amplifier and it is negligible because it is added to the large output signal.

The chopper amplifier exhibits several drawbacks. The main ones are: the limitation of frequency bandwidth (it is limited from the top by the frequency of the carrier signal), difficult realization of differential mode of amplification, and necessity of the signal filtration. For that reason the chopper technique is recently often substituted by auto-zero technique.

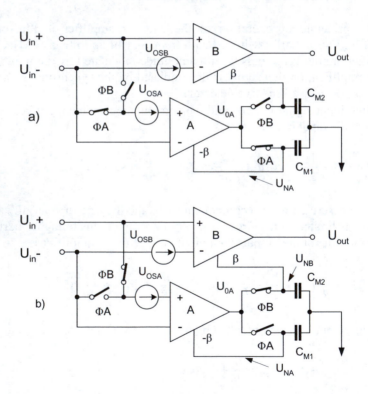

Figure 4.37. The operation principle of the auto-zero amplifier – model AD8551 of Analog Devices (a) step A of the operation, (b) step B (Analog Devices 2005)

The auto-zero amplifier presented in Fig. 4.37 consists of two amplifiers commuted by the switches ΦA and ΦB. The operation of the circuit is performed in two steps: A and B. During the first step of operation the switches ΦA are connected, in the second step – switches ΦB. In the first step (Fig. 4.37a) the amplifier B is connected to the input, while the input of the amplifier A is short-circuited. The amplifier A amplifies only the offset signal U_{OSA} and this amplified signal is connected to the feedback circuit. Thus the output signal is

$$U'_{OA} = K_u U_{OSA} - \beta U'_{OA} \qquad (4.35a)$$

and

$$U'_{OA} = \frac{K_u U_{OSA}}{1 + \beta} \qquad (4.35)$$

where K_u is the amplification of the amplifiers (the same for both amplifiers) and β is the amplification of the third input – also the same for both amplifiers.

The U'_{OA} signal charges the capacitor C_{MI} connected to the output. After finishing the step A the switches ΦB are connected (and switches ΦA are disconnected).

In the second step (Fig. 4.37b) at the output of the amplifier A there is a voltage

$$U''_{OA} = K_u U_{in} + K_u U_{OSA} - \beta U_{NA} \qquad (4.36)$$

The third component of the equation (4.36) is the voltage U_{NA} across the capacitor C_{MI} (the time constant of the capacitor circuit is relatively large during the B step and the switching process is sufficiently fast that the capacitor does not have time to discharge). From (4.35) and (4.36) we obtain

$$U''_{OA} = K_u U_{in} + K_u U_{OSA} - \beta \frac{K_u U_{OSA}}{1 + \beta} \qquad (4.37)$$

After simple calculations of the equation (4.37)

$$U''_{OA} = K_u \left(U_{in} + \frac{U_{OSA}}{1 + \beta} \right) \qquad (4.38)$$

In the second step the contribution of the zero drift component is decreased by a factor of $(1+\beta)$. The output signal in the step B is

$$U_{out} = K_u U_{in} + K_u U_{OSB} + \beta U_{NB} \qquad (4.39)$$

Because voltage $U_{NB} = U''_{OA}$ thus

$$U_{out} = K_u U_{in} + K_u U_{OSB} + \beta K_u \left(U_{in} + \frac{U_{OSA}}{1+\beta} \right) \qquad (4.40)$$

and

$$U_{out} = U_{in}\left(K_u + K_u \beta \right) + \frac{K_u \beta U_{OSA}}{1+\beta} + K_u U_{OSB} \qquad (4.41)$$

The amplifier circuit is designed in such way that $\beta >> 1$. Thus $K_U \beta >> K_u$ and the equation (4.41) can be simplified to

$$U_{out} \cong K_u \beta U_{in} + K_u \left(U_{OSA} + U_{OSB} \right) \qquad (4.42)$$

We can see that the input voltage is amplified by a factor of $K_u \beta$, while the zero drifts are amplified by the factor β-times smaller. The resultant zero drift component U_{OS} is

$$U_{OS} = \frac{U_{OSA} + U_{OSB}}{\beta} \qquad (4.43)$$

Because β is large the resultant zero component is reduced to the nV level. Thus these two steps of the auto-zero operation significantly reduce the zero drift component without limitation of the frequency bandwidth and with the possibility of the differential mode of amplification.

4.2.4. Amplifiers of very small AC signals

While in the case of amplification of small DC signals the main limitation is the zero drift, in the case of small AC signal this limitation is caused by the noise level. The noise consists of unlimited number of signals with various frequencies (white noise) and sometimes the noise itself is larger than the useful input signal. If we know the frequency of this input signal we can separate this signal from noises using a selective amplifier. The selective amplifiers (tuned to certain known frequency) are used in the measurements rather seldom, because it is very difficult to ensure stable amplification. Small deviation from the resonance frequency causes large variation of the amplification. For that reason for the amplification of small *AC* signals the *lock-in amplifiers* are used more often (Scofield 1994, SR 1000 2005).

In the lock-in amplifiers a phase sensitive detector is used as the selective element. This detector selects from the input signal only these components that have the same frequency as the reference voltage. Moreover it selects also only these signals that are in phase with the reference signal. When the ordinary selective amplifier exhibits the Q factor of about 50, the lock-in amplifier can exhibit this factor as large as 100 000 (it is possible to select the signals of 10 kHz in the frequency bandwidth 0.01 Hz). In this way it is possible to select from the noises signals of the level below the nV.

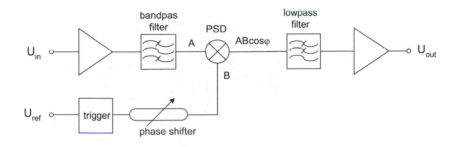

Figure 4.38. The operation principle of the lock-in amplifier (SR 1000 2005)

Figure 4.38 presents the operation principle of the lock-in amplifier. After preliminary amplification the input signal is connected to the phase sensitive detector, where it is rectified with respect to the frequency and phase of the reference signal. Next, this signal is filtered by low-pass filter.

In Fig. 4.39 there are presented typical applications of lock-in amplifier. One of the most important is the application in the bridge circuit amplifier for the strain-gauge sensors (Fig. 4.39a). Due to very small variations of the resistance of this sensor the output signal of the bridge circuit is also very small – only several μV. If we supply the bridge with AC voltage and we use the same voltage as the reference signal of the phase-sensitive detector then from the signal with noises we can separate only the signal of the frequency the same as supply voltage.

Similar principle is used for measuring very small resistances (Fig. 4.39b) (Gerstenhaber 1991). It is possible to measure very small voltage drop across this resistance by supplying this circuit with the current of the same frequency as the reference signal (from the same generator).

Using two phase sensitive detectors we can separate both components of the vector signal – in phase and shifted by 90°. In this way we can analyze the impedance components $Re(Z)$ and $Im(Z)$, determine the phase shift, perform spectral analysis, etc. (Fig. 4.39c).

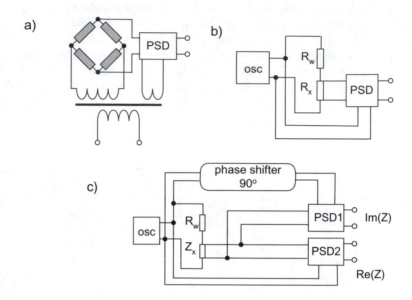

Figure 4.39. The examples of typical applications of the lock-in amplifier: a) measurement of small variations of the resistance – strain-gauge bridge, b) measurement of very small resistances – micro-ohm-meter) (Gerstenhaber 1991), c) measurement of both components of the vector signal

We can analyze the operation principle of the lock-in amplifier on the example of a bridge circuit used for the strain-gauge sensors presented in Fig. 4.40.

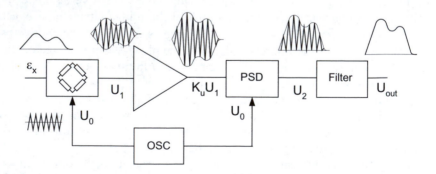

Figure 4.40. The block diagram of the bridge circuit for strain-gauge sensors

If the measured strain ε varies periodically with the frequency ω then the variation of the resistance of the sensor with the constant K is

$$\frac{\Delta R_x}{R_x} = \delta R_x = K\varepsilon_x = K\varepsilon_m \ sin \ \omega t \qquad (4.44)$$

The output voltage of the bridge circuit of the sensitivity factor S supplied by the voltage signal $U_0 = U_{om}sin\Omega t$ is

$$U_1 = S\delta R_x U_0 = SK\varepsilon_m U_{om} \ sin \ \omega t \ sin \ \Omega t \qquad (4.45)$$

After simple calculations we obtain

$$U_1 = U_{1m}\left[\cos(\Omega - \omega)t + \cos(\Omega + \omega)t\right] \qquad (4.46)$$

where $U_{1m} = SK\varepsilon_m U_{0m}$.

From the equation (4.46) we can see that the bridge circuit supplied by the AC voltage works as a modulator device – the magnitude of the output signal of the frequency the same as the supply voltage is modulated according to the variation of resistance (and stress). The spectral characteristics of these signals are presented in Fig. 4.41.

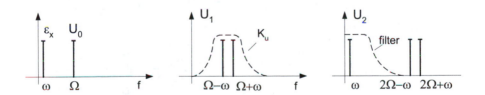

Figure 4.41. The spectral characteristics of the signals of the circuit presented in figure 4.40

The transfer characteristic of phase sensitive detector is

$$U_2 = K_u U_1 |cos \ \Omega t| \qquad (4.47)$$

or as a series

$$U_2 = \frac{2}{\pi}K_u U_1\left(1 - \frac{2}{3}cos \ \Omega t - \frac{2}{15}cos \ 4\Omega t + ...\right) \qquad (4.48)$$

thus

$$U_2 = \frac{2}{\pi} K_u U_{1m} \left[\sin \omega t + \frac{1}{3}\sin(2\Omega - \omega) t + \frac{1}{3}\sin(2\Omega + \omega) t + ... \right] \qquad (4.49)$$

The phase sensitive detector acts as a selective filter. Now it is only necessary to connect the low-pass filter to correctly recover the measured signal

$$U_{out} = \frac{2}{\pi} K_u U_{1m} \sin \omega t \qquad (4.50)$$

Figure 4.42 presents an example of the lock-in amplifier developed by Perkin Elemer Electronics (Signal Recovery) (SR 1002 2005). The device contains two phase sensitive circuits – one for the component in phase, and second for the shifted by 90° component. Internal or external source can be used as the reference signal.

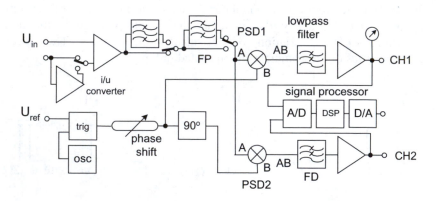

Figure 4.42. An example of the lock-in amplifier – model developed by Signal Recovery (SR 1002 2005)

The analogue lock-in amplifier exhibits several drawbacks: it is necessary to use a low-pass filter, the dynamics is rather poor, and the bandwidth is limited (about 1/5 of the carrier frequency). Therefore, recently there are also developed digital lock-in amplifiers (SR 1003 2005). In such a device both signals (measured and reference) are converted to the digital signals and the phase-sensitive detector is also a digital one.

As the measuring instrument the lock-in amplifier is usually a rather expensive and sophisticated device. Therefore it is reasonable to construct such devices using a monolithic phase sensitive modulator/demodulator (see Fig. 4.23). Figure 4.43 presents an example of the application of such monolithic circuit AD630 of Analog Devices as the lock-in amplifier used for the signal recovery. Figure 4.43b presents the oscilloscope traces of the signal before and after lock-in amplifier. It was possible to recover the signal measured in presence of noises 100 dB larger, which corresponds to noise greater from the useful signal by a factor of 10^5.

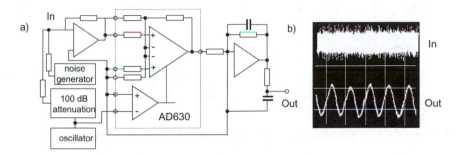

Figure 4.43. An example of the lock-in amplifier based on the AD630 device of Analog Devices (a) and the picture of the signals before and after amplification (b) (Analog Devices 2005)

4.5.5. The amplifiers of very large input resistance (electrometers)

The instrumentation amplifiers exhibit very large input resistance – even as large as 10^{12} Ω. But in certain applications such resistance is insufficiently large. For example, electrodes for measurements of *pH* exhibit internal resistance of about hundreds of $M\Omega$ and detection without current flow is necessary. The typical instrumentation amplifier is in this case not recommended. Similarly, stressed piezoelectric sensors generate electric charge, which can only be measured without the current flow.

Special kinds of amplifiers called *electrometers* or *charge amplifiers* are designed, which exhibit extremely large input resistance. For example, Keithley Company (Keithley 2005) developed the electrometers with the input resistance larger than hundreds of teraohms (*1 TΩ = 10^{12} Ω*) and the input current lower than several femtoampers (*1 fA = 10^{-15} A*). There are also instruments with the input resistance of 10 PΩ (*1 PΩ (penta) = $10^{15}\Omega$*) and input current 400 aA (*1a (atto) = 10^{-18}*). To put that into perspective consider that the voltage of about *10 V* causes in a typical insulator a current of several *pA*.

We can obtain very large input resistance by special technology (special kind of insulation and connections) and also by using special electrical circuits. Because as the source of very large resistance we can use the current source then we apply the current-voltage transducers for electrometers. For the circuit presented in Fig. 4.44a we can write that

$$U_{out} = -I_{in}R \qquad (4.51)$$

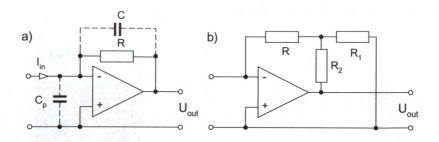

Figure 4.44. The typical circuits of the amplifiers with large input resistance

Thus the circuit presented in Fig. 4.44a operates as a current-voltage transducer with input resistance dependent on the resistance R. The time constant depends on RC and for large R it can be as large as several seconds. Therefore this circuit exhibits low-pass filtering characteristics with relatively small cut-off frequency. The application of very large $T\Omega$ resistance R can be inconvenient and therefore the modified amplifier circuit presented in Fig. 4.44b can be recommended (Pallas-Areny 1999). For this circuit we can write the following dependence

$$U_{out} = -I_{in}\left[R_2 + R\left(1+\frac{R_2}{R_1}\right)\right] \approx -I_{in}R\left(1+\frac{R_2}{R_1}\right) \qquad (4.52)$$

Thus we obtain multiplication of the input resistance by the ratio R_2/R_1.

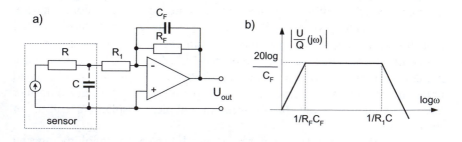

Figure 4.45. An example of the charge amplifier

For the circuit of the charge amplifier (Fig. 4.45) we can write that

$$U_{out} = \frac{Q}{C_F + (C + C_F)/K_u} \approx \frac{Q}{C_F} \tag{4.53}$$

We obtain the transducer of the charge with negligible influence of the cable capacitance C (when the gain of amplifier K_u is sufficiently large). The time constant of this transducer depends on the $R_F C_F$. For example for $R_F = 10\ G\Omega$ and $C_F = 100\ pF$ the low cut-off frequency is $0.16Hz$. The high cut-off frequency depends on $R_1 C$ and for example for $R_1 = 50\ \Omega$ and $C = 200\ pF$ it is $1.6\ MHz$.

4.2.6. The function amplifiers

By appropriate application of the feedback we can create various transfer characteristics of amplifier and in this way we can realize various mathematical functions as: integration, differentiation, logarithm, multiplication, division, addition, subtraction, etc.

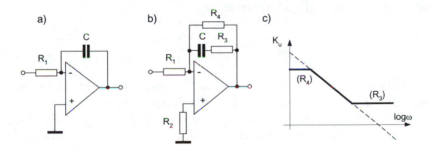

Figure 4.46. The integrating amplifier: a) principle of operation, b) practical circuit, c) frequency characteristic

Fig. 4.46 presents the circuit of the amplifier realizing the integration of the input signal. For ideal *integrator circuit* (Fig. 4.46a) we can write the following dependence

$$u_{out}(t) = -\frac{1}{R_1 C} \int_{t_0}^{t_0+T} u_{in}(t)dt + U_0 \tag{4.54}$$

thus the circuit realizes integration with the period of integration $T = R_1 C$ and U_0 is the voltage across the capacitor C before start of the integration. The presence of the zero voltage and associated zero drift is quite significant problem of the operation of integrator circuit and therefore usually the frequency bandwidth is limited by inserting additional resistor R_2 (Fig. 4.46b). We can set low and high cut-off frequencies by appropriate choice of resistance R_3 and R_4. Figure 4.46c presents the transfer characteristic without (dashed line) and with correction resistors.

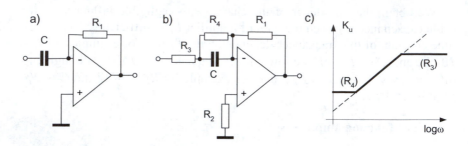

Figure 4.47. The differentiating amplifier: a) principle of operation, b) practical circuit, c) frequency characteristic

Similarly we can design the *differentiator circuit*. For an ideal differentiator circuit (Fig. 4.47a) there is

$$u_{out}(t) = -R_1 C \frac{du_{in}(t)}{dt} \qquad (4.55)$$

Without the limitation of frequency in the circuit of differentiator presented in Fig. 4.47a there is a risk of high frequency resonance – this can be prevented by inserting additional resistors R_3 and R_4 (Fig. 4.47b). Figure 4.47c presents the transfer characteristic without correcting resistors (dashed line) and with correction (solid line).

Fig. 4.48 presents the circuit of the *logarithmic amplifier*. The logarithmic amplifiers can be used for compression of the signals or linearization of the transfer characteristic (when the characteristic of the sensor is exponential). This amplifier is also used for *rms* conversion (see Fig. 4.15). In the design of such amplifier it is utilized the exponential dependence of the transistor collector current on the voltage U_{BE}

$$I_c \approx I_{es} e^{qU_{BE}/kT} \qquad (4.56)$$

where k is the Boltzman constant, q is the electron charge, T is temperature, and I_{ES} is the reverse current of base-emitter junction.

For the circuit presented in Fig. 4.48a we can write the following dependence

$$U_{out} = -\frac{kT}{q} \ln \frac{U_{in}}{RI_{es}} \tag{4.57}$$

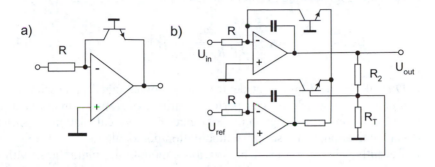

Figure 4.48. The logarithmic amplifier: a) principle of operation, b) practical circuit (Tran Tien Lang 1978)

The transfer characteristic $U_{out} = f(U_{in})$ strongly depends on the temperature (also current I_{es} depends on the temperature). Therefore, the practical circuit of the logarithmic amplifier is more complicated, as it is presented in Fig. 4.48b (Tran Tien Lang 1978). With correction the transfer characteristic is described as

$$U_{out} = -\frac{R_T + R_2}{R_T} \frac{kT}{q} \ln \frac{U_{in}}{U_{ref}} \tag{4.58}$$

It is possible to correct the temperature characteristic of the whole amplifier by appropriate choice of the temperature characteristic of the R_T element.

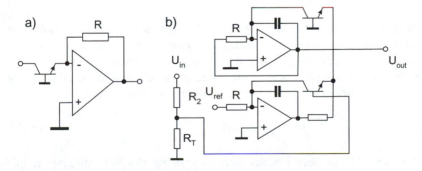

Figure 4.49. The exponential amplifier: a) principle of operation, b) practical circuit

By transformation of the logarithmic amplifier (Fig. 4.48) it is possible to obtain reverse characteristic of the amplifier. In this way the *exponential (antilog)* operation can be developed (Fig. 4.49). For the *antilog circuit* presented in Fig. 4.49 the transfer characteristic is described as

$$U_{out} = R_1 \frac{U_{ref}}{R} \exp\left(-\frac{q}{kT} \frac{R_T}{R_T + R_2} U_{in} \right) \qquad (4.59)$$

One of the most important devices in analogue signal processing is the *multiplier*, with which we can perform various operations: multiplication, division, square, root, trigonometric functions, *rms* calculation, electrical power calculation, phase sensitive modulation/demodulation.

Recently, there are available various monolithic multipliers with a multiplying error not larger than 0.1% and frequency bandwidth up to tens of MHz. In such circuits two techniques of multiplying are applied: *Gilbert transconductance multiplier* or *log/antilog operation*.

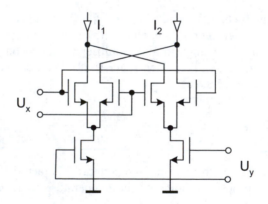

Figure 4.50. The example of transconductance Gilbert multiplier (Han 1998)

Designed by Gilbert in 1968 (Gilbert 1968, Gilbert 1972) the multiplier circuit is still (with small modifications) used in analogue semiconductor devices. Its main advantage is that it is easy to implement such a device into the integrated circuit (the same semiconductors as the rest of the circuit). For the Gilbert multiplier the following dependence is valid

$$I = I_1 - I_2 = K U_x U_y \qquad (4.60)$$

It is possible to obtain better accuracy using the log/antilog principle presented in Fig. 4.51. For this circuit we can write that

$$U_A = K_1 \lg \frac{U_x}{U_{ref}}, \quad U_B = K_2 \lg \frac{U_y}{U_{ref}} \tag{4.61}$$

and assuming that $K_1 = K_2 = K$

$$U_C = -(U_A + U_B) = -K \lg \frac{U_x U_y}{U_{ref}^2} \tag{4.62}$$

and also

$$U_{out} = U_{ref} 10^{-U_C/K_3} = \frac{U_x U_y}{U_{ref}} \tag{4.63}$$

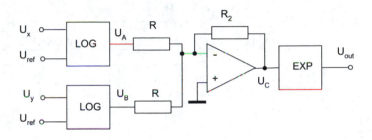

Figure 4.51. The log/antilog multiplier

The example of the typical multiplier (Analog Devices *model AD534*) is presented in Fig. 4.52.

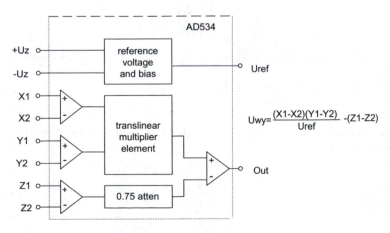

Figure 4.52. The typical functions of the integrated multiplier – the AD534 model of Analog Devices

The multiplier device enables calculation of the following equation

$$U_{out} = \frac{(X_1 - X_2)(Y_1 - Y_2)}{10V} + (Z_1 - Z_2)$$

(4.64)

By appropriate connection of the multiplier it is possible to realize various operations. Several examples proposed by Analog Device (Analog Devices 2005) are presented in Fig. 4.53.

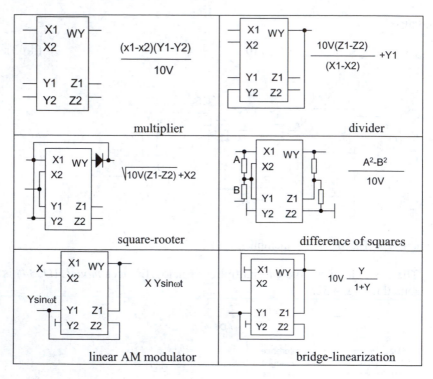

Figure 4.53 Various examples of the applications of a typical multiplier device (proposed by Analog Devices)

The multiplier based on the logarithmic function usually realizes the following equation

$$U_{out} = U_y \left(\frac{U_z}{U_x} \right)^M$$

(4.65)

Figure 4.54 presents the operational diagram of a typical log/antilog multiplier. The parameter M is set by connection of suitable resistors.

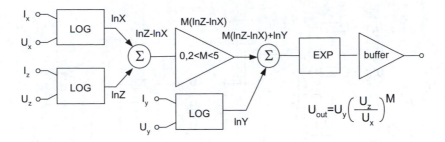

Figure 4.54. The operational diagram of typical log/anitlog multiplier – model AD538 of Analog Devices

Figure 4.55 presents the applications of such a multiplier for multiplication/division operation.

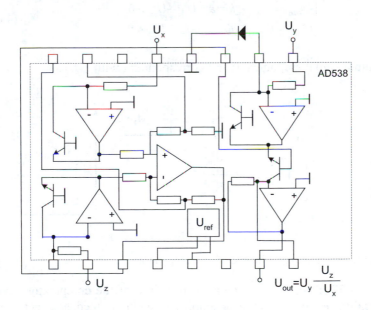

Figure 4.55. The example of application of the multiplier AD538 of Analog Devices as a multiplier/divider

Obviously, the operational amplifier allows performing the operations of addition or subtraction. Two examples of such devices are presented in Fig. 4.56.

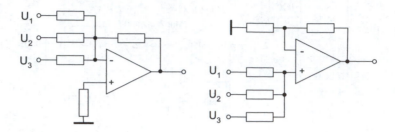

Figure 4.56. The operational amplifier used for addition of the signals

In the measurements often occurs the necessity of comparison of two signals. The devices called *comparators* realize the following function

$$\begin{cases} U_{out} = 1 & \text{for } \varepsilon = U_+ - U_- > 0 \\ U_{out} = 0 & \text{for } \varepsilon = U_+ - U_- < 0 \end{cases} \qquad (4.66)$$

where output signal are logical *0* or *1* corresponding with $\pm U_{out}$.

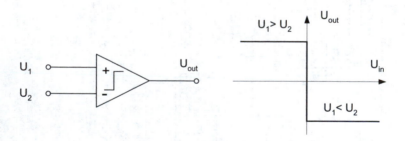

Figure 4.57. The operating principle of the comparator

Figure 4.57 illustrates the operating principle of the comparator device. Theoretically, every operational amplifier without feedback acts as a comparator. But in the typical operational amplifier's time of switching is relatively large – several μs. Therefore, there are available specially designed devices – comparators with switching time of several ns. If it is necessary to perform the comparison immediately, for example for detection of zero crossing by measurement of phase shift, then the comparator is very useful. For slow processes, for example monitoring of the level of the signal, ordinary operational amplifier is sufficient. Comparators play a very important role in digital signal processing.

4.3. NEGATIVE FEEDBACK IN THE MEASURING TECHNIQUE

As was demonstrated in the previous chapters the application of suitable feedback enables us to design various performances of the amplifier. In the case of measuring transducers the feedback helps in the improvement of the accuracy of conversion and enables us to influence the input/output resistances. Figure 4.58 presents examples of two main transducers: *current-voltage* converter and *voltage-current* converter.

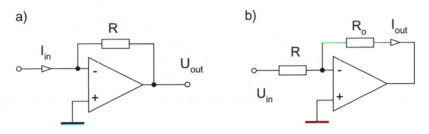

Figure 4.58. The realization of the converters: current-voltage (a) and voltage-current (b)

In the current-to-voltage converter (Fig. 4.58a) the output signal U_{out} is proportional to the input current according to the following equation

$$U_{out} = -RI_{in} \qquad (4.67)$$

while in the voltage-to-current converter (Fig. 4.58b) the output current is

$$I_{out} = \frac{U_{in}}{R} \qquad (4.68)$$

The conversion factor depends on the value of resistance R, which we can set very precisely.

In the measurements, the transducers of the current output are very important, because in the case of current output and the signal transmission the error introduced by the variation of the resistance of connection (change of voltage drop on the connection due to the variation of the ambient temperature) does not influence the output signal. There are various standards of current transducers – for example $I_{out} = 0 - 5\ mA$ or $I_{out} = 4 - 20\ mA$. The latter is very convenient because it is possible to distinguish zero level of the signal (*4 mA*) and the line defect (*0 mA*). An example of the transducer with the current output is presented in Fig. 4.59.

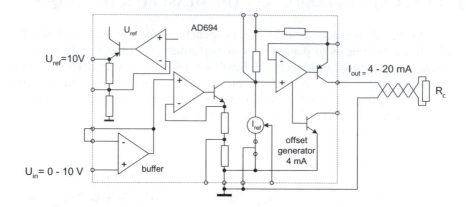

Figure 4.59. The transducer with the current output – AD694 transducer of Analog Devices

The transducer presented in Fig. 4.59 enables us to convert the voltage signal *0 – 2 V* or *0 – 10 V* to the current signal *0 – 20 mA* or *4 – 20 mA* with the nonlinearity error smaller than *0.002%*. The load resistance is limited by the supply voltage and is *200 – 2000 Ω* for the supply voltage *5 – 36 V*.

Generally, in the measuring transducers the feedback is very advantageous and it should be applied always if it is possible. Let us compare the performances of open-loop and feedback transducers (Fig. 4.60).

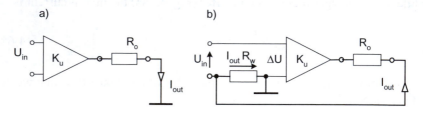

Figure 4.60. The voltage transducer: without feedback (a) and with feedback (b)

If the transducer operates without feedback (Fig. 4.60a) its conversion factor is

$$K' = \frac{I_{out}}{U_{in}} = K_u \frac{1}{R_o} \qquad (4.69)$$

Thus this factor directly depends on the gain factor of the amplifier. Usually, it is rather difficult to ensure stable gain, which is varying with the temperature, supply voltage or by the aging of the elements. If we apply the current feedback (Fig. 4.60b) then the conversion factor is

$$K = \frac{G}{1 + G\beta} = \frac{1}{\frac{1}{G} + \beta} \tag{4.70}$$

where G is the transmittance of the open circuit (gain of the amplifier) and β is the transmittance of the closed circuit (feedback).

After differentiation of (4.70) we obtain

$$\frac{dK}{K} = \frac{1}{1 + G\beta} \frac{dG}{G} - \frac{G\beta}{1 + G\beta} \frac{d\beta}{\beta} \tag{4.71}$$

Usually the feedback elements are stable and precise (in our example it is the resistance R), thus we can assume $d\beta/\beta \cong 0$. The equation (4.71) is

$$\frac{dK}{K} \cong \frac{1}{1 + G\beta} \frac{dG}{G} \tag{4.72}$$

Taking into account (4.70) and (4.72) we obtain

$$\frac{dK / K}{dG / G} = \frac{K}{G} \tag{4.73}$$

We can see that the influence of the variation of the gain on the accuracy of conversion is now significantly smaller – for sufficiently large $G\beta$ it is negligible. For example for $K = 5mA/10mV = 0.5 \ mA/mV$, $K_u=20 \ 000 \ V/V$ and $R_o = 2 \ k\Omega$ the conversion factor is $G = 10$. If the gain changes of about *20%* the resultant conversion factor changes only around *1%*.

Generally when $d\beta/\beta$ is not negligibly small, then the dependence (4.72) can be written in the form

$$\frac{dK}{K} = \sqrt{\left(\frac{dG}{G}\right)^2 \left(\frac{K}{G}\right)^2 + \left(\frac{d\beta}{\beta}\right)\left(1 - \frac{K}{G}\right)^2} \tag{4.74}$$

Thus after application of the feedback the accuracy of the transducer increases significantly. It should be noted that the feedback decreases only multiplicative errors, the additive errors (for example zero drift) do not decrease with feedback.

The feedback improves also the linearity of the transducer. The input signal of the amplifier is

$$\Delta x = x_{in} - \beta x_{out} \qquad (4.75)$$

and because

$$x_{out} = G\Delta x \qquad (4.76)$$

the input signal of the amplifier is decreased by $(1+G\beta)$

$$\Delta x = \frac{x_{in}}{1+G\beta} \qquad (4.77)$$

One of the sources of the nonlinearity is large range of input voltage of the amplifier (close to the saturation). If the input signal is small we use only linear part of the amplifier transfer characteristic. For the circuit presented in Fig. 4.60 the equations (4.75 – 4.77) are

$$\Delta U = Uin - I_{out} R_w ; \qquad I_{out} = \Delta U K_u \frac{1}{R_o + R_w} \qquad (4.78a)$$

$$\Delta U = \frac{U_{in}}{1 + K_u \dfrac{R_w}{R_w + R_o}} \qquad (4.78b)$$

The input signal of the amplifier ΔU is significantly smaller than the input signal U_{in}. of the whole transducer (for example if we process an input signal in the range of mV the input signal of the amplifier is in the range of μV).

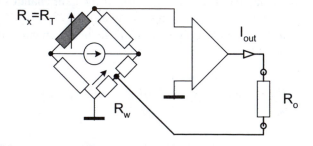

Figure 4.61. The transducer of the resistance with feedback and current output

Figure 4.61 presents the transducers of the resistance utilizing the bridge circuit (with thermoresitive temperature sensor R_T). The output current in the feedback resistor R_w causes the auto-balance of the bridge. The bridge circuit is in the balance state with a very small deviation ΔU signal in the input of amplifier (or very small unbalance ΔR_x). We described in Section 3.3.5 (see Figure 3.36) that the deflection type bridge is nonlinear and this nonlinearity increases with increase of the range. Thus if we decrease the range of the unbalance of the bridge we significantly decrease the nonlinearity error of the whole transducer.

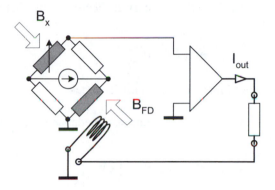

Figure 4.62. The transducer of the resistance (magnetic field) with feedback and the current output

Fig. 4.62 presents the transducer of magnetic field utilizing the magnetoresistive sensor. The output current in the feedback coil generates the feedback magnetic field B_{FD} which balances the bridge circuit. The magnetoresistive sensor operates as the zero magnetic field detector and even if it is nonlinear the whole transducer is linear (because we again use only a small linear part of the transfer characteristic).

It is recommendable if the transducer exhibits large input resistance, because the source of the signal is not loaded. Moreover, if the resistance of the source R_s is varying it does not influence the accuracy. The feedback enables significant increase of the input resistance. For the transducer presented in Fig. 4.60b we can write that

$$I_{in} = \frac{U_{in} - I_{out}R_w}{R_{in} + R_w + R_s} \qquad (4.79)$$

Taking into account the dependencies (4.78) we obtain

$$I_{in} = \frac{U_{in}}{R_{in} + R_w + R_s} \frac{1}{1 + K_u \beta} \tag{4.80}$$

Without the feedback (Fig. 4.60a) we have

$$I_{ino} = \frac{U_{in}}{R_{in} + R_s} \tag{4.81}$$

Neglecting the resistance R_w as rather small we can state that after applying of the feedback the input current decreases by factor of $(1+K_u\beta)$ and

$$R_{in} = (1 + G\beta)R_{ino} \tag{4.82}$$

where R_{ino} is the input impedance without feedback.

Similarly, we can prove that the output impedance of the transducer with current feedback is

$$R_{out} = R_{outo} + R_w(1 + K_u) \tag{4.83}$$

while the output impedance of the transducer with voltage feedback is

$$R_{out} = \frac{R_{outo}}{1 + K_u \beta} \tag{4.84}$$

By applying the current feedback we obtain the transducer with current output (large resistance – current source). By applying of the voltage feedback we obtain the transducer with voltage output (small resistance – voltage source).

Feedback helps also in improvement of the dynamic performances of the transducer. If the open circuit is inertial and is described by the following transmittance

$$G(s) = \frac{K_u}{1 + sT} \tag{4.85}$$

then the transmittance of such circuit with feedback is

$$K(s) = \frac{G(s)}{1 + \beta G(s)} = \frac{K_u}{1 + \beta K_u} \frac{1}{1 + s\dfrac{T}{1 + \beta K_u}} \tag{4.86}$$

We see that the time constant T decreases by a factor of $(1+\beta Ku)$ (the sensitivity also decreased by the $(1+\beta Ku)$ factor). Fig. 4.63 presents the comparison of the response for the step function.

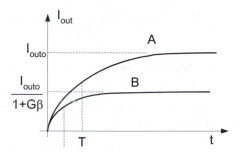

Figure 4.63. The dynamic characteristic of the transducer inertial type in the time domain, (A – without feedback, B – with feedback)

Also in the case of the oscillation type of the transducer we obtain improvement of the performance after applying the feedback. Without the feedback the transmittance is

$$G(s) = \frac{K_u \omega_o^2}{\omega_o^2 + 2b\omega_o s + s^2} \tag{4.87}$$

where ω_o is the resonance frequency and b is the damping coefficient of the oscillations.

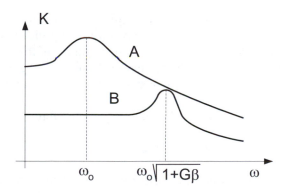

Figure 4.64. The dynamic characteristic of the transducer of oscillation type in the frequency (A – without feedback, B – with feedback)

After applying of the feedback the transmittance is

$$K(s) = \frac{K_u \omega_o^2}{\left(\omega_o \sqrt{1+K_u\beta}\right)^2 + 2\left(\omega_o \sqrt{1+K_u\beta}\right)\left(\frac{b}{\sqrt{1+K_u\beta}}\right)s + s^2} \tag{4.88}$$

We see that with the feedback the resonance frequency increases by $\sqrt{1+K_u\beta}$ while damping decreases by a factor of $\sqrt{1+K_u\beta}$. The comparison of the frequency characteristics for the circuits with and without the feedback is presented in Fig. 4.64.

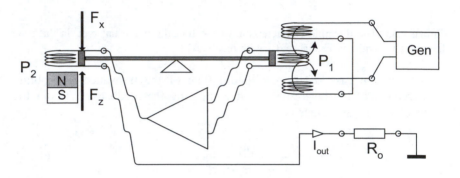

Figure 4.65. The transducer of the force with the feedback and the current output

Figure 4.65 presents the force transducer. Measured force F_x causes the deflection of the bar and moves the displacement sensor P_1 from the state of balance. The output signal of this sensor after amplification is connected to the coil of electromagnet P_2. The force of repulsion of this coil moves the bar back in order to obtain again the state of balance (and zero signal from the sensor P_1). Therefore this transducer is also called the current weight.

The output current creates the balancing repulsion force

$$F_z = BzdlI_{out} = k_1 I_{out} \tag{4.89}$$

where B is the induction of the electromagnet, d an l are the dimensions of the coil and z is a number of turns.

Thus the output current is proportional to the measured force

$$I_{out} = kF_x \tag{4.90}$$

We see that the feedback improves almost all features of the transducer: accuracy, linearity, input resistance, output resistance and dynamics. However, we should also ask about drawbacks. The main downside of the feedback application is that the feedback causes the risk of instability of the whole transducer. Therefore, in some cases it is necessary to introduce special corrections.

Figure 4.66 presents the results of experimental testing of the dynamics of the force transducer presented in Fig. 4.65. The output signals were tested after the step change of the measured force. We can see (case a) that the transducer without correction tends to oscillate (high frequency, small amplitude oscillations) and for an inappropriate choice of the correction it is unstable (large oscillations – case c). Introduction of the correction removed the oscillations but the same time caused the increase of inertia (case b). But as is demonstrated in Fig. 4.65d it is possible to design the correction in such way as to obtain excellent dynamics of the transducer.

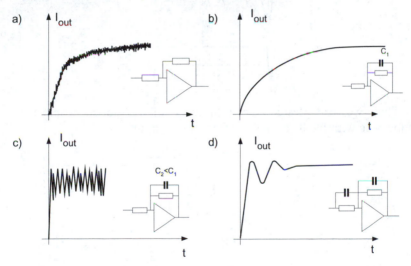

Figure 4.66. The experimental results of the force transducer for testing of the dynamics for various correcting elements

In this chapter have been presented the advantages of negative feedback used in transducers. So far, the transducers of voltage (Fig. 4.60b), resistance (Fig. 4.61) or force (Fig. 4.65) with the current output signal were presented. Figure 4.67 presents another type of transducer with feedback. This transducer converts the resistance, capacitance or inductance to the frequency signal.

There is a certain group of bridge circuits, in which the condition of balance depends on the frequency of the supplying signal. For example, the balance condition for the bridge circuit presented in Fig. 4.67a is

$$\omega = \sqrt{\frac{1}{C_2 C_3 (R_2 R_3 - R_4 R_x)}} \qquad (4.91)$$

If we use the oscillator circuits with voltage dependent on the frequency then we obtain a transducer of resistance with feedback. For the circuits presented in Fig. 4.67b and 4.67c the balance conditions are described by the equations

$$\omega_b = \sqrt{\frac{1}{C_2 C_3 \left(R_2 R_3 - \dfrac{L_x}{C4} \right)}} \; ;$$

$$\omega_c = \sqrt{\frac{1}{R_2 R_3 (C_2 C_3 - C_4 C_x)}} \qquad (4.92)$$

Thus we can design the transducer of resistance, capacitance or inductance with feedback and the frequency output.

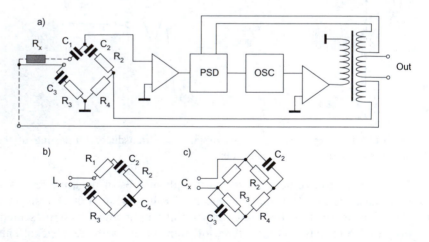

Figure 4.67. The RLC to frequency transducer with the feedback circuit and the frequency output signal: resistance transducer (a), bridge circuit for inductance transducer (b) and for capacitance transducer (c)

4.4. THE IMPROVEMENT OF THE QUALITY OF THE ANALOGUE SIGNALS

4.4.1. The noises and interferences of the analogue signals

The measurement signals are usually accompanied by some noises and interferences, sometimes of the level comparable to the level of the measured signal. As the noises and interferences we assume all signals other then the measured signal – the noises are the stochastic signals with indefinite frequency and magnitude (*the white noise* is the signal with theoretically all frequency components). The interferences are the signals coming from various external sources; very often these signals are of the main frequency *50 Hz* and harmonics of this frequency.

The typical interference signals are generated by the electric power lines, electrical machines, lighting equipment, commutating devices, radio communication transmitters, atmospheric discharges or cosmic noises. There are also internal sources of noises – resistors and semiconductor devices.

In the previous chapters various methods of rejection of noises or interferences have been described:
- application of the differential input of amplifiers (common mode signals rejection) or application of differential sensors (for example to eliminate the external temperature influence);
- application of the galvanic separations of the circuits (isolation amplifiers);
- application of the phase sensitive detectors – in lock-in amplifiers to rejection of the noises;
- elimination of the temperature zero drift by application of the auto-zero function;
- correct grounding and shielding of various parts of the circuit, for example application of the *Wagner earth* in the *AC* bridge circuits.

Modern signal recovering techniques enable us to eliminate the noises and interferences significantly larger than the measured signal. But we can look at the problem of noises and interferences from a other point of view. By using incorrect connections of the signal or by applying the incorrect grounding we can deteriorate the quality of the measuring signal.

One of the most difficult interferences to repair is the signal deteriorated by the zero drifts. The zero drift can be caused by several sources: the connections of two metals (thermoelectric voltage), instable contacts, vibrations of various parts of the circuit, bad quality insulation, piezoelectric effects, electrochemical effects, etc. The basis of the zero drift is very often technological (quality of contacts, uniformity of materials) and it can be amplified by the temperature differences of various parts of the circuit. For that reason it is important not to accept the formation of the temperature differences (for example non-uniform heating of various parts). Sometimes, conversion of the *DC* into the *AC* signals can be helpful (carrier amplifiers, chopper amplifiers, etc.).

One of the basic sources of the noises is the resistor *R*. The *thermal Johnson noise* U_{nT} is generated due to the chaotic thermal movement of the charges. This noise can be described by the *Nyquist relation*

$$U_{nT} = \sqrt{4kTR\Delta f} \qquad (4.93)$$

where *k* is the Boltzman constant ($k = 1.38 \ 10^{-23}$ *Ws/K*), *T* is the temperature, and *Δf* is the frequency bandwidth.

The thermal noises can be reduced by the decrease of the resistances used, by the limitation of the bandwidth and of course by the control of the temperature.

Another important source of noises is the semiconductor junction. When the current is on the potential barrier a number of charges *q* are randomly crossing this barrier and these random current fluctuations are the source of the *shot noises* I_{ss} described by the *Schottky relation*

$$I_{ss} = \sqrt{2qI\Delta f} \qquad (4.94)$$

where *q* is the electric charge ($q = 1.6 \ 10^{-19}$ *C*).

In the low frequency range there are *1/f type noises* (sometimes called *low frequency noise, flicker noise* or *excess noise*). There are many sources of these noises, most of them unknown. Such noises are inversely proportional to the frequency, and the power spectral density S(*f*) of this noise is:

$$S(f) = \frac{E_f^2}{f^\alpha} \qquad (4.95)$$

where E_f is the *rms* voltage of noises and *α* coefficient is $\alpha = 0.8 - 1.3$.

Because the noise level depends on the frequency these noises are not the white noises (noises containing every frequency of signals) and are called *pink noise* – random noise having the same amount of energy in each octave.

Due to the random character of the noises usually they are described not by the voltage level U_n but by the spectral density S(*f*)

$$S(f) = \frac{U_n^2}{\Delta f} = \left(\frac{U_n}{\sqrt{\Delta f}}\right)^2 \qquad (4.96)$$

The $U_n/\sqrt{\Delta f}$ value is called the *spectral density of noises* and it means the *rms* value of the voltage signal in relation to the square root from the frequency bandwidth. Often the noises are characterized by the *SNR factor – signal to noise ratio.*

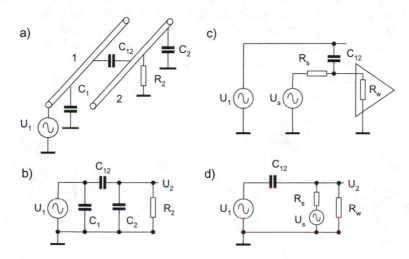

Figure 4.68. The interferences connected to the amplifier by the capacitive coupling: the access to the adjacent wire (a,b) and access to the amplifier input (c,d)

The external interferences can penetrate the measuring circuit by the capacitive, inductive or conductive coupling. Fig. 4.68 presents two examples of the interferences connected to the measuring circuit by the *capacitive coupling*. In the case presented in Fig. 4.68a the wire 2 infiltrates the signal from the wire 1 by the inter-wire capacitance C_{12}. The equivalent circuit is presented in Fig. 4.68b. For the large frequency (larger than $\omega = 1/R_2\,(C_{12}+C_2)$ the capacitance voltage divider is created and the interference signal in the wire 2 is

$$U_2 = \frac{C_{12}}{C_2 + C_{12}} U_1 \qquad (4.97)$$

For lower frequencies the interference signal depends on the frequency f and the resistance R_2 (Pallas-Areny 1999)

$$U_2 = j\omega R_2 C_{12} U_1 \qquad (4.98)$$

In the case presented in Fig. 4.68c the amplifier connected to the measured signal U_s is additionally connected to the signal U_1 by the coupling capacitance C_{12}. The equivalent circuit is presented in Fig. 4.68d.

A typical example of capacitance coupling is the penetration of the interferences by the inter-turns capacitance of the separation transformer. The researcher performing experiments would expect that the connection of the galvanic separation (by use of the separation transformer) makes the circuit free from the *50 Hz* interferences. To effectively obtain such rejection of interferences it is necessary to introduce between turns the copper grounded electrostatic shield (this way we eliminate the capacitive coupling of signals).

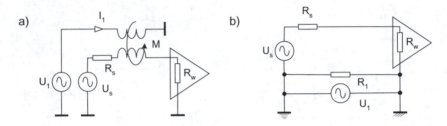

Figure 4.69. The interferences connected to the amplifier by the inductive (a) or conductive (b) coupling

A typical example of the *conductive coupling* of interferences is the current in the common wire: supplying or grounding. Especially in the case of connection of two various grounding points the inter-ground difference of potentials (voltage U_1 in the Fig. 4.69b) can be dangerous.

Fig. 4.69a presents the inductive coupling when the current in adjacent wires can cause an additional interference voltage to be induced

$$U_L = \omega M I_1 \tag{4.99}$$

A typical example of *inductive coupling* is the penetration of the measuring circuit by the voltages induced by the external sources of electromagnetic fields, from radio transmitters or mobile telephony communication. Therefore, such interferences are often called as *RFI – radio frequency interferences*.

There are several universal methods and tools of rejection or limitation of the interference. The capacitive coupling can be reduced by the application of the electrostatic shield. A conducting plate or foil grounded at one point can be used as the electrostatic shield (Fig. 4.70). After introduction of this shield most of the currents coming from the interference source U_1 are shorted to the ground by the shield and do not penetrate the measuring circuit (Fig. 4.70b).

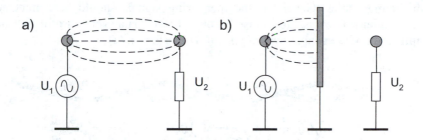

Figure 4.70. The reduction of the capacitive coupling by the electrostatic shield

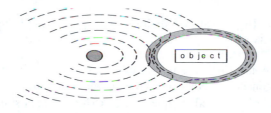

Figure 4.71. The reduction of the magnetic coupling by the introduction of the magnetic shield

Figure 4.71 presents the method of reduction of magnetic coupling by the application of a magnetic shield. The magnetic shield is prepared from high-permeability magnetic material. The lines of magnetic field are closed in the shield and do not penetrate the area of the measuring circuit. To obtain effective shielding the material of the shield should be properly chosen – for low magnetic fields it is necessary to use different material than for high magnetic field since the magnetic permeability strongly depends on the level of magnetic field. Also for *DC* magnetic field different magnetic materials should be used than for the *AC* magnetic fields. Therefore, sometimes the magnetic shield is composed of several shields prepared from various materials.

Although the techniques of shielding seem to be relatively simple, for correct application of shielding and grounding it is necessary to have extensive knowledge, experience and even intuition. There are many references, including books on this subject (Agilent 1441 2005, Morrison 1991, Morrison 1998, Ott 1988, Rich 2005, Voseghi 2000).

4.4.2. The connection of the measuring signal to the amplifier

Many of the interferences penetrate the measuring circuits through the connecting cables. Generally, the measuring signals should be connected using shielded wires. In transmission of the data very useful is the application of the simple *twisted pair of wires* (Fig. 4.72).

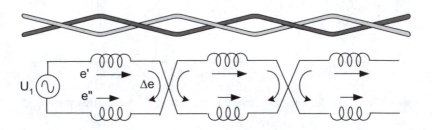

Figure 4.72. The twisted pair of wires and the principle of reduction of interferences in such connection

In the twisted pair the interferences are reduced because the voltages *e'* and *e"* induced in adjacent wires compensate each other and potential remaining of induced voltages *Δe* exhibits opposite direction in the neighbouring loops of the twisted wires.

For transmission of the digital data more and more important are fiber-optic cables. The optical system is immune to the electromagnetic interferences. The transmission of the data is extremely fast, theoretically the speed of frequency *THz* is possible and in practice the transmission of hundreds *MHz* is achieved.

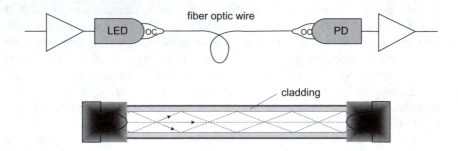

Figure 4.73. The fiber-optic cable as the communication wire for transmission of the measuring data

The principle of data transmission using the fiber-optic connection is presented in Fig. 4.73. The electrical signal is converted to the light by means of the LED or laser diode, then the optical system *OC – optical coupler* is formed. The optical signal should be converted again to the electrical form by means of a photo-detector device, before it is fed to the receiver. The effect of internal reflection of the light (the multimode transmission) is utilized for transmission of the signal through the fiber-optic wire. The core of the fiber-optic cable is surrounded by a special glass material called *cladding*. This part of fiber-optic cable exhibits the reflection coefficient ensuring that the light does not leave the wire.

Fig. 4.74 presents the results of experimental investigation of the shielding effect of various connecting cables (Ott 1988). These cables have been inserted to the external 50 kHz magnetic field.

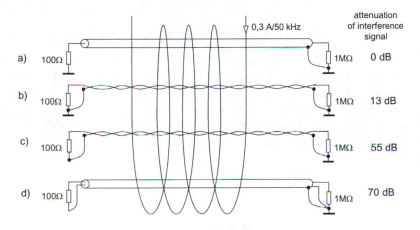

Figure 4.74. The attenuation of the magnetic interference signal depending on the method of the connection of the cables (Ott 1976)

The simple single wire in the coaxial shield grounded at the end practically did not attenuate the interference (Fig. 4.74a). Similarly the twisted wire grounded at both ends exhibited poor attenuation of interferences (Fig. 4.74b). The same twisted pair cable but grounded at the receiver end attenuated the interferences much better (Fig. 4.74c). Satisfying attenuation of interferences exhibited double wire cable with a coaxial shield grounded at the receiver end (Fig. 4.74 d).

There is no one simple receipt for the grounding, shielding and cabling. Practically, it is always recommended to perform a series of experiments to find the best solution – sometimes a simple and cheap twisted pair of wires can exhibit better performances than expensive shielded wire.

Generally we should avoid the connection of the signal source and the receiver in such a way that the shield of the cable is a current-carrying part of the circuit[1] (Fig. 4.75) *(Do not allow shield current to exist and do not allow the shield to be at a voltage with respect to the reference potential)* (Rich 2005).

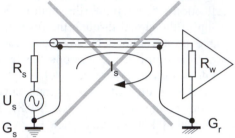

Figure 4.75. Not recommendable connection of the signal source and the receiver allowing the existence of the shield current

Especially dangerous is the connection of the cable shield at both ends to grounds (Fig. 4.75). When these grounds have various potentials G_s and G_r there is an equalizing current I_s in the shield. This current generates additional interference signal, which can penetrate the measuring circuit by the inductive coupling. For that reason *the shield should be grounded at only one end*.

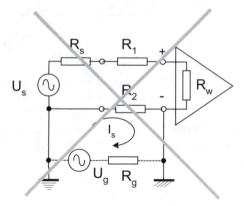

Figure 4.76. Incorrect connection of the signal source to the amplifier

---

[1] However, there are exceptions to this rule. For example, in the experiment of Ott (Ott 1988) presented above the best results of attenuation of magnetic interferences exhibited the cable with both ends of the shield connected to the source of signal and to receiver respectively. The explanation of this result is that both currents (in the shield and in the wire) generated mutually compensating magnetic fields.

Similarly, it is not recommended to connect the signal source to the amplifier of different grounding potentials (Fig. 4.76). In such case this potential difference U_g implies that in the R_2 resistance flows the equalizing current I_s. This current generates the conductive interference as the voltage U_z at the terminals of the amplifier

$$U_z = U_g \frac{R_2}{R_2 + R_g} \qquad (4.100)$$

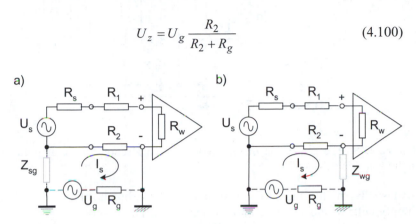

Figure 4.77. The connection of the signal to the amplifier with floating ground of the signal source (a) and floating ground of the amplifier input (b)

Fig. 4.77a presents the connection of the signal source with isolated ground *(floating ground)*. The source is connected to the ground by large $(10^8 - 10^{10}\ \Omega)$ impedance of leakage Z_{sg}. In the input of amplifier the interference signal is

$$U_z = U_g \frac{R_2}{R_g + R_2 + R_{sg}} \approx U_g \frac{R_2}{R_{sg}} \qquad (4.101)$$

The interference signal is significantly attenuated because the impedance Z_{sg} is large. If it is not possible to separate the signal source from the ground we can use the amplifier with floating point (connected to the ground by the leakage impedance Z_{wg} – Fig. 4.77b). Modern digital voltmeters generally have the floating ground terminal – practically the ground terminal is not introduced at the front panel of the instrument.

In modern amplifiers or voltmeters a special kind of shielding is used – a double shield. The first shield "ground shield" is connected to the ground, but the second shield "guard shield" is with floating ground – it is connected to the ground and to the "minus" terminal by large leakage impedances Z_2 and Z_3 (Fig. 4.78). The equalizing current I_s is now in a closed loop outside the input signal connection.

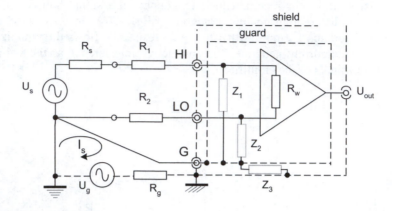

Figure 4.78. The connection of the grounded signal source to the amplifier with additional shield "guard"

Usually, in the measuring instruments the grounds of the digital part, supply voltage and analogue part are separated. The first one (digital part and supply voltage) is called *power GND* (*power ground*) while the second is called *LL GND* (*low level ground*). The input terminals are usually with floating ground and are called *HI* (*high*) and *LO* (*low*). In high accuracy instruments the terminal *GUARD* is additionally introduced.

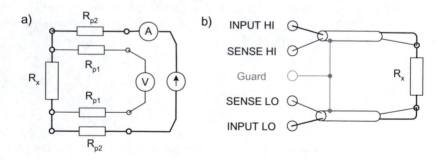

Figure 4.79. The connection of the measured low resistance to the four-terminal multimeter (the terminal guard in the circuit presented in Fig. 4.79b is optional)

When small resistance R_x is connected to the measuring device as the rule the *four-wire connection* is used – with separate current and voltage terminals (Fig. 4.79). The current terminals are usually larger and more massive and the connections wires R_{p2} are outside the measuring circuit. Because the voltage drop measurement is practically performed without current (large input resistance of the amplifier) the resistances of connection R_{p1} can be neglected. In digital multimeters the voltage terminals are indicated as "*Sense Hi*" and "*Sense LO*" while the current terminals are indicated as "*Input Hi*" and "*Input LO*".

Fig. 4.80 presents the application of the toroidal choke for attenuation of the interferences. This method is used in computer connections and it is realized in such a way, that the signal cable is wound on the toroidal ferrite yoke.

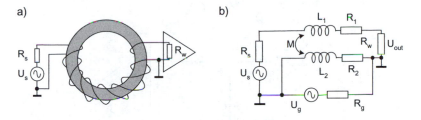

Figure 4.80. A toroidal longitudinal choke used for attenuation of the interferences (Northrtop 1997)

Fig. 4.80b presents the equivalent circuit of such a toroid connection. The interferences are represented by the source U_g. The ratio of the signals to the interference is

$$\frac{U_{out}}{U_g} = \frac{R_2}{R_2 + R_g} \frac{1}{1 + \dfrac{\omega L_2}{R_g + R_2}} \tag{4.102}$$

The interference is attenuated and additionally a low-pass filter $L_2/(R_g+R_2)$ is formed. For higher frequencies (above 1 MHz) instead of toroidal yoke sometimes the ferrite ring (ferrite bead) thread on the cable is used.

More difficult to attenuate are the low frequency interferences (industrial frequency interferences). Usually, various kinds of filters are used in order to suppress such interferences.

It can be important to match the noise to the amplifier. The noises of the amplifier are represented by the voltage e_n and current i_n noises. The noise factor F is the ratio of the amplifier noises to the thermal noises

$$F = \frac{e_n^2 + i_n^2 R_s^2}{e_T^2} = \frac{e_n^2 - i_n^2 R_s^2}{4kTR_s} \tag{4.103}$$

The optimal value of R_s which makes F minimal can be calculated as

$$\frac{\partial F}{\partial R_s} = 0 \qquad (4.104)$$

From (4.103) and (4.104) the optimal source resistance is

$$R_{sopt} = \frac{e_n}{i_n} \qquad (4.105)$$

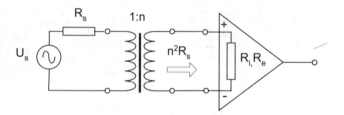

Figure 4.81. The noise matching by transformer coupling

Usually the R_{sopt} is relatively large, even up to several $M\Omega$, while the typical voltage sources exhibit rather small resistance. Therefore, matching the resistance by using a transformer can result in reduction of the noises. If the ratio of turns of the transformer is n the primary resistance R_s is reflected at the secondary winding as

$$R_s' = n^2 R_s \qquad (4.106)$$

Thus the optimal transformer turns ratio is

$$n_{opt} = \sqrt{\frac{R_{sopt}}{R_s}} = \sqrt{\frac{1}{R_s}\frac{e_n}{i_n}} \qquad (4.107)$$

The transformer inserted between the source and the amplifier matches the resistances, but also plays the role of isolation device (and sometimes as the symmetrization device – converting the single-ended source to the differential input).

4.4.3. The analogue filtering of the signals

The interferences can be rejected by the filter devices. By using *lowpass LP filters* we can reject the high frequency interferences (for example radio frequency interferences). And *vice versa* by using the *highpass HP filter* we can eliminate the industrial frequency interferences (if our signal is in the bandwidth above these frequencies). Also, we can separate the useful signal from the interference signals by using the *bandpass BP filter*. There also exist *bandreject* filters.

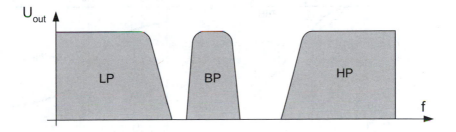

Figure 4.82. Typical filters and their frequency characteristics: LP – low-pass filter, BP – band-pass filter, HP – high-pass filter

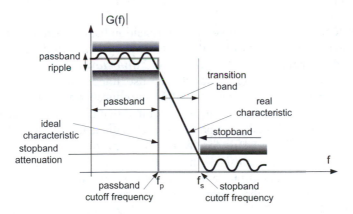

Figure 4.83. Amplitude response of real low-pass filter

The ideal filter should pass the signals in assumed frequency bandwidth and stop the signals of other frequencies (Fig. 4.83). The real analogue filter does not exhibit flat frequency characteristics in the passband and the transition between the passband and stopband is not vertical (there is a finite transition frequency bandwidth).

The performance of the filter is described by the amplitude and phase frequency characteristics (Fig. 4.84). The ideal amplitude characteristic should be flat up to passband cut-off frequency (attenuation 0 dB). For real characteristic we assume that the passband is for the attenuation 0 – 3dB. As the stopband we assume the frequency band where the attenuation is larger than the assumed value (for example 100 dB).

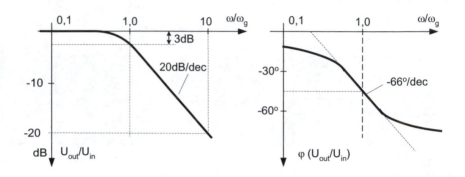

Figure 4.84. An example of the amplitude and phase frequency characteristics of the lowpass filter

The slope of the characteristic in the transition band depends on the order of the filter – for the first order filter it is *20 dB/decade* (*6 dB/octave*[1]), for the second order filter it is *40 dB/decade* (*12 dB/octave*) etc.

The phase frequency characteristic is also important because incorrect phase performances mean that the filter introduces signal distortions. To obtain correct phase performances it is required that the phase varies linearly with the frequency. Usually the phase characteristic is not linear in the whole bandwidth (see Fig. 4.84). Moreover it is very difficult to ensure both performances simultaneously: the filters with excellent amplitude characteristic exhibit poor phase linearity and *vice versa*.

Fig. 4.85 presents the examples of realization of simple passive RC filters. The low-pass RC filter presented in Fig. 4.85a is described by the equation

$$G(j\omega) = \frac{1}{1 + j\omega RC} \qquad (4.108)$$

---

[1] Octave it is the ratio of frequency equal to 1:2, while the decade corresponds with the ratio of frequency 1:10.

or

$$G(\omega) = \frac{1}{\sqrt{1 + (\omega RC)^2}} \qquad (4.109)$$

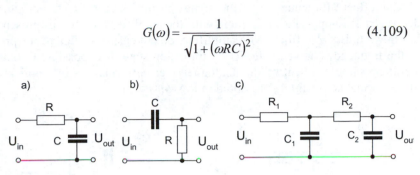

Figure 4.85. The passive RC filters: a) low-pass filter, b) high-pass filter, c) second order cascade low-pass filter

The cut-off frequency of such filter is $\omega_g = 1/RC$. If we connect two filters in the cascade (Fig. 4.85c) then the frequency characteristic is described in the form

$$G(j\omega) = \frac{1}{1 + j\omega(R_1 C_1 + R_2 C_2 + R_1 C_2) - \omega^2 R_1 C_1 R_2 C_2} \qquad (4.110)$$

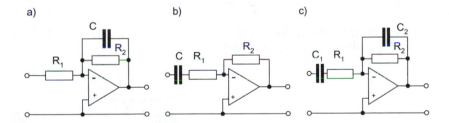

Figure 4.86. The RC filters supported by the amplifier: a) low-pass filter, b) high-pass filter, c) band-pass filter

By connecting several filters in series we obtain an increase of the order of the filter and the same the steepness of the characteristic in the transition band. But it is not reasonable to connect many filters in the cascade (more than three) because each of the filters significantly attenuates also the useful signal. Moreover, the cascade type filter is not a simple sum of single filters because each following filters loads the previous one, thus influencing the frequency characteristic (the dependence (4.108) was derived under assumption that the source resistance is negligibly small and the load resistance is large).

The effect of attenuation of the signal by the passive RC filters can be decreased by supporting such filters with the amplifier circuit – as presented in Fig. 4.86. But also filters with amplifiers do not ensure sufficient steepness of the frequency characteristic in the transition state. It is possible to obtain significant improvement of the analogue filter performances by application of the special active filters presented in Fig. 4.87.

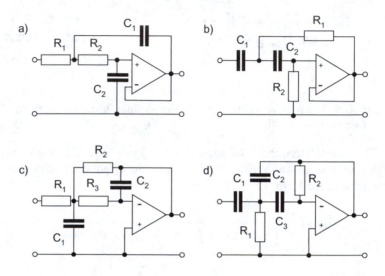

Figure 4.87. The active analogue RC filters: ab) Sallen-Key filters, cd) MFB filters, ac) low-pass filters, bd) high-pass filters

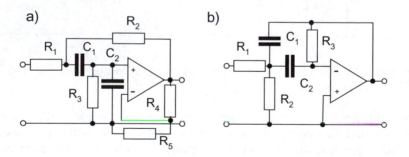

Figure 4.88. The active analogue RC band-pass filters: a) Sallen-Key filter, b) MFB filter

The active filters use the operational amplifiers and the RC elements in the feedback circuit. Fig. 4.87 presents the typical active filters – *Sallen-*

Key filters (called also *filters with voltage controlled source - VCS*) (Fig. 4.87ab) and MFB filters (*multi- feedback filters*) (Fig. 4.87cd). Note that the low-pass filters and high-pass filters are obtained simply by swapping around the R and C elements. The band-pass filters are obtained by the combination of low-pass and high-pass filters – Fig. 4.88.

For the Sallen-Key lowpass filter the transfer characteristic can be described by the following s-operator function

$$G(s) = \frac{\dfrac{1}{R_1 R_2 C_1 C_2}}{s^2 + \dfrac{1}{R_1 C_2}s + \dfrac{1}{R_1 R_2 C_1 C_2}} = \frac{\omega_p^2}{s^2 + \dfrac{\omega_p}{Q} + \omega_p^2} \qquad (4.111)$$

where

$$\omega_p = \sqrt{\frac{1}{R_1 R_2 C_1 C_2}} \; ; \qquad Q = \sqrt{\frac{R_1 C_2}{R_2 C_1}}$$

For the special case when $R_1 = R_2 = R$ the capacitances can be calculated from the following conditions (Jamal 2003)

$$C = \frac{1}{2\pi R f_p} \; ; \qquad C_1 = 2QC \; ; \qquad C_2 = C/2Q \qquad (4.112)$$

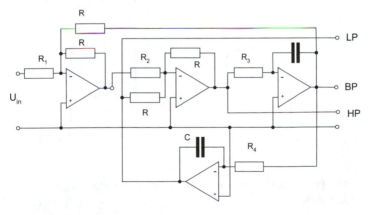

Figure 4.89. The universal active second order filter that provides simultaneous lowpass LP, highpass HP and bandpass BP outputs

Figure 4.89 presents the example of the universal active filter enabling to obtain simultaneous three types of filters: low-pass, high-pass and band-pass filters.

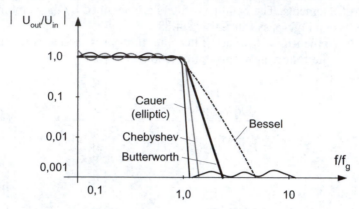

Figure 4.90. The amplitude characteristics of various filter designs

There is no one best design of filter. Figure 4.90 presents the amplitude characteristics of various filters. The most popular is the *Butterworth filter*, which has the most flat characteristic in the passband. The *elliptic Cauer filter* (and also *Chebyshev*) ensure the steepest amplitude characteristic in the transition band, but the flatness of the characteristic in the passband is poor (there are several ripples in this band). The *Bessel filter* exhibits the best phase characteristic but the steepness of the amplitude characteristic is inferior.

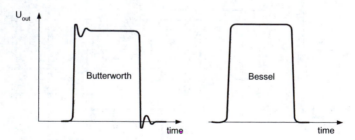

Figure 4.91. The response after step function of the input signal of two filters: Butterworth and Bessel

The phase distortions are visible in the time domain response of the filter. Fig. 4.91 presents the response after the step function of the input signal of two filters. The Bessel filter exhibits excellent performances taking into account the phase distortion.

The design of the analogue filters requires quite complicated calculations (Huelsman 1993, Thede 2004, Schaumann 2001, Winder 2002, Volkenburg 1995). Fortunately, on the market there are available various design procedures enabling fast design of the filter with desired performances. For example, the *Matlab* software offers various tools for the design of the filters (Lutovac 2000). Also, in the *LabVIEW* platform there are ready to use design procedures of the filters.

Usually, the filter is described using *s*-operators. A simple RC filter (equation 4.106) can be described in the form

$$G(s) = \frac{1}{1 + sRC} \qquad (4.113)$$

In the normalized form it is assumed that $R = 1\ \Omega$, $C = 1\ F$, $\omega_p = 1\ rad/s$ and the dependence (4.113) can be rewritten as

$$G(s) = \frac{1}{1 + s} \qquad (4.114)$$

Similarly the fourth order filter is described by the equation

$$G(s) = \frac{1}{a + bs + cs^2 + ds^3 + s^4} \qquad (4.115)$$

Table 4.3. Butterworth filter denominator polynomials

Order	Polynomials
1	$s + 1$
2	$s^2 + \sqrt{2}s + 1$
3	$s^3 + 2s^2 + 2s + 1$
4	$s^4 + 2.61s^3 + 3.414s^2 + 2.61s + 1$
5	$s^5 + 3.236s^4 + 5.236s^3 + 5.236s^2 + 3.236s + 1$

All most popular filters are described as the polynomials representing the denominator in equation (4.113). These polynomials calculated for Butterworth, Chebyshew, Cauer or Bessel are completed as tables in various publications. Table 4.3 presents the Butterworth denominator polynomials.

The active filters presented in Fig. 4.87 represent the second order filters. The simplest way to obtain higher order of the filter is to connect the filters in a cascade form – for example, to obtain the fourth order filter usually there are connected two filters of the second order, and to obtain the fifth order filter additional first order filter is connected. Table 4.4 presents the denominator polynomials derived under the assumption that we dispose to connect the first and second order filters.

Table 4.4. Butterworth filter denominator polynomials for the filter composed from the first and second order components

Order	Polynomials
1	$s+1$
2	$1+1.414s+s^2$
3	$(1+s)(1+s+s^2)$
4	$(1+1.848s+s^2)(1+0.765s+s^2)$
5	$(1+s)(1+1.618s+s^2)(1+0.618s+s^2)$

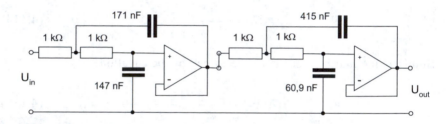

Figure 4.92. The 4th-order Butterworth low-pass filter with the cut-off frequency *1 kHz* - after (Maxim 2002)

Let us design the low-pass Butterworth filter of the fourth order with the cut-off frequency *1000 Hz* (Maxim 1795 2002) (Fig.4.92). The characteristic of the Sallen-Key filter can be described using the following equation

$$G(s) = \frac{G_1 G_2}{G_1 G_2 + (G_1 + G_2)C_2 s + C_1 C_2 s^2} \qquad (4.116)$$

where resistances R are presented as conductances G (for simplifying further calculations). With normalized resistors 1Ω the equation (4.116) can be rewritten in a form

$$G(s) = \frac{1}{1+(2C_2)s+(C_1 C_2)s^2} \qquad (4.117)$$

According to the polynomial $(s^2+1.8478s+1)$ $(s^2+0.7654s+1)$ and the equation (4.117) in the first filter $C_1C_2 = 1$ and $2C2 = 1.8478$ thus the capacitances in the first filter are $C_2 = 0.9239\ F$ and $C_1 = 1.08\ F$. Similarly in the second filter $C_1C_2 = 1$ and $2C2 = 0.7654$ thus the capacitances in the first filter are $C_2 = 0.3827\ F$ and $C_1 = 2.61\ F$.

If we assume that the resistors are *1 kΩ* (multiplied by factor 1000) the capacitances should be divided by 1000. To obtain the cut-off frequency as *1kHz* (instead of *1 rad/s*) we should divide the capacitances by $2\pi\cdot1000$. As result we obtain the circuit of the filter presented in Fig. 4.92.

In the microscale (for example in the integrated circuit technology) it is rather difficult to obtain precise values of the *RC* elements. Therefore, instead of resistors there are developed filter circuits with *switched capacitors* because in the integrated circuit technology it is relatively easy to design the switch element.

Consider the circuit presented in Fig. 4.93a. If the capacitor is switched with the period *T* (first time to the input u_1 the second time to the output u_2) the charge transfer is

$$\Delta Q = C(u_1 - u_2) \tag{4.118}$$

and the equivalent current is

$$I = \frac{\Delta Q}{T} = \frac{C(u_1 - u_2)}{T} \tag{4.119}$$

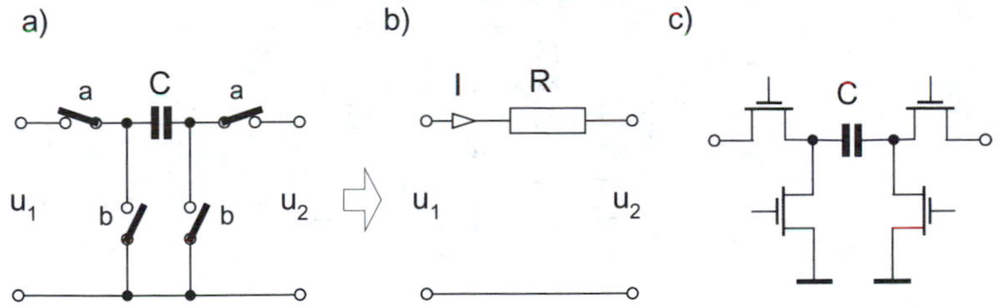

Figure 4.93. The switched capacitor as the simulator of the resistance: a) the principle of switching, b) the equivalent circuit, c) the semiconductor realization

The circuit with a switched capacitor is equivalent to the circuit with the resistor *R*

$$R = \frac{u_1 - u_2}{I} = \frac{T}{C} = \frac{1}{fC} \tag{4.120}$$

Thus it is possible to design the filters with the R element substituted by the switched capacitor. The important advantage of this solution is the possibility of the tuning of filter by the change of the frequency of switching. Fig. 4.94 presents the design of the switched capacitor filter.

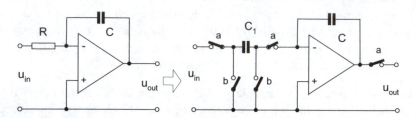

Figure 4.94. The RC filter (a) and equivalent filter with switched capacitor (b)

Fig. 4.95 presents the *state variable* universal filter circuit designed for the programmable active filters with the switched capacitors. This filter can operate as the low-pass filter LP, high-pass filter HP, band-pass filter BP and additionally band-reject filter BR. The cut-off frequency can be changed by the change of the resistors R_F of two integrating circuits (or the switching capacitor C_1 simulating this resistor as it is presented in Fig. 4.94)

$$\omega_p = \frac{1}{R_F C} = \frac{1}{f C_1 C} \qquad (4.121)$$

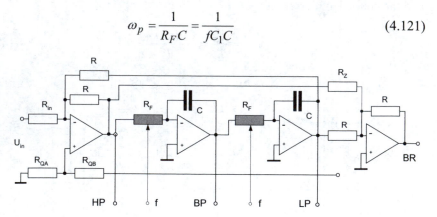

Figure 4.95. The state variable filter with the possibility of application of the frequency tuning (switching capacitors instead of resistors R_F) (Jamal 2003)

Figure 4.96 presents the integrated filter circuit developed by the *National Semiconductor* - model *LMF100*. This circuit utilizes the operation principle of the state variable filter presented in Fig. 4.95.

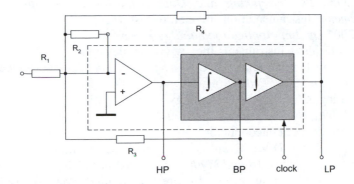

Figure 4.96. The functional circuit of the high performance dual switched capacitor filter model LMF100 of National Semiconductor (LMF 100 1999)

The integrated circuit of the filter with switched capacitors enable to design the second order filter (high-pass, low-pass or band-pass). The external resistors can be used to the setting of the band-pass gain or low-pass gain. The center frequency is tuned by the external clock frequency.

REFERENCES

Agilent 1441 2005 *Reducing Noise in Switching*, Agilent Application Note 1441-2

Analog Devices 2005 Analog Devices Inc. www.analog.com

Burr Brown 2005 Burr Brown (Texas Instruments), www.burr-brown.com

Coughlin R.F., Driscoll F.F. 2000 *Operational Amplifiers and Linear Integrated Circuits*, Prentice Hall

Franco S. 2001 *Design with Operational Amplifiers and Analog Integrated Circuits*, McGraw-Hill

Gerstenhaber N., Murphy M. 1991 *Synchronous System Measures* $\mu\Omega$, Analog Devices – Application Note AN 306

Gilbert B. 1968 *A Precision Four-quadrant Multiplier with Subnanosecond response*, IEEE J. Solid-State Circuits, SC-3, pp.365-373

Gilbert B. 1972 *Four-quadrant Multiplier Circuit*, US Patent 3 689 752

Han G., Sanchez-Simencio E. 1998 *CMOS Transconductance Multipiers,* IEEE Trans. Circuits and Systems II, 45, pp. 1550-1562

Huelsman L.P. 1993 *Active and Passive Filter Design*, McGraw-Hill

Jamal R., Steer R. 2003 *Filters*, Chapter 22 in Electrical Measurements, Signal Processing and Displays, CRC Press

James K. 2000, *PC Interfacing and Data Acquisition: Techniques for Measurements*, Instrumentation and Control, Newnes

Jung W.G. 2004 *Op Amp Application handbook*, Newnes

Keithley 2005 Keithley Instruments Inc., www.keithley.com

Kitchin C., Counts L. 1986 *RMS to DC Conversion Guide*, Analog Devices

Lai E. 2004 *Digital Signal Processing*, Newnes

Linear Technology 2005 Linear Technology Corp., www.linear-tech.com

LMF 100 1999, *High Performance Dual Switched Capacitor Filter*, Datasheet National Semiconductor

Lutovac M.D., Tosic D.V., Evan B.L. 2000 *Filter Design for Signal Processing using MATLAB and Mathematica*, Prentice Hall

Maxim 1795 2002 *Analog Filter Design Demystified*, Maxim (Dallas Semiconductors) Technical Note No. 1795

Morrison R. 1998 *Grounding and Shielding Techniques in Instrumentation*, John Wiley

Morrison R. 1991 *Noise and other Interfering Signals*, John Wiley

Nonvolatile Electronics 2005 Nonvolatile Electronics Corp., www.nve.com

Northrop R.B. 1997 *Introduction to Instrumentation and Measurements*, CRC Press,

Ott H.W. 1988 *Noise Reduction in Electronic System,* John Wiley & Sons

Pallas-Areny R., Webster J.G. 1991 *Sensors and Signal Conditioning*, John Wiley & Sons

Pallas Areny R., Webster J.G. 1999 *Analog Signal Processing*, John Wiley & Sons

Park J., Mackay S. 2003, *Practical Data Acquisition for Instrumentation and Control*, Newnes

Rich A. 2005 *Shielding and Guarding*, Analog Devices Application Note AN-347

Schaumann R. 2001 *Design of Analog Filters*, Oxford University Press

Scofield J.H. 1994 *A frequency domain description of a lock-in amplifier*, American Journal of Physics, 62, pp. 122-133

Signal Recovery 2005 Signal Recovery, www.signalrecovery.com

Smith S.W. 2003 Digital Signal Processing, Newnes

SR 1000 2005 *What is a Lock-in Amplifier?*, Signal Recovery Technical Note TN 1000

SR 1001 2005 *Specifying Lock-in Amplifiers*, Signal Recovery Technical Note TN 1001

SR 1002 2005 *The Analog Lock-in Amplifier*, Signal Recovery Technical Note TN 1002

SR 1003 2005 *The Digital Lock-in Amplifier*, Signal Recovery Technical Note TN 1003

Stanley W.D. 2001 *Operational Amplifiers and Linear Integrated Circuits*, Prentice Hall

Texas Instruments 2005 Texas Instruments, www.ti.com
Thede L. 2004 *Practical Analog and Digital Filter Design*, Artech House
Tran Tien Lang 1987 *Electronics of Measuring Systems*, John Wiley & Sons
Vaseghi S.V. 2000 *Advanced Digital Signal Processing and Noise Reduction*, John Wiley & Sons
Volkenburg M.E. 1995 *Analog Filter Design*, Oxford University Press
Winder S. 2002 *Analog and Digital Filter Design*, Newnes

5

Digital Processing of the Measurement Signals

5.1. ANALOGUE-TO-DIGITAL CONVERTERS

5.1.1. Sampling, quantization and coding of signals

The technical world is becoming more and more digital because digital signals are very convenient for information processing. However, most physical phenomena are analogue and the sensors measure analogue quantities. For that reason, the digital signal processing is often realized in the following sequence: conversion of the analogue signal to digital form ⇒ digital signal processing ⇒ conversion of the digital signal back to the analogue one. The conversion is realized by the *analogue-to-digital converters ADC* while the reverse process is realized by *digital to analogue converters DAC*.

The analogue signals are of *continuous time* – the value of such signal is determined in every instant of time. An example of the analogue signal is presented in Fig. 5.1a. The conversion of the analogue signal $x(t)$ to the digital form is realized in such a way that in assumed moment of time the value of the signal $x(n)$ is determined and represented by a number. We can say that the digital signal is determined in *discrete time*, which means that the value of the signal is known only in selected moments. Usually the discrete time is realized by collecting the samples of the analogue signal at the constant interval called the *period of sampling* T_s (Fig. 5.1b).

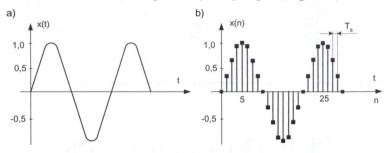

Figure 5.1. The analogue signal (a) and its conversion to the discrete one (b)

The process of collection of the samples is called the *sampling process* of analogue signals. The frequency $f_s=1/T_s$ is called the *sampling frequency* and it is described in *Hz* or *SPS* – *samples per second*. The process of determination of the digital value of the samples is called the *quantization* of the signals. The sampling is the digitization of the time, while the quantization is the digitization of the signal value.

As the result of sampling the time on the axis x is substituted by the number (index) n and every sample is described by its index n. The analogue signal described by the equation $x(t) = X_m sin \omega t$ is converted to the signal $x(n)$ = X_n (where X_m is the magnitude of analogue signal while the X_n value is the value of the signal of the number (index) n).

The conversion from the index n to the time t is evident because index n indicates the time with the period $T_s = 1/f_s$. For example, if we are sampling the signal of the frequency *50 Hz* and we would like to obtain the discrete signal represented by *64* samples per the period of signal[1] the sampling frequency should be $f_s = 3200 \ Hz$ (and period of sampling is $T_s = 312.5 \ \mu s$). Thus the $n = 50$ corresponds with the time *50 × 312.5 µs = 15.625 ms*. If we would like to have *128* samples per period of the measured signal then the sampling frequency should be two-times larger (*6400 Hz* in our case).

The analogue sinusoidal signal of the frequency f_a is described by the equation

$$x(t) = X_m \sin 2\pi f_a t \qquad (5.1)$$

---

[1] It is advantageous to have 2^n samples per period because Fast Fourier Transform requires such number of samples – this subject is discussed later.

and is represented by one spectral line of the frequency f_a (Fig. 5.2a). After sampling with the period T_s the same signal is described as

$$x(n) = X_m \sin 2\pi f_a n T_s \qquad (5.2)$$

Because the sinusoid is identical with the period 2π $(sin\varphi = sin\,(\varphi \pm 2k\pi)$ the dependence (5.2) can be rewritten in the form

$$x(n) = X_m \sin 2\pi f_a n T_s = X_m \sin(2\pi f_a n T_s \pm 2k\pi) \qquad (5.3)$$

and introducing the value $m = k/n$ we obtain

$$x(n) = X_m \sin 2\pi\left(f_a \pm \frac{k}{n} f_s \right)nT_s = X_m \sin 2\pi(f_a \pm mf_s)nT_s \qquad (5.4)$$

Comparing the equations (5.3) and (5.4) we see that after sampling of the signal of frequency f_a appear the components $f_a \pm mf_s$ (Fig.5.2b).

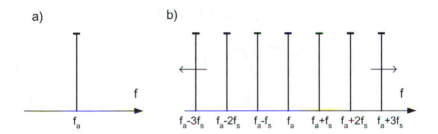

a) b)

f_a $f_a\text{-}3f_s$ $f_a\text{-}2f_s$ $f_a\text{-}f_s$ f_a $f_a\text{+}f_s$ $f_a\text{+}2f_s$ $f_a\text{+}3f_s$

Figure 5.2. The spectrum of the sinusoidal signal (a) and its replication after sampling (b)

After sampling of the sinusoidal signal the spectrum is represented by the infinite number of lines repeated with the multiple of $\pm f_s$. The spectral lines are repeated and instead of one signal of frequency f_a we have a lot of signals with frequencies $f_a \pm mf_s$ (Fig. 5.2b).

Similarly, if instead of one sinusoidal signal we have the signals within a bandwidth w (Fig. 5.3a) after sampling we obtain the multiplication of this bandwidths with the frequency f_s (Fig. 5.3b). We obtain a lot of signals of the frequencies $w \pm mf_s$.

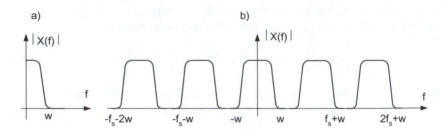

Figure 5.3. The signal of the bandwidth w (a) and its replication after sampling (b)

The signals presented in Fig. 5.3 were sampled with the frequency $f_s >$ $2w$. Thus in the frequency bandwidth $0 < f < w$ the signals before (Fig. 5.3a) and after sampling (5.3b) are the same – it is possible to remove the other signals of the frequency $f > w$ with a filter. But if the sampling frequency is smaller than $2w$ the duplicated signals interferes mutually and in the bandwidth around the frequency of sampling f_s distorted signals appear. This effect is called *aliasing* – Fig. 5.4.

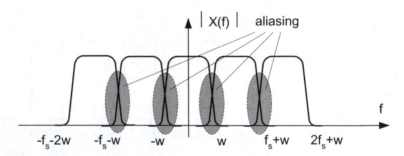

Figure 5.4. The interference of the signals (aliasing) when the sampling frequency is too small

To avoid this ambiguous aliasing effect *the sampling frequency should be at least two times larger then the highest frequency of the sampled signal* (two times larger than the bandwidth w). This rule is known as *the Shannon theorem.*[1] In other words the Shannon theorem is: *the signal is properly*

---

[1]   Before the Shannon the sampling theorem was originated by British mathematicians Whittaker and Ferrar. Independently similar theorem was introduced by a Russian scientist Kotel'nikow. Therefore the Shannon theorem is sometimes also called as the WKS sampling theorem (WKS – Whittaker, Kotel'nikow, Shannon).

sampled only if it does not contain a frequency component above one-half of the sampling rate.

Let us consider a case when an analogue acoustic signal with the bandwidth *20 Hz – 20 kHz* is converted to a digital one. The sampling frequency usually used in CD technology for such signals is f_s = *44 kHz*. According to the Shannon theorem the sampling frequency is sufficiently high (more then two times larger than 20 kHz). However, if in the processed acoustic signal there is a signal of the frequency 45 kHz this signal is normally inaudible (beyond the audibility of the human ear). But according to equation (5.4) after sampling this signal appears as $f_a - f_s$ = *45 kHz – 44 kHz = 1 kHz*. Thus, after sampling a new distorted audible signal *1 kHz* appears due to the aliasing.

To avoid this aliasing effect before the analogue-to-digital converter there should be introduced a special *anti-alias lowpass filter* with the cut-off frequency equal to the *Nyquist frequency*. The Nyquist frequency f_N according to the Shannon theorem is half of the sampling rate $f_N = f_s/2$.

The cut-off frequency of the anti-alias filter depends on the dynamics of the signal[1]. As was discussed in Section 4.7.3 the typical slope of the Mth-order filter is $M \times 6$ *dB/octave*. If our sampled signal exhibits the dynamics of *60 dB* then to limit this signal to the bandwidth *w* it is necessary to use a tenth order filter, which is rather difficult in practical realization. We can see that for large dynamics of the signal the filter should exhibit very large steepness of the frequency characteristic in the transition band. Therefore as the anti-alias filter often elliptical (Cauer) filters with large steepness of the frequency characteristic are used. But high-order filters with large steepness introduce phase distortion, which in the case of acoustic signals is unacceptable.

Fig. 5.5 presents the principle of application of the anti-alias filter. According to the Shannon theorem the sampling frequency f_s should be two times larger than the bandwidth *w*. Such a choice of sampling frequency creates a risk that the signals of frequency $f_a > w$ can generate the signals f_a-f_s in the bandwidth after sampling. For that reason it is safer to set the sampling frequency f_s two times larger than the frequency when the anti-alias filter sufficiently attenuates the signals (thus the Nyquist frequency $f_s/2$ is slightly larger than the bandwidth *w*).

---

[1] Take into account that as the bandwidth of the amplifier we assume the frequency range where the amplitude of the signal does not drop more than 3dB. Thus even outside the bandwidth there are signals with quite large amplitude.

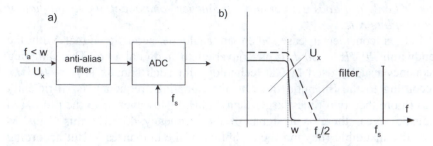

Figure 5.5. The sampling of the signal with the anti-alias filter at the input

The anti-alias filter is analogue. The digital filters could exhibit much better performances, but for application of such filters it is necessary to convert first the signal into digital (which means the alias problem related to the sampling operation without application of the anti-alias filter).

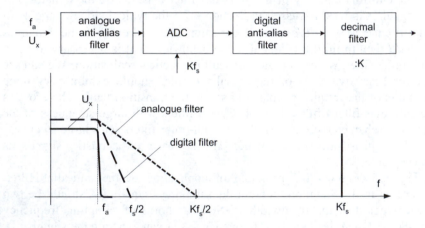

Figure 5.6. The principle of the oversampling technique

Higher sampling frequency means less critical requirements of the filter performances. Such conclusion results in the technique of sampling called *oversampling* technique (Fig. 5.6). This method is currently applied in high quality sound processing. For example in SACD system introduced by Sony (*SACD – Super Audio Compact Disc*) the sampling frequency is *2.82 MHz* which means the oversampling factor $K = 64$. In *DVD Audio* system introduced by Technics the sampling frequency is *192 kHz* and the oversampling factor is $K = 4$.

By applying the oversampling we can use the analogue anti-alias filter of lower order. After conversion to the digital signals we can use much better digital anti-alias filter and then the decimal filter recovering the lower sampling rate. The profit related to the application of the cheaper and less complicated anti-alias filter is at the expense of the necessity of application of the analogue-to-digital converter of higher sampling speed, thus the most costly one.

Seemingly the profit of applying of the oversampling is not so obvious (with the exception that we use a simpler anti-alias filter). By application of an increase of a sampling frequency we enlarge the bandwidth and significant part of this bandwidth is without the signal. But as will be proved later, the noises are distributed in the whole bandwidth and by increase of the bandwidth we diminish the noise in the useful bandwidth. And decrease of noises in the useful bandwidth is crucial for analogue-to-digital conversion because the dynamics and resolution of the AD conversion is much better.

Let us consider another case when we process the signal in the bandwidth *45 MHz – 55 MHz*. Applying the sampling frequency *110 MHz* (according to the Shannon theorem) seems to be extravagance. In such a case we can modify the Shannon rule: *the sampling frequency should be at least two times larger than the bandwidth and not the largest frequency signal*. In our case of the signals in bandwidth *45 MHz – 55 MHz* it is sufficient to use sampling frequency *20 MHz* instead of *110 MHz*. This technique is called the *undersampling technique* (or sometimes band-pass sampling). Of course, by using the undersampling technique we apply a band-pass anti-alias filter instead of a low-pass filter.

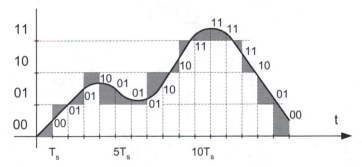

Figure 5.7. The quantization of the continuous signal with 2-bit resolution (the error of quantization is indicated with the dark color)

In the quantization process to each sample there is a digital value assigned, most often in the binary code. Fig. 5.7 presents the quantization with 2-bit resolution. In 2-bit quantization the converted value can be represented by four possible levels: *00, 01, 10* and *11*. The value of the continuous signal is rounded to the nearest possible level of quantization – thus the maximal value of the quantization error is half of a quant. In our case of 2-bit quantization this error is equal to 12.5% of full value. It is obvious that the larger is the digital word representing the quantized value (as more bits are represented this value) the better is the quality of quantization (lower quantization error and larger quantization dynamics)[1]. Table 5.1 presents the performances depending on the number of bits of various typical analogue-to-digital converters.

Table 5.1. The performances of the quantization process depending on the number of bits N (determined under assumption, that the range of the conversion is $0-2V$)

Number of bits N	Number of the quantization levels 2^N	Value of a quantum q	resolution %FS	*rms* noises $q/\sqrt{12}$	dynamics dB
8	256	8 mV	0.39	2.3 mV	48
10	1024	2 mV	0.098	580 µV	60
12	4096	0.5 mV	0.024	144 µV	72
16	65 536	31 µV	0.0015	8.9 µV	96
24	16 777 216	120 nV	0.000006	34. nV	144

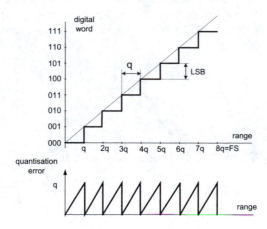

Figure 5.8. The characteristic of quantization of the 3-bit ADC

---

[1] But the larger is the number of bits the more expensive is the analogue-to-digital converter.

Fig. 5.8 presents the example of the conversion with a 3-bit converter. The *LSB* (*LSB – least significant bit*) is the abbreviation assigned to the smallest quantity of converted value and for N-bit converter it is equal to the resolution $1/2^N$. On the other hand the smallest quantity of the measured value is one quantum q determined as the smallest part of the *FS* value (*FS – full scale*)

$$q = \frac{FS}{2^N}$$

(5.5)

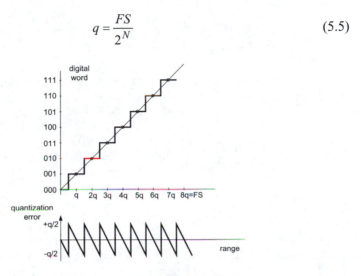

Figure 5.9. The characteristic of quantization shifted by q/2 value

As the percentage value the resolution can be determined as $1/2^N$ 100% and for the 8-bit converter the resolution is $100/2^8 = 100/256 = 0.39\%$. From Fig. 5.8 we can see that the quantization error is varying between 0 and q value. Usually the characteristic of the quantization is shifted by the $q/2$ value thus the error of quantization is varying between $-q/2$ and $+q/2$ (Fig.5.9).

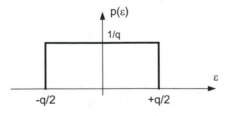

Figure 5.10. The probability distribution of the error of quantization

According to the characteristic presented in Fig. 5.9 the error of quantization ε is $\pm q/2$ and the probability distribution $p(\varepsilon)$ is uniform for all values of errors between $-q/2$ and $+q/2$ (Fig. 5.10).

The mean square value (*rms value*) of the error is

$$\varepsilon_{rms} = \sqrt{\int_{-q/2}^{q/2} \varepsilon^2 p(\varepsilon)d\varepsilon} = \sqrt{\frac{1}{q} \int_{-q/2}^{q/2} \varepsilon^2 d\varepsilon} = \frac{q}{\sqrt{12}} \tag{5.6}$$

The ε_{rms} value is often described as the noise of quantization. The signal to noise ratio SNR is

$$SNR = 20\log\frac{RMS\ SIGNAL}{RMS\ NOISE} = 20\log\frac{2^N \frac{q}{2\sqrt{2}}}{q/\sqrt{12}} = 20\left(\log 2^N - \log\frac{2}{\sqrt{6}}\right) \tag{5.7}$$

$$SNR = 6.02N + 1.76\ dB \tag{5.8}$$

The relation (5.8) is valid only if the noise is determined in bandwidth from *DC* to $f_s/2$. If the signal bandwidth w is less than $f_s/2$ then the expression (5.8) should be corrected to the form

$$SNR = 6.02N + 1.76 + 10\log\left(\frac{f_s}{2w}\right) \tag{5.9}$$

The expression (5.9) reflects the effect of noise reduction due to oversampling – for given signal bandwidth doubling of sampling frequency increases the *SNR* ratio by *3dB*.

The noises level is important for the dynamics of conversion. This dynamics can be calculated as the ratio of the signal $2^N q$ to the resolution of quantization q

$$dynamics = 20\log\frac{2^N q}{q} = 6.02N \tag{5.10}$$

The formula (5.10) is often expressed as "*six dB per one bit*". For example, in acoustic signal processing it is assumed that the bandwidth is *20 kHz* while dynamics is *100 dB*. Thus the sampling frequency should be *40 kHz* (in CD technique it is *44 kHz*) and to obtain the dynamics *100 dB* the number of bits should be: *100/6.02=16.6*. Thus to obtain correct dynamics of the audio signals the converter should be a 16-bit one.

As the result of quantization the value of the sampled signals is represented by the binary code. There are various systems of number encoding – generally we use two formats of the number: *fixed point number* (sometimes called integer number) and *floating point number* (called also real number).

Table 5.2. Various formats of the fixed point numbers

Decimal	unsigned integer	offset binary	sign and magnitude	two's complement
7	0111	1110	0111	0111
6	0110	1101	0110	0110
5	0101	1100	0101	0101
4	0100	1011	0100	0100
3	0011	1010	0011	0011
2	0010	1001	0010	0010
1	0001	1000	0001	0001
0	0000	0111	1000 or 0000	0000
-1		0110	1001	1111
-2		0101	1010	1110
-3		0100	1011	1101
-4		0011	1100	1100
-5		0010	1101	1011
-6		0001	1110	1010
-7		0000	1111	1001

In the *fixed point format* every bit is in fixed position, starting from the largest one (*MSB – most significant bit*) and ending by the smallest one (*LSB – least significant bit*). In natural binary code called *unsigned integer* every bit represents the digit 2^N. Thus the digit of the range R is represented as by the dependence

$$x = R\left(a_1 2^{-1} + a_2 2^{-2} + ... + a_n 2^{in}\right)$$ (5.11)

For $R = 1$ the number *0101* is corresponding to the

$$x = 0 \cdot 0.5 + 1 \cdot 0.25 + 0 \cdot 0.125 + 1 \cdot 0.0625 = 0.3125$$

The unsigned binary format cannot represent negative numbers. This problem can be solved by the *offset binary* format where the decimal value is shifted to obtain the negative number. The digit in this format is described by the equation

$$x = R\left(a_1 2^{-1} + a_2 2^{-2} + ... + a_n 2^{in} - 0.5\right) \qquad (5.12)$$

In the case of the 4-bit format the decimal number is shifted by 7 as is illustrated in Table 5.2.

Another format also enabling to represent the negative number is the *format sign and magnitude*. In this format the first left bit is reserved for the sign (zero for positive number and one for negative one). These two formats (binary offset and sign and magnitude) are difficult to implement in operational unit. Moreover in sign and amplitude format there are two representations of decimal zero.

The most popular is *format two's complement* that is easy to implement in the computer arithmetic unit. In this format the positive numbers are represented similarly to the unsigned integer format and the sign and magnitude format. Also, similarly as in the sign and magnitude format, the first bit is reserved for sign. For negative numbers the following algorithm is used: the decimal number is taken as the absolute value ⇒ next this number is convert to binary format ⇒ all bits are complemented: ones become zero, zero becomes one ⇒ a 1 is added to this number. For example *-5* is converted in following way: *-5* ⇒ *0101* ⇒ *1010* ⇒ *1011*. The most important advantage of the format two's complement is that the arithmetic unit in the same way adds positive and negative numbers (by subtracting it automatically counts in two's complement).

Many limitations of the fixed point numbers (especially in the case of large numbers) can be avoided in *floating point format*. Floating point format is similar to the scientific notation of numbers: *mantissa M* is multiplied by 2^E, where *E* is *exponent*. Additionally whole number is multiplied by $(-1)^S$ where *S* is the sign bit

$$x = (-1)^S \times M \times 2^E \qquad (5.13)$$

The most popular is the ANSI/IEEE 754-1985 standard where in a 32-bit representation of the number the first bit is a sign bit, next 8 bits are assigned to the exponent and last 23 bits are assigned to the mantissa (Table 5.3) according to the formula

$$x = (-1)^S \times 2^{E-127} \times M \qquad (5.14)$$

Table 5.3. The representation of floating point number according to the ANSI/IEEE 754-1985 standard

bit	31	30 29 28 27 26 25 24 23	22 21 20 ...... 2 1 0
	S	2^7 2^6 2^5 2^4 2^3 2^2 2^1 2^0	2^{-1} 2^{-2} 2^{-3} .... 2^{-21} 2^{-22} 2^{-23}
	sign	MSB exponent E LSB	MSB mantissa M LSB

The mantissa is represented by the following notation

$$M = 1 + m_{22} 2^{-1} + m_{21} 2^{-2} + ... + m_1 2^{-22} + m_o 2^{-23}$$

For example the number: *1 00000101 01110000000000000000000* corresponds to *(-1)×1.4375×2^{-122} = -2.70363×10^{-37}*.

The floating point format enables representation of the numbers with better dynamics but with worse resolution.

Of course it is possible in every moment to convert the binary numbers into decimal, hexadecimal or other format. But if the signal is being further processed digitally the binary format is the most convenient to use.

Although modern analogue-to-digital converters are very fast they need certain time to perform sampling and quantization process. Therefore, the AD converters are usually preceded by a special circuit holding the processed signal for the time necessary for the conversion. These circuits are called *SH – sample-and-hold circuits*.

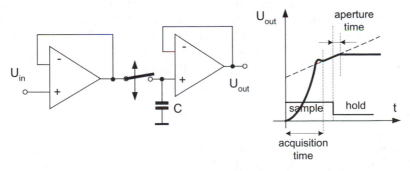

Figure 5.11. The simple sample-hold circuit and its time characteristic

An example of the SH circuit is presented in Fig. 5.11. After closing of the switch the capacitor *C* is charged to the voltage value equal to the input voltage. After disconnection of the switch the capacitor C stores (holds) the voltage. In the holding time the conversion (processing) of the signal is performed. The working cycle of the SH circuit consists of three parts: sampling time, short transient time when the holding value is fixed and holding time.

The sampling time can be as short as possible, only to equalize the input voltage and the capacitor voltage. This time can be extended and the changes of the voltage on the capacitor can follow-up the input voltage. Such circuits are called *track-and-hold circuits*.

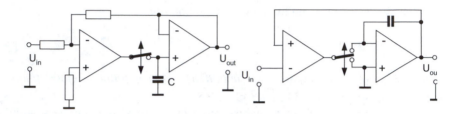

Figure 5.12. Two examples of the sample-and-hold circuits with feedback

The simple circuit presented in Fig. 5.11 is often substituted by slightly more complicated circuits with feedback. Two examples of the circuits with feedback are presented in Fig. 5.12. The SH circuits with feedback operate slower than the simple circuits, but the accuracy of signal processing is better.

The sample-and-hold circuits are indispensable parts of many digital processors, among them analogue-to-digital and digital-to-analogue converters. In the latter case they help in smoothing of the signal and elimination of the pulse interferences. On the market, there are also available amplifiers with SH circuit – *SHA – sample-and-hold amplifiers*. The typical times of sampling are of about *1 µs* and the *aperture time*[1] is not larger than several *ps*. There are also very fast sample-and-hold circuits with sampling time of about *10 ns* and aperture time less than *1 ps*.

5.1.2. Analogue-to-digital converters ADC

Many various AD converters have been designed and developed (Candy 1991, Goeshele 1994, Jespers 2001, Norsworthy 1996, van de Plasche 2003, Schreier 2004). However, currently on the market there are only a few main types of them: successive approximations register SAR, pipeline, delta-

---

[1] Aperture time is the time between hold command and disconnection of the signal from the hold capacitor (Fig. 5.11)

sigma, flash and integrating converters. Fig 5.13 presents the comparison of the main performances of these AD converters.

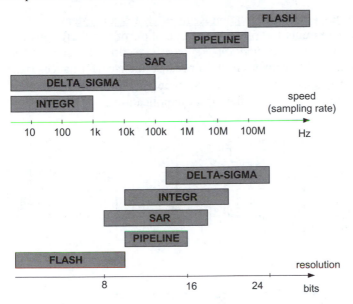

Figure 5.13. The comparison of the performances of the main AD converters

Figure 5.13 presents the comparison of two important parameters of the AD converters: the sampling frequency (speed) and number of bits (resolution). We can see that there is no one universal AD converter – the converters of high speed are of the poor resolution and *vice versa* – accurate (large number of bits) converters are rather slow. The most commonly used are the SAR (Successive Approximation Register) and Delta-Sigma converters. SAR converters are very accurate, operate with relatively high accuracy (16-bit) and wide range of speed – up to 1 MSPS[1]. For higher speed, up to 100 MSPS the pipeline converters are recommendable. For converting of very fast signals the direct flash converters are used.

The Delta-Sigma converters (16-bit and 24-bit) are used when high accuracy and resolution are required. Recently, these converters are still in significant progress. They gradually substitute the integrating (dual-slope) converters more often used in the past (Maxim 2102 2003). The integrating converters are mainly used for conversion of DC signals because their conversion time is relatively long *10 – 150 ms* (for comparison the

---

[1] *MSPS – mega samples per second.*

conversion time of the SAR converters is *3 – 30 μs*, while the flash converters need only *10 ns*).

5.1.2.a. Successive Approximation Register –SAR converters

Fig. 5.14 presents the principle of operation of the SAR converter. The *SAR* (*Successive Approximation Register*) is currently one of the most commonly used AD converters. It is because their performances (resolution 16- or 18-bit, speed 0.5 – 5 MSPS, time of conversion 1 μs for 16-bit converter) are acceptable for the most of applications.

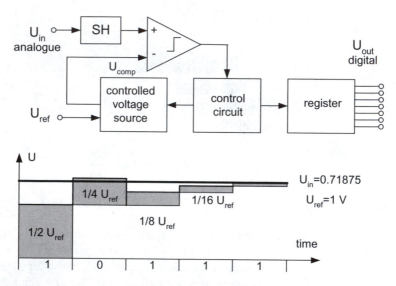

Figure 5.14. The principle of operation of SAR converter

The principle of operation of the SAR device resembles the weighting on the beam scale. Successively the standard voltages in sequence: $U/2$, $U/4$, $U/8...U/2^N$ are connected to the comparator. These voltages are compared with converted U_{in} voltage. If the connected standard voltage is smaller than the converted voltage in the register this increment is accepted and the register sends to the output *1* signal. If the connected standard voltage exceeds the converted voltage the increment is not accepted and register sends to the output *0* signal.

Fig. 5.15 presents the example of the SAR converter – model AD7667 of Analog Devices (AD 2005). The standard voltages are obtained using the array of 16 binary weighted capacitors. During the acquisition phase all switches are connected to analogue input U_{in} and the capacitors are charged (they are used as the sampling capacitors). In the conversion phase the capacitors are disconnected from the U_{in} and connected to the reference ground. This way the captured voltage is applied to the comparator input. Next, the switches connect successively the capacitor array to the standard voltage U_{ref}. Every connection changes the (U_{in}-U_{ref}) voltage according to binary weighted voltage steps $U_{ref}/2$, $U_{ref}/4$....$U_{ref}/65536$. This difference is connected to the comparator input. The control logic unit toggles switches as the comparator is balanced. As this process is completed the control logic sends the code to the digital output.

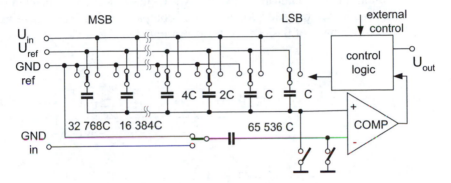

Figure 5.15. The principle of operation of the PulSAR converter of the Analog Devices (model AD7667)

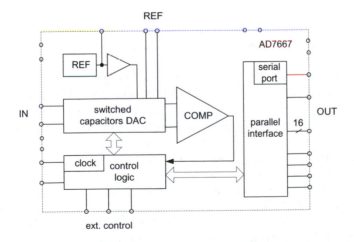

Figure 5.16. Functional block diagram of the AD7667 PulSAR converter of Analog Devices

The main advantages of the presented converter are its simple design and low consumption of power – only one comparator is used for the conversion. Fig. 5.16 presents the functional diagram of this converter. The 16-bit device enables conversion of the $0 - 2.5$ V voltage to the digital output (serial or parallel) with uncertainty *0.004%FS*, dynamics *96 dB* and sampling rate *800 kSPS*. The power dissipation is only *130 µW*.

5.1.2.b. Flash converters

In the *flash converters* instead of successively connecting weighted binary voltages to one comparator (as in SAR devices) there are connected at the same time binary weighted voltages to 2^N comparators (each representing one bit). The example of flash converter is presented in Fig. 5.17.

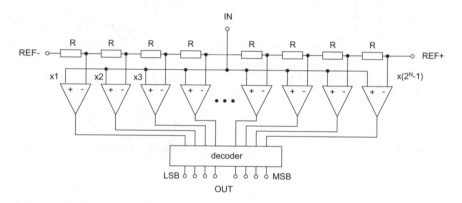

Figure 5.17. An example of the flash type AD converter

In the case of the 8-bit converter it is necessary to connect 255 resistors to the 255 comparators (in the case of 16-bit converters it would be 65 535 comparators!). No wonder that the flash converters are designed as at most 8-bit converters. The main advantage of the flash converters is that the conversion is performed in one step. Therefore the time of conversion is very small (less than *1ns*) and the sampling rate above *1 GSPS* is possible. The main drawback of the flash converter is its poor resolution (number of bits) and large power dissipation (due to great number of comparators).

As an example of flash converter we can consider the *MAX108* model of Maxim (Maxim 2005). It is an 8-bit converter with a sampling rate up to *1.5 GSPS* and conversion time *0.7 ns*. The uncertainty of this converter is *0.25 LSB* and the power consumption is *5 W*.

It is possible to decrease the number of converters in the half-flash type converter – presented in Fig. 5.18. In such a converter the sampling is performed in two sub-ranges. The first 4-bit flash converter processes roughly the first four bites. The converted voltage is subtracted from the input voltage (from the track-and-hold circuit) and this voltage difference is converted by the second fine 4-bit flash converter. Due to this solution the number of converters in 8-bit device is diminished to 30 (from the original 255).

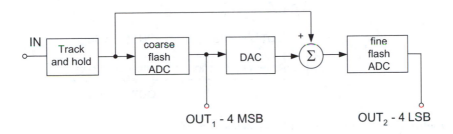

Figure 5.18. An example of the half-flash type AD converter

As an example of the half-flash converter we can consider TCL5540 converter of Texas Instruments (TI 2005). This converter enables 8-bit conversion with the sampling rate 80 MSPS and conversion time 9 ns. The uncertainty of this converter is 1 LSB.

5.1.2.c. Pipeline converters

Pipeline converters operate similarly to the half-flash converter – the input signal is processed in several stages (these converters are sometimes also called as "*subranging*"). The main differences between half-flash and pipeline converters are as follows: in a half-flash converter there are two stages while in pipeline converters there can be several stages; after each stage there are inserted amplifiers for improving the resolution of the next stage; between the stages there are inserted track-and-hold circuits and it is possible to process the signal by the first stage at the same time as the last stages process the previous sampling result (the signal is pipelined through the converter).

An example of two-stage pipeline converter is presented in Fig. 5.19. The input signal after TH circuit is converted to digital signal by ADC1 converter – 6 most significant bits. The remaining signal is again converted to a digital one by DAC1 circuit and it is subtracted from the input signal. This residual analogue signal is amplified to obtain better resolution in the next stage. The signal is converted again to a digital signal by ADC2 converter – 7 least significant bits. The important is the error correction logic circuit. In a 12-bit converter both converting stages, 6 bits and 7 bits, have common 1 bit. This overlapped additional bit is used for the eventual error correction.

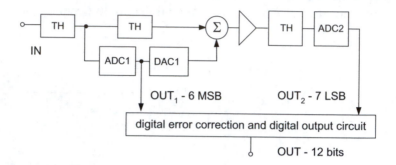

Figure 5.19. The example of 12-bit pipeline converter

The multistage operation enables to perform the conversion with relatively high resolution 14 – 18 bits and sampling rate up to 100 MSPS. In comparison with flash converters a much smaller number of comparators is required – for example four-stage 16-bit converter requires only 60 comparators. Figure 5.20 presents the three-stage pipeline converter of Analog Devices (model AD6645). It enables the conversion with *14-bit* resolution and sample rate *105 MSPS*. The time of conversion is *10 ns*, power consumption *1.5 W* and uncertainty *1.5 LSB*.

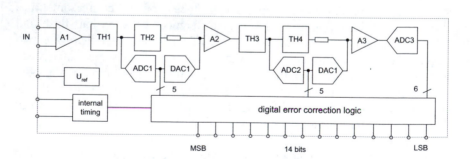

Figure 5.20. The functional block diagram of AD6645 pipeline converter of Analog Devices

5.1.2.d. Delta-Sigma AD converters

The *delta-sigma ΔΣ converters* called also *1-bit converters* or *bitstream converters*[1] utilize the oversampling technique. Due to many advantages (most of all the best resolution – even up to 24-bit) these converters are currently very intensively developed (Candy 1991, Norsworthy 1996, Schreier 2004). The principle of operation of such converters is presented in Fig. 5.21.

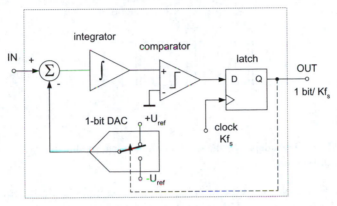

Figure 5.21. The block diagram of the delta-sigma converter

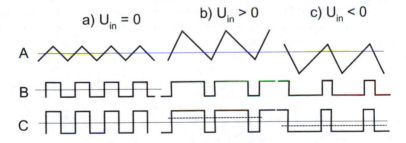

Figure 5.22. The principle of operation of the delta-sigma converter: A – output signal of the integrator, B – output signal of the comparator, C – output signal of the DAC 1-bit converter in the feedback

In delta-sigma conversion the delta modulation is used (hence the name of this device). In delta modulation the width of the impulse is proportional

---

[1] or sigma-delta converters.

to the value of converted signal. As the 1-bit ADC quantizer operates the comparator and latch switched with the frequency Kf_s forced by the clock (K is the oversampling factor). The output voltage is converted again to analogue form by 1-bit DAC. The adder in the input compares the input value and the output signal.

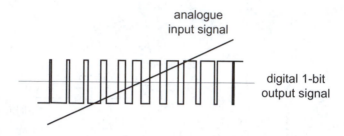

Figure 5.23 The dependence of the digital output signal on the analogue input signal in a delta-sigma converter.

Due to feedback the average value of output signal should be equal to the value of the input signal. If the input signal increases the integrating circuit need more time to obtain the zero value, the width of the impulse decreases and the average value of the output signal increases – as it is illustrated in Fig. 5.22 case b). Figure 5.23 presents the dependence of the digital output signal on the analogue input signal and Fig. 5.24 presents the same dependence for sine input signal.

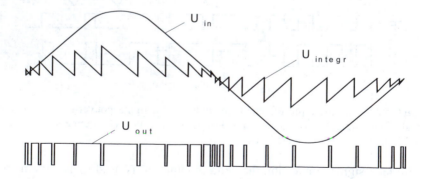

Figure 5.24. The integrator and output signal of the delta-sigma converter as the dependence of the input signal value

The important advantage of the delta-sigma converter is the noise suppression. In the previous chapter (Eq. 5.10) it was shown that the increase of the output signal of 1 bit results in increase of dynamics of 6 dB. This conclusion can be inverted – an increase of the dynamics (*SNR – signal to noise ratio*) of 6 dB would give the possibility of increasing the resolution by one bit. Thus the SNR of about 140 dB enables us to obtain a 24-bit converter (see Table 5.1).

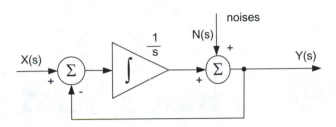

Figure 5.25. The equivalent circuit of the delta-sigma converter

Figure 5.25 presents the equivalent circuit of the delta-sigma converter with the source of noises. The input value is

$$Y(s) = [X(s) - Y(s)]\frac{1}{s}$$

(5.15)

For *N(s) = 0* we can describe the transmittance of the converter as

$$\frac{Y(s)}{X(s)} = \frac{1}{s+1}$$

(5.16)

Expression (5.16) is the transmittance of the low-pass filter. If the *X(s) = 0* we can write that

$$Y(s) = -Y(s)\frac{1}{s} + N(s)$$

(5.17)

and transmittance for the noise source is

$$\frac{Y(s)}{N(s)} = \frac{s}{1+s}$$

(5.18)

The circuit operates for the noises as a high-pass filter and reduces the noises for low frequency (Fig. 5.26c). This feature is called *noise shaping*. Thus the delta-sigma converter suppresses the noises in two ways. Due to oversampling the noises are decreased, because the noises energy is distributed in the larger bandwidth (Fig. 5.26b). And additionally the noises are attenuated, because the signal is filtered as low-pass while the noises are filtered as high-pass (Fig. 5.26c)

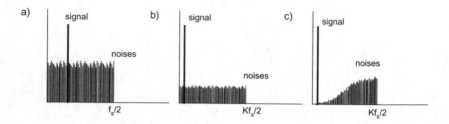

Figure 5.26. The noise (a) suppression due to oversampling (b) and noise shaping (c)

To obtain a noise suppression of about 40 dB it is necessary to apply a oversampling factor equal to 64 (Fig. 5.27). Further noise suppression is possible by increasing the order of the modulator. From the graph presented in Fig. 5.27 we can see that to obtain a 24-bit converter (140 dB dynamics) we should apply a third order modulator. Figure 5.28 presents the circuit of the second order delta-sigma converter.

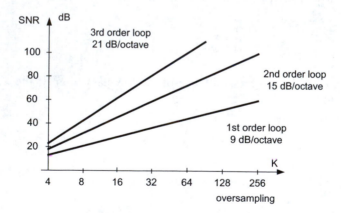

Figure 5.27. The dependence of SNR on the order of delta-sigma modulator and the oversampling factor

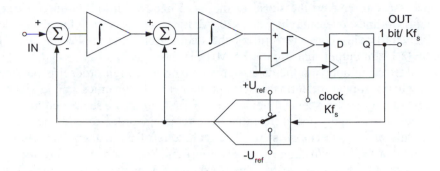

Figure 5.28. The second order delta-sigma converter

Depending on the application various manufacturers propose various delta-sigma converter design. For example, Analog Devices developed a 24-bit resolution AD7713 model of delta-sigma converter with an auto-zero function. Such converter enables us to process the signals in bandwidth *200 Hz* with uncertainty better than *0.0015%*.

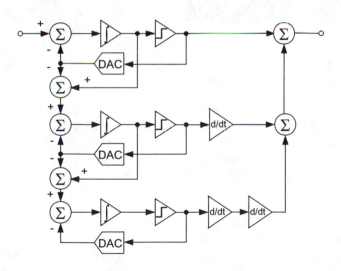

Figure 5.29. The MASH type multistage delta-sigma converter

To obtain large dynamics Analog Devices developed the fourth order delta-sigma converter with dynamics 120 dB in the 1 kHz bandwidth. If it is necessary to convert the signal of the large bandwidth the 16-bit converter with 8-times oversampling and 20 MHz clock was developed (model AD9260). This converter utilizes the decimation filter processing the 10 MHz/12-bit output signal to the 2.5 MHz/16-bit signal.

Instead of applying the delta-sigma converter of high order it is possible to obtain similar performances (improvement of dynamics and SNR) by cascade connecting several converters of first order. In such a circuit it is necessary to apply the differentiating circuits in order to add the output signals of the subsequent steps. The technique of multistage converting is called MASH (*Multistage Noise Shaping*) and such converters are used in high quality audio devices to obtain excellent dynamics (for example such technique introduced Technics). Fig. 5.29 presents the circuit of the MASH type converter.

Figure 5.30. The sinusoidal signal processed by one-bit delta-sigma converter and multi-bit converter (Wolfson Microelectronics 2000)

Apart from one-bit sampling there are also multi-bit delta-sigma converters (Geerts 2002). In such converters instead of one-bit quantizator the flash converter can be used. Figure 5.30 presents the output signal of a four-bit delta-sigma converter. In the multi-bit converter it is possible to improve the dynamics and decrease the noise level.

5.1.2.e. The integrating AD converters

The integrating converters are often realized as the *dual slope converters*. The principle of operation of dual slope converter is presented in Fig. 5.31. The integrating circuit is connected to the comparator that detects the zero-level of the integrator signal. This comparator controls the logic gate connecting the clock generator to the counter.

The dual slope converter operates in two half-cycles. In the first one the input voltage is connected to the integrating circuit for the fixed time T_1. At the same time the clock oscillator of frequency f_{cl} is connected to the counter.

The first half-cycle is finished when the counter indicates assumed value, for example $N_1 = 1000$. The voltage at the output of the integrating circuit increases with a fixed slope to the value

$$U_{int} = \frac{1}{RC}\int_0^{t_1} U_x dt = \frac{U_x}{RC}\frac{1}{f_{cl}}N_1 \qquad (5.19)$$

a) b)

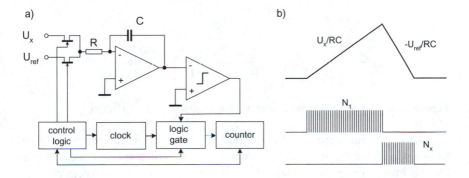

Figure 5.31. Principle of operation of the dual slope integration ADC

In the second half-cycle the reference voltage of the reverse polarization is connected to the integrating circuit and the counter starts counting the clock oscillator pulses. The voltage at the integrator output is decreased to the moment when the comparator detects zero. The zero state is when the following condition is fulfilled

$$\frac{U_x}{RC}\frac{1}{f_{cl}}N_1 = \frac{U_{ref}}{RC}\frac{1}{f_{cl}}N_x \qquad (5.20)$$

and the number of counted pulses is

$$N_x = \frac{N_1}{U_{ref}}U_x \qquad (5.21)$$

Thus the final state of the counter depends on the N_1 value (this we can fix very precisely), on the reference voltage value U_{ref} and of course on the converted voltage value U_x. The value indicated by the counter does not depend on the *RC* value and the frequency of clock oscillator.

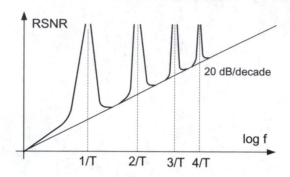

Figure 5.32. The noise rejection ration in the integrating AD converter

The important feature of the integrating converters is the rejection of AC noises. Consider the case that the measured *DC* voltage U_x is accompanied by the interference *AC* voltage $U_{int} = U_m sin(\omega t + \varphi)$. After integration we obtain

$$U_{int} = \frac{1}{T}\int_0^T [U_x + U_m \sin(\omega t + \varphi)]\, dt = U_x - \frac{U_m}{\omega T}[\cos(\omega T + \varphi) - \cos\varphi] \quad (5.21)$$

We can see that if the integration period T is fixed in such a way that $T = 2\pi/\omega$ then the second term (AC interferences) is equal to zero. The noise rejection ratio *RSNR* is (Tran Tien Lang 1987)

$$RSNR = \frac{noise}{error} = 20\log\frac{\omega T}{\cos(\omega T + \varphi) - \cos\varphi} \quad (5.22)$$

Figure 5.32 presents the dependence of the RSNR factor on the frequency. The integration converter behaves like a selective filter rejecting not only the component of the frequency $f = 1/T$ but also the harmonics of this signal. Usually, the value of T is fixed to be equal to *20 ms*, which enables rejection of the *50 Hz* signal and its harmonics. In practical circuits the *T* period is synchronized with the frequency of the supply *AC* voltage.

The relatively long time of integration is a drawback of the dual slope converter. This problem can be overcome by applying the multislope converter (Fig. 5.33).

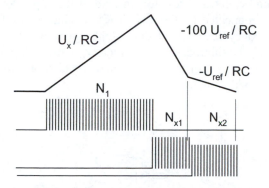

Figure 5.33. Principle of operation of a three-fold-slope converter

There are three-fold-slope and quad-slope devices. In the three-fold-slope device the second cycle (of dual slope device) is divided to the two steps. In the second step the reference voltage is connected to the integrator with smaller R resistance (for example 100 times smaller). This way the time necessary to decrease the output voltage of the integrator is 100 times shorter. After the integrator output voltage reaches a defined threshold voltage it is again connected the R resistor for precise detection of the zero state. Thus after these three phases the following relationship is realized

$$\frac{U_x}{RC}\frac{1}{f_{cl}}N_1 = \frac{U_{ref}}{RC}\frac{1}{f_{cl}}100N_{x1} + \frac{U_{ref}}{RC}\frac{1}{f_{cl}}N_{x2} \qquad (5.23)$$

and

$$U_x = \frac{U_{ref}}{N_1}\left(100N_{x1} + N_{x2}\right) \qquad (5.24)$$

The multi-slope integrating technique offers improvement of the conversion speed (or resolution in the same time) at the expense of more complexity and the need to apply two precise resistors.

Another problem appearing in the integrating converters is a zero drift. The minimization of this effect is possible in quad-slope converters, where an additional cycle is performed for the short-circuited input, which enables us to introduce the required correction. Another method is the application of the auto-zero function. An example of such a converter is presented in Fig. 5.34. The conversion time is divided into four cycles – in the first one for the short-circuited input and connected resistor, instead of the capacitor (switch K4), the capacitor C_o (connected by the switch K5) is charged to the offset voltage. In the two next cycles (typical dual slope operation) this voltage across the capacitor C_o is subtracted automatically, introducing the zero correction. In an additional fourth cycle the capacitor is short-circuited in order to remove the charged voltage.

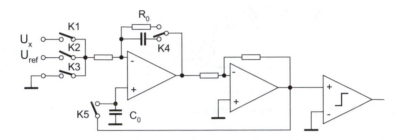

Figure 5.34. The integrating converter with the auto-zero correction (Tran Tien Lang 1987)

The integrating converters are typically used as the end part of DC digital voltmeters. Therefore they are usually equipped with a digital display – as example the converters of Maxim or Intersil can be considered. Currently, the tendency could be observed in substituting the integrating converters with cheaper delta-sigma converters. The main drawback of the integrating converter (apart from the long time of conversion) is the necessity of application of the expensive, high quality capacitors. Although there is no capacity C in the equation 5.21, the accuracy of the converter depends on the quality of this capacitor (the effect of memorizing the residual voltage).

Typical integrating converters operate as 12-bit or 15-bit (3 ½ or 4 ½ digit displays). The 18-bit integrating converter of Maxim (model MAX132) exhibits an uncertainty of 0.006%.

5.1.3. The main specifications of analogue-to-digital converters

As two most important parameters of the ADC usually the sampling frequency and the number of bits are considered. Table 5.4 presents the comparison of analogue-to-digital converters of various manufacturers.

Table 5.4. The comparison of the market available analogue-to-digital converters

Part	Type	Sample rate	Bits	Manufacturer	Approx.. Price $
ADC180	integration	2048 ms	26	Thaler	210
ADS1256	delta-sigma	300 kSPS	24	Texas	9
AD7714	delta-sigma	1 kSPS	24	AD	9
AD1556	delta-sigma	16 kSPS	24	AD	27
MAX132	integration	63 ms	18	Maxim	8
AD7678	SAR	100 kSPS	18	AD	27
AD7674	SAR	800 kSPS	18	AD	30
AD10676	pipeline	80 MSPS	16	AD	900
ADS8412	SAR	2MSPS	16	AD	23
MAX1200	pipeline	1 MSPS	15	Maxim	20
ADS5500	pipeline	125 MSPS	14	AD	95
AD9410	pipeline	210MSPS	10	AD	200
AD9480	pipeline	250MSPS	8	AD	200
HI1276	flash	500 MSPS	8	Intersil	300
MAX105	flash	800 MSPS	6	Maxim	36
MAX108	flash	1.5 GSPS	8	Maxim	9

The former parameter is important if the frequency of the processed signal (frequency bandwidth) is taken into account, while the latter one informs us about the resolution. It does not mean that the number of bits (length of digital word) automatically describes the accuracy. Similarly, the sampling frequency does not automatically mean that the investigated frequency can be at least half of the sampling frequency. Only the knowledge about other parameters as errors, level of noises, dynamic performances give us full information about the performance of an analogue-to-digital converter.

It can be seen from the data presented in Table 5.4 that the most expensive are the converters of the high resolution (16 bits) and large speed *100 MSPS*. These performances are possible to obtain using the pipeline converter. In the region of the best resolution (24 bits) the delta sigma converters dominate. The SAR converters are appreciated due to their accuracy and practically all data acquisition boards utilize these converters.

Similarly as in the case of analogue signal processing the digital signal processing can be influenced by the zero drift of the amplifier (Fig. 5.35a) (especially the temperature zero drift) and the gain error (Fig. 5.35b).

Figure 5.36 illustrates the main errors of linearity. *The integral nonlinearity INL* is the deviation of the values of the actual transfer function from a straight line (Fig. 5.35a). The *differential nonlinearity DNL* is the incorrect quantization resulting in not equal quanta. If the elementary quant is LSB the DNL is deviation from the ideal 1 LSB code (Fig.5.36b). The special case of the large differential nonlinearity is the *missing code error*. This error occurs when the quantization step is larger than 2 LSB. For example due to large DNL the number 100 (in Fig. 5.36b) is not indicated during the conversion. Because this error is dangerous for accuracy of conversion many of manufactured converters are described as "*no missing code*".

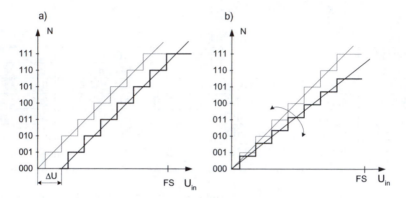

Figure 5.35 The error of zero drift (a) and the gain error

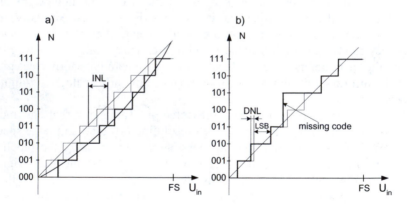

Figure 5.36. The transfer characteristic of the 3-bit converter with the integral nonlinearity error (a) and the differential nonlinearity error (b)

The performances of the analogue-to-digital converters are described by many parameters (Maxim 644 2000) presented in data sheets. Let us present the most important of these parameters.

SNR is the described earlier *signal to noise ratio* (usually it is the ratio of amplitude of the signal to the amplitude of the noises but also the ratio of *rms* values is used). SNR of analogue-to-digital converters are described by the relationships (5.8) and (5.9).

SINAD (*signal to noise and distortion ratio*) is defined as the ratio of *rms* value of the sine wave to the *rms* value of noises plus all harmonics of the signal.

THD (*total harmonic distortion*) is the ratio of *rms* sum of the harmonics to the fundamental component.

IMD (intermodulation distortion) appears when the input signal contains two signals of similar magnitude and frequencies f_1 and f_2. After the sampling process there can be generated components of the frequencies $f_1 - f_2, f_1 + f_2, 2 f_1 - f_2$ etc. IMD is defined as the ratio of the *rms* of intermodulation components to the signal without distortion.

SFDR (spurious free dynamic range) is defined as the ratio of *rms* value of fundamental signal component to the *rms* value of the largest spurious component (mainly spurious pulses).

Transient response is the response of the converter after the step unit change of the input signal.

FPBW (full power bandwidth) is defined as the point of the frequency characteristic where the amplitude of the digitized conversion result is decreased by 3 dB.

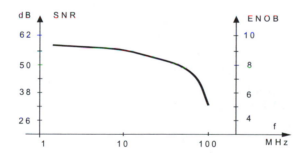

Figure 5.37. The example of the dependence of the SNR and ENOB factor on the frequency of input signal (for the flash converter)

Special importance is related to the *ENOB (effective number of bits)* factor. In an ideal analogue-to-digital converter there is only the quantization error. But in the real converters with the increased frequency additional noises and distortion can be quite significant.

In Fig. 5.37 the ENOB dependence for the flash converter is presented. Although the data sheet reveals that the number of bits is equal to *10* we can see that for the frequency of *100 MHZ* due to the increase of SNR the resolution is effectively decreased to *5*. For the sinusoidal input signal the ENOB can be determined from the relation

$$ENOB = \frac{SINAD - 1.76}{6.02} \qquad (5.25)$$

5.2. DIGITAL-TO-ANALOGUE CONVERTERS

5.2.1. The reconstruction of the analogue signal

The *digital-to-analogue converters DAC* are mainly used for the recovery of original analogue signals from the digital code. Hence, this process is sometimes called the reconstruction of the analogue signal. Each digital value of the code is related to the defined value of the analogue signal resulting from the partition of the full range to the number of quantity – as it is illustrated in Fig. 5.38.

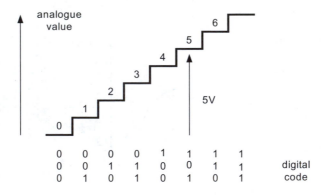

Figure 5.38. The conversion of the digital code to the analogue value (for LSB equal to 1 V)

After conversion of the series of digital data we obtain a series of pulses with the amplitudes proportional do the digital values of the signal in the moments of sampling (Fig. 5.39a).

In the simplest case we can complete the lack of the signal between the pulses by the holding the magnitude of the pulse until the delivery (generation) of the next pulse. This process is called *ZOH – zero order hold* – or *staircase reconstruction* (Fig. 5.39b).

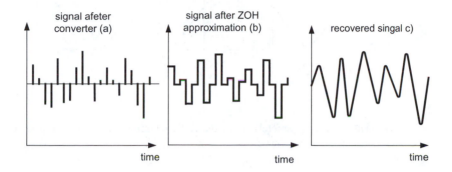

Figure 5.39. The reconstruction of the analogue signal

For the reconstruction of the signal the best would be to apply the ideal low-pass filter. If we use the zero order hold we realize following relationship

$$x(t) = \begin{cases} 1 & for\ 0 \le t \le T_s \\ 0 & for\ other\ moments \end{cases} \tag{5.26}$$

The function (5.26) in the frequency domain is described as

$$X(j\omega) = \int_{-\infty}^{\infty} x(t)e^{-j\omega t}\,dt = T_s e^{-j\omega T_s/2} \frac{\sin(\omega T_s/2)}{\omega T_s/2} \tag{5.27}$$

The function (5.27) is presented in Fig. 5.40. From Fig. 5.40 we can see that the ZOH technique results in the dependence of the signal on the frequency – it decreases with the increase frequency and disturbs the signal (as compared to the flat horizontal characteristic of the ideal low-pass filter presented in Fig. 5.40). For that reason, at the output of the digital-to-analogue converter a correcting filter is sometimes inserted, which increases the signal value with the frequency – as the inverse of the *sinx/x* relationship (Fig. 5.41).

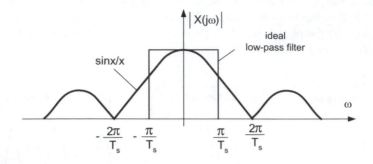

Figure 5.40. The transfer characteristic of the ZOH technique converting the series of the pulses into the staircase curve

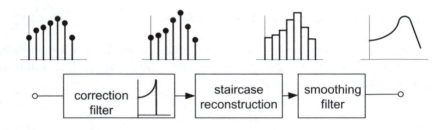

Figure 5.41. The reconstruction of the analogue signal from the series of pulses

In the case of reconstruction of the signal the number of the samples can be important. In the example presented in Fig. 5.42a the signal after conversion is represented by 10 pulses (samples) per period and the staircase curve is near the original. Theoretically the Shannon rule is fulfilled and the signal should be correctly reconstructed by the low-pass filtering of the non-desirable harmonics.

The small number of the quantization levels can be inconvenient during the sampling and signal reconstruction. To increase the levels of quantization sometimes the signal is supplemented by the additional noise signal of the value not larger than 1 LSB. This technique (called *dithering*) results in increasing of the noises (see right hand side of the Fig. 5.42) but it helps in better reconstruction of the signal. The additional noise (usually the white noise) is possible to reject by applying the filter.

a)

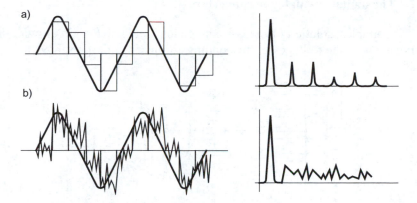

b)

Figure 5.42. The sinusoidal signal after staircase reconstruction (a) and supplemented by the signal (b) (on the right hand side the spectral characteristics of these signals are presented)

Figure 5.43 presents the typical structure of the circuit designed for conversion of the digital signal to the analogue one. At the input there is inserted a *register circuit* (*latch circuit*), which is required to save the signal for the time necessary for conversion of the last digit (*the settling time*). The input register plays the same role as in the case of analogue-to-digital conversion the sample-and-hold circuit. An analogue signal is generated as the sum of the component signals corresponding to appropriate levels of quantization. At the output the filter circuit and eventually the amplifier are inserted.

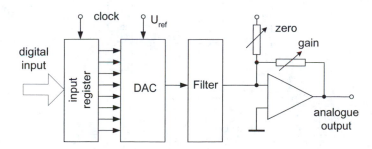

Figure 5.43. The structure of typical circuit of conversion of digital signal into the analogue one

5.2.2 The digital-to-analogue converters - DAC

The simplest solution of the DA conversion would be the circuit with the summation of the voltages corresponding to binary code presented in Fig. 5.44.

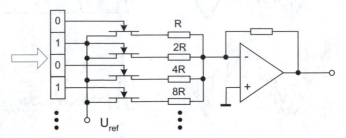

Figure 5.44. The digital to analogue converter with weighted resistors

The converter presented in Fig. 5.44 is rather difficult to manufacture because it requires precise resistors with a very wide range of values. For example for 20-bit conversion it is necessary to use a precise resistor equal to *500 MΩ* for the most significant bit, when the LSB is represented by *1 kΩ* resistor. Technologically simpler is to use the same value of resistors, although it means that the number of these resistors can be huge.

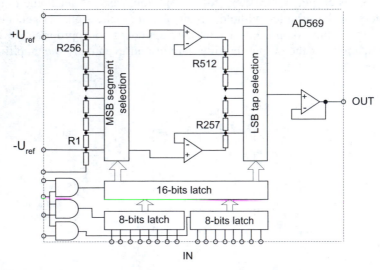

Figure 5.45. The functional block diagram of the string digital-to-analogue converter AD569 of Analog Devices

The converter built from single-valued resistors requires 256 resistors for 8-bit conversion and as much as 65 536 of them for 16-bit conversion. In practical circuits the voltage divider can be composed of two dividers – one for coarse conversion (the first 8 bits) and the second for fine conversion (last 8-bits). In such design in order to achieve 16-bit conversion only 512 resistors are required. Although such number of resistors seems to be great these converters (called *segmented converters* or *string converters*) are available on the market – as an example we can point to the AD converter model AD569 developed by Analog Devices (Fig. 5.45). The main advantage of the string converter is relatively large speed and very good linearity of conversion. The AD569 converter presented in Fig. 5.45 enables 16-bit conversion with nonlinearity less than 0.01% and settling time 3 μs.

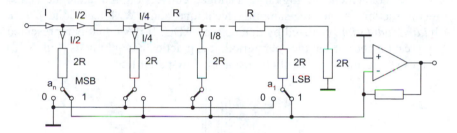

Figure 5.46. The R-2R digital to analogue converter with ladder network

Further simplification of the converter circuit is possible in the *R-2R converter* presented in Fig. 5.46. In this case also the resistors of the same value R are used (*2R* can be composed from two resistors). At each node the current splits into halves. The resulting output voltage is proportional to the total current summed at the inverting input of the amplifier. It is advantageous that the whole network is consuming the same current from the supply source independently of the positions of the switches.

In the *R-2R* converter it is not required to have precise value of the resistors – it is only necessary to have the resistors with precisely the same value of each resistance. Instead of resistors it is possible to switch the current sources (Fig. 5.47).

The advantage of the converter with switched currents (called a *current steering converter*) is relatively high speed (up to *500 MSPS*). For example, the 8-bit converter model DAC08 of Analog Devices converts data with an update rate up to *12 MSPS* and settling time *85 ns*.

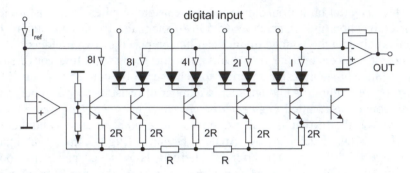

Figure 5.47. Digital-to-analogue converter with weighted currents

Very convenient for digital-to-analogue conversion and analogue signal reconstruction is the oversampling technique and PWM technique *(Pulse Width Modulation)* realized by delta-sigma ADC. If the signal is represented by great number of samples per period, it is practically continuous (Fig. 5.48) and can be reconstructed only by using the low-pass filter.

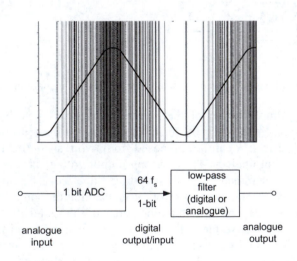

Figure 5.48. The reconstruction of the analogue signal after oversampling after PWM modulation

The conventional audio CD technique uses PCM *(Pulse Code Modulation)* with 44.1 kHz sampling frequency and 16 – 24 bit resolution. Therefore using oversampling technique it is necessary to use decimation filter to recover such digital signal from the 1-bit bitstream (as presented in Fig. 5.49a).

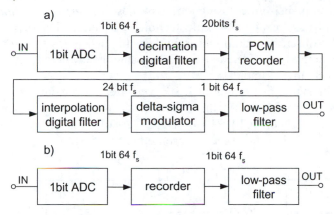

Figure 5.49 Conventional (a) and DSD (b) techniques of audio signal recording and reproduction

In the new DSD *(Direct Stream Digital)* technique introduced by SONY in its SACD audio system the 1-bit stream of the oversampling frequency 2.82 MHz is directly recorded on the DVD type disc profiting high density of this disc and enhanced speed of data transmission.

Signal converted by applying the oversampling technique often needs to be converted again to "ordinary" sampling form with sampling frequency decreased to f_s and the same time the resolution increased to multi-bit form (for example 20-bit resolution as it is presented in Fig. 5.49a). To perform such operation special filters called *decimation filters* can be used.

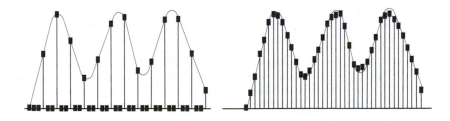

Figure 5.50. The increase of number of samples by insertion of intermediate samples between existing ones

Sometimes to improve the possibility of signal reconstruction the increase of the number of samples in one period of signal is recommended. Such an operation can be performed using various kinds of *interpolation filters*. Fig. 5.50 presents the relatively simple technique of increasing of number of samples by inserting additional samples. In first step the zero value additional samples are added and next after using the lowpass filter we obtain the signal with more samples.

The delta-sigma oversampling DSD converters are mainly used in high quality sound processing. As the example, let us consider the delta-sigma converters DSD1700, PCM1710 or DAC1220 of Burr-Brown. The block diagram of PCM1710 converter is presented in Fig. 5.51.

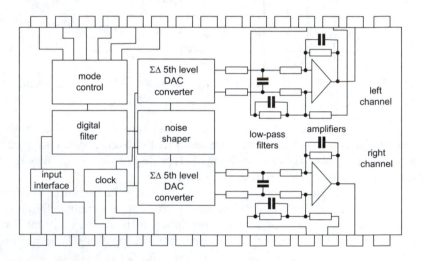

Figure 5.51. 5th level delta sigma DA converter model PCM1710 of Burr Brown

Presented in Fig. 5.51 the two-channel (stereo) converter consists of 5th level delta sigma converters and 4th order filter. At the input 8× interpolation filter realizes the oversampling (and also other functions such as digital attenuation, double speed dubbing, de-emphasis, etc). The converter enables us to convert the 16-bis or 20-bit signal to analogue one with *98 dB* dynamic range, *110 dB SNR* (in DSD1700 dynamic range is *110 dB*, THD is *0.001%* and frequency response is *100 kHz*). The DAC120 20-bit converter utilizes 2nd order delta sigma converter and converts the signal with linearity error less than *0.0015%.*

5.2.3. The main specifications of digital to analogue converters

In the data sheets describing the performances of digital-to-analogue converters similar parameters are used as in the case of analogue-to-digital converters. Fig. 5.52 presents the *integral nonlinearity error INL* and *differential nonlinearity error*.

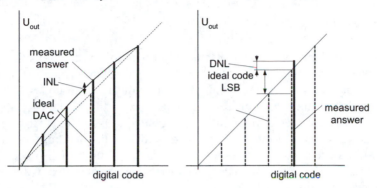

Figure 5.52. Integral nonlinearity error INL and differential nonlinearity error DNL of DA converters

The integral nonlinearity is the difference between a real transfer function and an idealized straight line. The differential nonlinearity is the difference between ideal step equal to 1 LSB code and the real step.

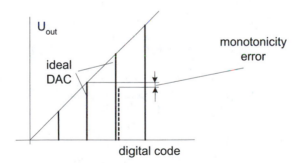

Figure 5.53. The error of monotonicity

The *error of monotonicity* is corresponding to the differential nonlinearity error. The DAC is monotonic if the analogue output always increases as the input code increases. Because this error deteriorates the DAC performances the manufacturers often mention "guaranteed monotonicity". An example of the monotonicity error is presented in Fig. 5.53.

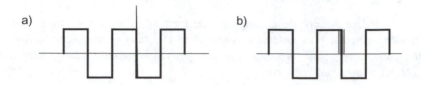

Figure 5.54. Errors of glitches (a) and jitters (b)

As the converter performs the switching operations in the transition state can appear short spikes called *glitches*. These pulses are dangerous because they are not observable on the standard (poor quality) oscilloscopes, although they can disturb the signal processing. Therefore at the output of converter often is inserted a special filter called a *deglitcher*. One of the techniques to eliminate glitches is to use sample-and-hold circuits holding the signal during the switch process. Another error related to the switch process is a jitter (Fig. 5.54b). The *jitter error* means the unrepeatability of the pulse slope, pulse duration or pulse phase.

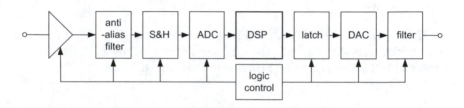

Figure 5.55. A typical cycle of the digital signal processing DSP

The *digital signal processing DSP* offers many unique possibilities not available in the analogue signal processing (Antoniou 2005, Deziel 2000, Khan 2005, Lai 2004, Lyons 2004, Mitra 2002, Oppenheim 1999, Proakis 1995, Rorabough 2005, Smith 2003, Stranneby 2001). The most popular application of digital signal processing techniques is Fast Fourier Transform FFT and Digital Filtering. Currently a large area of digital signal processing application is image processing. After digital signal is processed sometimes it is necessary to come back to the original analogue form. As example can be presented an audio application where the last step is an analogue loudspeaker. Therefore often the signal is converted into digital one; next it is processed and then again is converted into analogue signal as presented in Fig. 5.55.

5.3. METHODS AND TOOLS OF DIGITAL SIGNAL PROCESSING

5.3.1. The main terms of digital signal processing

The *digital signal processing DSP* requires the knowledge of several new specific mathematical methods – most of the methods used in analogue signal analysis correspond to special equivalents in digital signal analysis. For example the Fourier Transform in the analogue technique is equivalent to the Discrete Fourier Transform DFT, analogue convolution is equivalent to the digital one, Laplace *s*-operators are sometimes substituted by the *z*-transform. In this chapter the main term of DSP technique are collected or reminded.

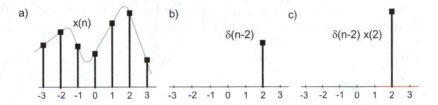

Figure 5.56. The discrete function (a), shifted unit impulse (b) and the selected impulse *x(k)* (c)

One bit can be represented by one impulse. If this impulse exhibits a short duration time it can be represented by the *Dirac delta function* denoted by δ(t). The delta function is a normalized impulse, that is the sample has a value of one

$$\delta(n)=\begin{cases}1 & n=0\\0 & n\neq 0\end{cases} \tag{5.28}$$

This impulse can be shifted (Fig. 5.56b) and this operation is denoted as *δ(n-k)*

$$\delta(n-k)=\begin{cases}1 & n=k\\0 & n\neq k\end{cases} \tag{5.29}$$

Thus the impulse of the discrete function of the value *x(k)* can be described as

$$x(k) = x(n)\delta(n-k) \qquad (5.30)$$

To one discrete function we can add another discrete function $x(n) + y(n)$, we can also multiply this function by the other one. We can also change the order of operations. These operations are possible if the system is *linear time invariant LTI*. The system is *linear* if the additive relation called as the *superposition principle* is valid. The superposition principle is described as

$$f(x_1 + x_2) = f(x_1) + f(x_2) \qquad (5.31)$$

Thus if $y_1(n)$ is the system response to the input signal $x_1(n)$ and $y_2(n)$ is the system response to the input signal $x_2(n)$ and

$$x(n) = a_1 x_1(n) + a_2 x_2(n) \qquad (5.32)$$

the output of a linear system is

$$y(n) = a_1 y_1(n) + a_2 y_2(n) \qquad (5.33)$$

The superposition is very important in DSP. Let us assume that the input signal *x(n)* can be *decomposed*, which means that it can be broken into two or more additive components $x_1(n)$, $x_2(n)$..... We can determine the output signal component of each input signal $y_1(n)$, $y_2(n)$... Next the output signal can be *synthesized* as the sum of each component. The synthesized output signal is identical as calculated directly *y(n)=f(x(n))*. Thus if the system is complicated we can analyze it as superposition of simpler components.

The system is *time invariant* (stationary) if the delay (shift in the time domain) of the input signal causes appropriate delay of the output signal. Thus if $x(n)=x_1(n-n_o)$ the response is $y(n)=y_1(n-n_o)$.

For the analysis of the discrete signal it is required if the system is casual. In the *casual system* the output signal depends only on the previous or present values of the input signals. Thus if input samples are *x(n)* for $n < n_o$ the output signal does not depend on the samples $n > n_o$.

The discrete signal is composed of the series of impulses with the magnitude proportional to the sampled signal *f(t)* and with the period T_s

$$y(nT_s) = f(nT_s)\delta(t - nT_s) \qquad (5.34)$$

and

$$y(t) = \sum_{n=-\infty}^{\infty} f(t)\delta(t - nT_s)$$ (5.35)

If the input signal $x(n)$ and the output signal (response) $y(n)$ are related by the function $F[x(n)]$, thus $y(n) = F[x(n)]$ we can write the relationship (5.35) in the following form

$$y(n) = F\left[\sum_{k=-\infty}^{\infty} x(k)\, \delta(n-k) \right] = \sum_{k=-\infty}^{\infty} x(k) F[\delta(n-k)]$$ (5.36)

or

$$y(n) = \sum_{k=-\infty}^{\infty} x(k) h(n-k)$$ (5.37)

The function $h(n)$ is called *impulse response* and it is crucial for the digital filters analysis[1]. A linear time invariant digital filter is completely characterized by its impulse response. Impulse response is the output signal of the linear system when the input signal is a delta function. The equation (5.37) we can write as

$$y(n) = \sum_{k=-\infty}^{\infty} x(k) h(n-k) = x(n) * h(n)$$ (5.38)

or

$$y(n) = \sum_{k=-\infty}^{\infty} x(n-k) h(k) = h(n) * x(n)$$ (5.39)

The term *convolution* denoted by the symbol * plays an important role in discrete systems, especially in the digital filters analysis. The convolution is the relationship between input signal and output signal of LTI systems and enables us to determine the response signal if the impulse response $h(n)$ is known.

---

[1] The impulse response is sometimes called the convolution *kernel* or simply the kernel.

The operation of the convolution $y(n) = \sum h(k)x(n-k)$ consists of the following steps:
a) flipping left-for-right the second signal $h(n)$
b) shift this signal by n samples
c) multiply this signal by the first signal: $h(k)x(n-k)$
d) sum all multiplied results.
 A nice illustration of the convolution calculation is the "convolution machine" proposed by Steven Smith (Smith 2003) and presented in Fig. 5.57.

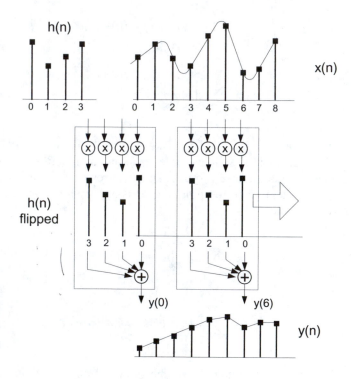

Figure 5.57. The "convolution machine" as a tool for calculation of the convolution (Smith 2003)

The algorithm of calculation of the convolution presented in Fig. 5.57 works as follows. Let us assume that we are determining the impulse $y(6)$ of the response. The impulses of the input function $x(n)$ are multiplied by the flipped impulse response $h(n)$. The sum of these multiplications is the value of the $y(6)$ sample of the response. If we would like to determine the next impulse of response, i.e. *y(7)*, it is necessary to repeat these operations for an impulse response shifted by one.

The convolution has the following properties

$$y(n) = x(n) * h(n) = h(n) * x(n)$$

$$[w(n) + x(n)] * h(n) = [w(n) * h(n)] + [x(n) * h(n)] \qquad (5.40)$$

$$[w(n) * x(n) * h(n)] = w(n) * [x(n) * h(n)]$$

The inverse operation to the convolution is deconvolution. The *deconvolution* operation is the calculation of the *x(n)* signal from the convoluted result *y(n)* knowing the *h(n)* function (see equation 5.38). The deconvolution is much easier to perform in the frequency domain than in the time domain – this problem is described in more detail later (Section 5.4.3).

Another important term is the *correlation function*, used for the comparison of two signals *x₁(n)* and *x₂(n)*. The correlation function is described by following relation

$$r_{12}(k) = \frac{1}{N} \sum_{n=0}^{N-1} x_1(n) \, x_2(n+k) \qquad (5.41)$$

The correlation is a mathematical operation very similar to the convolution (compare the equations 5.37 and 5.41). If we compare two signals we determine the *cross-correlation*, and if we compare the signal with itself we determine the *autocorrelation*.

As was mentioned earlier the Fourier Transform corresponds in the digital domain to Discrete Fourier Transform. Thus, the Fourier Series for the analogue signal *x(t)* is equivalent to the *Discrete Fourier Series DFS* determined for the discrete signals *x(n)*

$$x(t) = \sum_{n=-\infty}^{\infty} c_n e^{jn\omega_0 t} \qquad \text{is equivalent to} \qquad x(n) = x(nT_s) = \sum_{k=0}^{N-1} c_k e^{jn\omega_s} \qquad (5.42)$$

where $c_k = \dfrac{1}{N} \displaystyle\sum_{n=0}^{N-1} x(n) e^{-j2\pi k n / N}$; $\omega_s = 2\pi k / N$.

Similarly the Fourier Transform in the digital domain is represented by the *Discrete Fourier Transform – DFT* described by the equations

$$X(j\omega) = \int_{-\infty}^{+\infty} x(t) e^{-j\omega t}\, dt \quad \text{is equivalent to} \quad X(k) = \sum_{n=0}^{N-1} x(n) e^{-jn\omega_s} \quad (5.43)$$

and the Inverse Fourier Transform is represented by the *Inverse Discrete Fourier Transform - IDFT*

$$x(t) = \frac{1}{2\pi} \int_{-\infty}^{\infty} X(j\omega) e^{j\omega t}\, dt \quad \text{is equivalent to} \quad x(n) = \frac{1}{N} \sum_{k=0}^{N-1} X(k) e^{jn\omega_s} \quad (5.44)$$

In the Discrete Fourier Transform N samples collected with the sampling frequency f_s

$$x(kT_s) = x(0),\, x(T_s),\, x(2T_s),....,x((N-1)T_s) \qquad (5.45)$$

are transformed to the N components discrete series in the frequency domain

$$X\left(\frac{nf_s}{N}\right) = X(0),\, X\left(\frac{f_s}{N}\right),\, X\left(\frac{2f_s}{N}\right),...., X\left(\frac{(N-1)f_s}{N}\right) \qquad (5.46)$$

The series $X(nf_s/N)$ is represented by the complex values – the real and imaginary part (or by magnitude and phase in polar notation). The spectrum of the DFT is represented by the spectral lines with the period of f_s/N. The DFT is sometimes described in the form

$$X(k) = \sum_{n=0}^{N-1} x(n) W_N^{kn} \qquad (5.47)$$

where coefficient $W_N = exp(-j2\pi/N)$.

The Fourier transform allows the conversion of the signal from the time domain to the frequency domain and *vice versa*. Below, are presented the main properties of the discrete Fourier transform.

If $x_1(n)$ is represented by the Fourier transform $X_1(k)$ and accordingly $x_2(n)$ by $X_2(k)$ then

$$x(n) = ax_1(n) + bx_2(n) \Leftrightarrow X(k) = aX_1(k) + bX_2(k) \qquad (5.48a)$$

The relation (5.48a) describes the *linearity* properties of the DFT.
The DFT is *periodic* with a period of N even if the $x(n)$ is non-periodic.

$$X(k) = \sum_{n=0}^{N-1} x(n)e^{-j2\pi \, kn/N} = \sum_{n=0}^{N-1} x(n)e^{-j2\pi \, kn/N} e^{-j2\pi \, Nn/N} = X(k+N) \quad (5.48b)$$

If the input signal is real then the real part of the DFT is an even function and the imaginary part of DFT is an odd function. This kind of symmetry is sometimes called *hermitian symmetry*.

$$x(n) = x^*(n) \Leftrightarrow X(k) = X^*(-k) \qquad (5.48c)$$

The relationship (5.48d) describes the *symmetry properties* of DFT. If $x(n)$ is an even function then $X(k)$ is also even. If $x(n)$ is an odd function then $X(k)$ is also odd. Furthermore, if $x(n)$ is real and even than $X(k)$ is real and even. And if $x(n)$ is real and odd then $X(k)$ is imaginary and odd.

$$x(n) = x(-n) \Leftrightarrow X(k) = X(-k) \qquad (5.48d)$$

The relationship (5.48e) is very important in DSP, because it means that the convolution operation can be performed by DFT of both components, then by multiplication of the results, and finally, by the inverse transform to the time sequence (*circular convolution*). And inversely we can perform the deconvolution of $x(n)$ by transforming it to $X(k)$ and by dividing it by one component $X_1(k)$.

$$x(n) = x_1(n) * x_2(n) \Leftrightarrow X(k) = X_1(k) X_2(k) \qquad (5.48e)$$

The relationship (5.48f) known as the *Parceval theorem* states that the energy of the signal in time domain is the same as the energy in the frequency domain. Thus the time domain representation of the signal is fully transformable to the frequency domain if the system is linear time invariant.

$$\sum_{n=0}^{N-1}|x(n)|^2 = \frac{1}{N}\sum_{k=0}^{N-1}|X(k)|^2 \tag{5.48f}$$

The *shift m in the time domain* (*time delay*) is equivalent to the multiplication in the frequency domain by the component *exp(-jω m)*. Thus, the phase component of the complex representation is increased by *ω m*.

$$x(n-m) \Leftrightarrow W_N^{km} X(k) \tag{5.48g}$$

The *shift M in the frequency domain* is equivalent to the multiplication of the signal in time domain by the component *exp(-jω M)*.

$$X(k-M) \Leftrightarrow W_N^{Mn} x(n) \tag{5.48h}$$

In the analysis of the discrete signal (and especially in the analysis of the digital filters) the *z-Transform* is very useful. The discrete signal described by the relation (5.30) i.e. *x(n)* = *Σx(k)δ(n-k)* can be rewritten in *z*-domain as

$$X(z) = \sum_{k=0}^{N} x(k)z^{-k} \tag{5.49}$$

Thus the *z*-transform denotes the shift of the signal in the time domain by *k*. For example, z^{-1} is equivalent to the delay of the signal by one sample

$$Z[x(n-1)] = z^{-1}X(z) \tag{5.50}$$

while z^{-m} means the delay by *m* samples

$$Z\{x(n-m)\} = z^{-m}X(z) \tag{5.51}$$

For analysis of the filter response we use the convolution relationship

$$y(n) = \sum_{k=0}^{\infty} h(k) x(n-k) = h(n) * x(n) \qquad (5.52)$$

where the impulse response is

$$y(n) = \sum_{k=0}^{\infty} h(k) \delta(n-k) = h(n) \qquad (5.53)$$

The z-transform of the impulse response is

$$H(z) = \sum_{k=0}^{\infty} h_k z^{-k} \qquad (5.54)$$

and

$$Y(z) = H(z)X(z) \qquad (5.55)$$

The convolution can be realized as multiplication in the z-domain. The transmittance $H(z)$ (called also *transfer function*) explicitly describes the properties of the casual system.

Especially important is the analysis of the stability conditions. If the system is described by the transmittance

$$H(z) = \frac{Y(z)}{X(z)} = \frac{a_1 z^{-1} + a_2 z^{-2} + a_3 z^{-2} \ldots}{1 + b_1 z^{-1} + b_2 z^{-2} + b_3 z^{-3}} \qquad (5.56)$$

we can analyze the *zeros* (values of the z when the numerator is equal to zero) and *poles* (the values of z when the denominator is equal to zero). We can rewrite the relation (5.56) in a form

$$H(z) = \frac{(z - z_1)(z - z_2)(z - z_3)\ldots}{(z - p_1)(z - p_2)(z - p_3)\ldots} \qquad (5.57)$$

The zeros (z_1, z_2, z_3...) and poles (p_1, p_2, p_3...) are complex numbers. If we analyze the position of the poles in the z-plane we can test the conditions of stability. The casual system is stable if the poles are located inside the unit circle $|z|=1$ in the z-plane (Fig. 5.58).

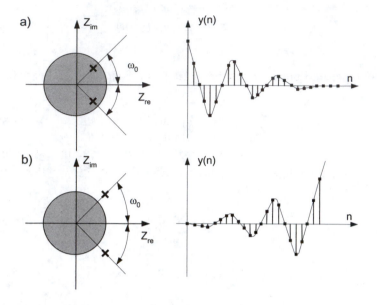

Figure 5.58. The system is stable because poles are inside the unit circle (a) and the system in unstable with the poles outside of the circle (b)

The relationship of the DFT in a form

$$X(j\omega) = \sum_{n=-\infty}^{\infty} x(nT_s)e^{-j\omega nT_s} \tag{5.58}$$

can be rewritten in the *s*-domain (after applying of the Laplace transform) as

$$X(s) = \sum_{n=-\infty}^{\infty} x(nT_s)e^{-nsT_s} \tag{5.59}$$

By substitution in equation (5.50) of the component e^{sT_s} by z i.e. $z = e^{sT_s}$ the DFT in the *z*-domain is

$$X(z) = \sum_{n=-\infty}^{\infty} x(nT_s)z^{-n} \tag{5.60}$$

The relationship between the z-domain and the ω-domain is

$$z = e^{j\omega T_s} \tag{5.61}$$

5.3.2. The Discrete Fourier Transform DFT and Fast Fourier Transform FFT

We can determine the discrete Fourier transform for selected N samples of the signal. As the result of the discrete transform of the signal $x(n)$ we obtain two arrays of $N/2+1$ samples: a real part $X_{re}(k)$ and an imaginary part $X_{im}(k)$. Often the result of the transform is presented as the absolute value $|X(k)|$ and the phase value $X_\Phi(k)$

$$|X(k)| = \sqrt{X_{re}^2(k) + X_{im}^2(k)} \; ; \quad X_\phi(k) = arctg\frac{X_{im}(k)}{X_{re}(k)} \tag{5.62}$$

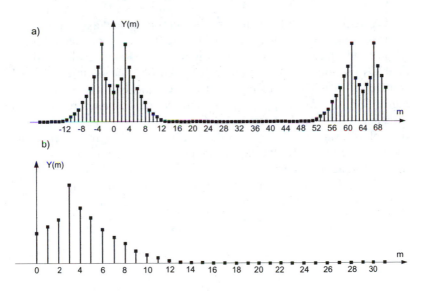

Figure 5.59. An example of the results of DFT analysis for 64 samples of the signal (a) and first 32 samples representing the analyzed signal (b)

Figure 5.59 presents the example of the Fourier analysis of 64 samples of the signal. The result is periodical with period $N/2$. Due to symmetry of obtained results only first $0 - (N/2)$ samples are useful because the remaining samples are meaningless (also the samples of negative frequencies do not introduce new information). Figure 5.59b presents the useful result of DFT analysis.

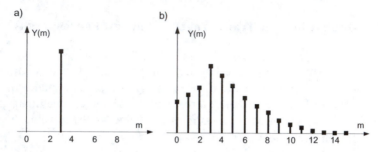

Figure 5.60. The result of the synchronous spectral analysis (a) and the result of the analysis of the same signal when the sampling frequency does not correspond to the frequency of the signal (asynchronous analysis)

Figure 5.60 presents the result of synchronous and asynchronous spectral analysis. Synchronous analysis is when the frequency of spectral lines f_s /N is equal to the frequency of analyzed harmonics. From results of the spectral analysis presented in Fig. 5.60 we can see that the samples of the Fourier transform do not have to represent exactly the harmonics of the analyzed signal. Fourier transform is represented by spectral lines resulting from the division of the sampling frequency f_s to N samples. If the f_s /N is the multiple of analyzed signal (*synchronous analysis*) then this signal is represent by the spectral line of exactly the same frequency (Fig. 5.60a). But if the analysis is asynchronous then the analyzed signal is represented by several spectral lines around the frequency nearest to the frequency of the signal. This effect of spectral line broadening is called as *leakage* (Fig. 5.60b).

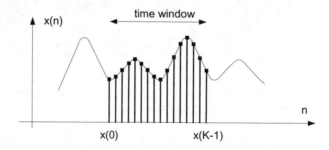

Figure 5.61. The selection of the samples for Fourier analysis

We choose selected series of the samples for the analysis – we analyze only the samples contained in the time window (Fig. 5.61). It means that we select samples by multiplying the signal by the window function. For example the rectangular window function (presented in Fig. 5.61) is in the form:

$$w(n) = \begin{cases} 1 & for \quad n_0 \le n \le n_0 + K \\ 0 & for \quad other \quad samples \end{cases} \tag{5.63}$$

where n_0 is the starting point of the window and K is the number of samples.

The n_0 value can be the arbitrary chosen but often it is selected by trigger function of spectrum analyzer as the equal to zero value (starting point of the periodic function). Taking into account the properties of the most frequently used algorithm of the DFT analysis (*FFT – Fast Fourier Transform*) usually $K = 2^n$.

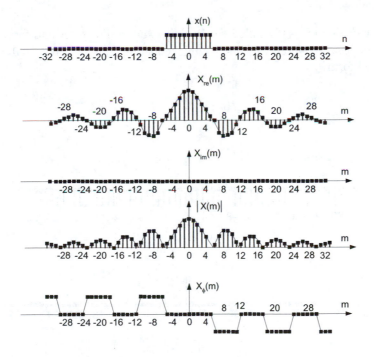

Figure 5.62. The Fourier analysis results of the series of impulses of the same value in the rectangular time window (Lyons 004)

The larger the number of samples the better the quality of the Fourier analysis, but a large number of samples requires more computing power, hence more time. Therefore, it is assumed that $K = 1024$ samples is sufficient to obtain a satisfying result of the analysis. It would be better if the analysis is synchronous but we can not guarantee such selection of f_s (usually we do not know the frequency components of the analyzed signal).

Figure 5.62 presents the Fourier analysis results of the series of impulses of the same value in the rectangular time window. The spectrum of the series of the absolute values shows the tails formed from side lobes, which are not expected. The appearance of the side lobes results from the theory because the Fourier transform of the rectangle signal is described by the equation

$$X(m) = \frac{\sin(\pi mK / N)}{\sin(\pi m / N)} \qquad (5.64)$$

where N is the number of samples, and K is the number of samples in the time window.

The equation (5.64) is called the *Dirichlet kernel*. For $K=N$ and for small values of $\pi m/N$ (when *sin* $x \approx x$) this equation can be presented in a simplified form

$$X(m) = \frac{\sin \pi m}{\pi m} \qquad (5.65)$$

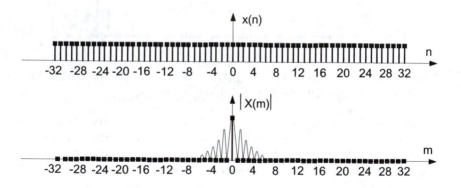

Figure 5.63. The Fourier analysis results of the series of impulses of the same values and $N=K$ (Lyons 2004)

The width of the side lobes is equal to N/K – therefore if $N = K$ the spectrum is represented just by a single spectral line. Such case is presented in Fig. 5.63 where the time window contains all samples, thus $N = K$.

The leakage of the spectral lines is well illustrated in Fig. 5.64 where the signal $x = cos(2\pi nk/N)$ is analyzed. In such a case the Fourier transform is described as

$$X(m) = \frac{1}{2} \frac{sin[\pi(k-m)]}{sin[\pi(k-m)/N]} \qquad (5.66)$$

If the analysis is synchronous then the m samples correspond exactly to the zero values of side lobes and only one spectral line is presented (Fig. 5.64b). For asynchronous analysis the positions of the m samples can occur in various places of the side lobes and additional leakage lines can appear (Fig. 5.64c).

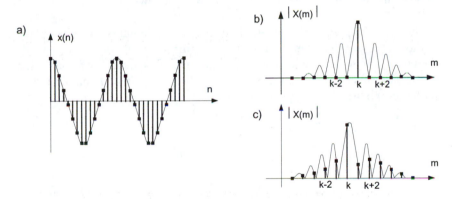

Figure 5.64. The Fourier transform of the cosine function (a) in the case of synchronous (b) and asynchronous (c) analysis

The side lobes appear due to the sharp border of the rectangle window. We can reduce the amplitude of the side lobes using other shapes of the window, when the border is smoother. Fig. 5.65 presents the effect of applying of the Hanning window described by the relation: $w(n)=0.5-0.5cos(2\pi n/N)$.

We can see that after application of the Hanning widow the side lobes have reduced amplitude but at the expense of widening the main window. There are various kinds of windows: Hanning, Hamming, Chebyshev,

Keiser, etc. and by choosing a correct window it is possible to improve the quality of the spectral analysis.

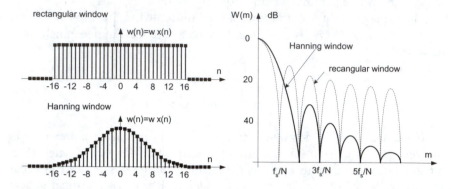

Figure 5.65. The effect of applying of the Hanning window

Figure 5.66 presents the example of the spectral analysis results obtained without and with the Hamming window. After use of the Hamming window the tails of the spectral line decrease in the case of asynchronous analysis, but in the case of synchronous analysis the application of the Hamming window increases the broadband of the spectral line.

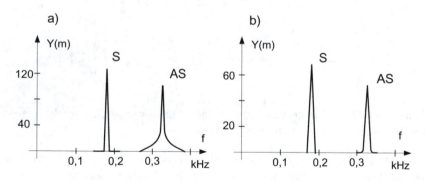

Figure 5.66. The example of the results of spectral analysis (synchronous S and asynchronous AS) without (a) and with Hamming window (b) (Smith 2003)

The more samples of the Fourier analysis, the better resolution of the result. This effect is illustrated in Fig. 5.67.

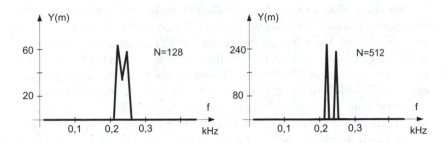

Figure 5.67. The examples of the spectral analysis performed for two numbers of samples (Smith 2003)

Sometimes we have the number of samples in the window less than 2^N – for example *50*. We can artificially increase of the number of samples by supplementing these samples by next samples equal to zero as demonstrated in Fig. 5.68

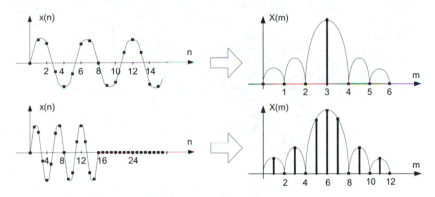

Figure 5.68. The results of Fourier analysis of the 16-sample sinusoidal signal (a) and the results of the analysis after supplementing this signal by 16 zero value samples (b) (Lyons 2004)

From the results presented in Fig. 5.68 we can conclude that after supplementing our samples by additional zero samples we improved the resolution of the analysis. For 16-samples analysis we obtained 7 spectral lines and the side lobes were not detectable. For 16 samples + 16 zero samples analysis we obtained 13 spectral lines with lines representing side lobes.

The computation of the N-point Fourier transform requires N^2 complex multiplication operations. Thus, the calculation of the Fourier transform is not a trivial task even for fast processors, because the time of computations practically excludes this analysis in the real time systems. The turning point was the invention of Cooley and Tukey (Cooley 1965). They proposed a special algorithm allowing faster Fourier transform calculation. This algorithm with some modifications is used currently in the Fourier analysis and is called *FFT – Fast Fourier Transform*. By applying the FFT analysis it is possible to decrease the number of multiplication from 2^N to $0.5\ Nlog_2N$. For example to realize the 1024-sample DFT it is necessary to perform *1 048 576* multiplication operations and *1 047 552* addition operations, while the FFT analysis requires in such a case only *5 120* multiplications (around 200 times less than in the DFT) and *10 240* additions (around 100 times less).

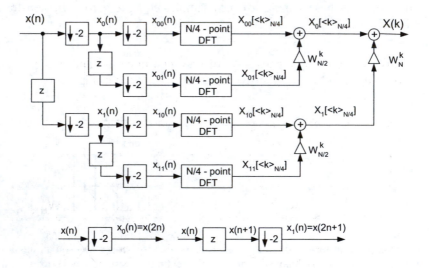

Figure 5.69. The principle of the decomposition of the DFT procedure into two stages (Mitra 2002)

The FFT is a family of algorithms that profits from the symmetry and periodicity of Fourier transform to increase the computational efficiency by diminishing the number of operations. The DFT equation is divided into two parts, even and odd sequence

$$X(k) = \sum_{n=0}^{N/2-1} x(2n)W_N^{2nk} + W_N^k \sum_{n=0}^{N/2-1} x(2n+1)W_N^{2nk} \tag{5.67}$$

The factor W_N^{kn}, known as the *twiddle factor,* appears in both parts of the equation and it is sufficient to be computed just once. Moreover, there are only four different values of this factor and there is no need to compute them so many times. The computation of N-point transform is a rather difficult task, therefore the sequences described by Eq. (5.67) are decomposed into several sub-sequences finishing on the two-point DFT (Fig. 5.69). The flow-graph of this algorithm is presented in Fig. 5.70.

The calculation of two-point transform is relatively simple and for example the first transform from Fig. 5.70 can be expressed as

$$X_{00}(k) = \sum_{n=0}^{1} x_{00}(n)\, W_2^{nk} = x(0) + W_2^k x(4) \tag{5.68}$$

This two-point DFT consisting of one multiplication and two additions can be expressed by the butterfly flow-graph presented in Fig. 5.70.

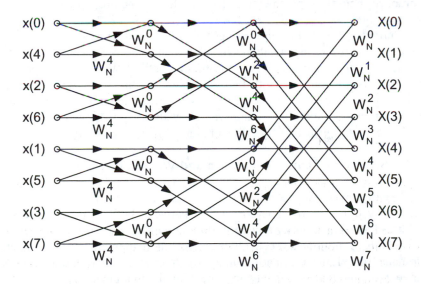

Figure 5.70. The flow-graph of the decimation-in-time 8-point FFT algorithm (Mitra 2003)

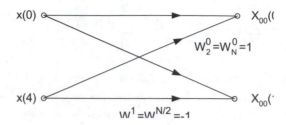

Figure 5.71 The butterfly operation of the two-point DFT

The decomposition of the data sequence can be performed in the time domain (*decimation-in-time*) or in the frequency domain (*decimation-in frequency*). Figure 5.70 presents the example of the flow-graph of the decimation-in-time FFT algorithm for N=8. By using such an algorithm it is possible to determine the 8-point DFT by applying 24 multiplication operations.

The FFT algorithms are described in detail elsewhere (Bracewell 1999, Brighan 1988, James 2002, Mitra 2002, Sneddon 1995). Most of those algorithms are of the basic radix-2, therefore it is required to provide 2^N samples. If there is a smaller number of samples it is recommended to complete them by supplementing the samples by the appropriate number of zero samples.

Currently, tools for computing the FFT (and *Inverse Fast Fourier Tansform – IFFT*) are available in ready-to-use form in many computing platforms (for example in MatLab or LabVIEW), many measuring instruments are equipped with FFT (for example digital oscilloscopes). The calculation of the FFT is possible using scientific calculators or spreadsheets (for example MS Excel). It is important that the FFT exhibits all features of DFT, because it is just a very efficient algorithm and not another transform.

5.3.3. Short-time Fourier Transform and Wavelet transform

Usually, we analyze the signals in the time domain *x(t)* or in the frequency domain *X(f)*. The conversion between both domains is possible with the aid of the Fourier transform. In many cases, especially in the case of short transition signals, such dualism (the time domain or the frequency domain analysis) is a limitation of the signal analysis. It would be useful to have the possibility of analyzing the signal simultaneously as the time varying signal and as the frequency spectrum. Figure 5.72 presents the example of such time-frequency analysis of a human voice.

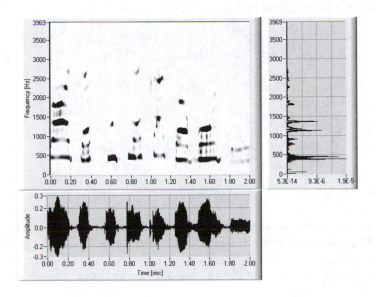

Figure 5.72. The result of analysis of human voice performed with Short Time Fourier Transform (Rak 2004)

The demand of simultaneous observation of the signal in time and frequency domain is realized for a long time in music notation, where the vertical position of the note on the stave denotes its frequency (pitch) and the horizontal position the instant of time at which the note appears (see Fig. 5.108).

There is another important limitation of the Fourier transform. It is required that the transform be *reversible*. This means that we should be able to reconstruct the primary time varying signal from the frequency spectrum – for example using the Inverse Fourier Transform. But Fourier transform is correct only for *stationary signals* – signals not varying in the time or the frequency. We only obtain the information on which frequency component exists in the signal and not how it varies.

The solution to this short time, non-stationary signals analysis is the development of the Fourier Transform called the time-dependent Fourier transform or *Short-time Fourier Transform STFT*. In this transform the time window of the analyzed signal is shifted in time – the analyzed signal is multiplied by shifted time window function (*Moving Window Method – MWM*). Thus, the non-stationary signal can be further reconstructed by the results of STFT distributed in time.

The short Fourier time is expressed by the equations

$$X(t,f) = \int_{-\infty}^{\infty} x(\tau)w(\tau - t)e^{-j\omega\tau} d\tau \tag{5.69}$$

for a continuous signal, and

$$X(n,k) = \sum_{m=0}^{N-1} w(m)x(n-m)e^{-jm(2\pi k / N)} \tag{5.70}$$

for discrete signals.

The function $w(t)$ or $w(m)$ is the function of the time window. The results of the time/frequency analysis should be presented as a 3D picture. Usually it is presented as the frequency/time $F(t)$ 2D picture called a spectrogram – sometimes supplemented by the $x(t)$ and $X(F)$ dependence – as presented in Fig. 5.72.

The main problem of the STFT results directly from the *Heisenberg theorem on uncertainty* (uncertainty principle). This principle states that it is not possible to know exactly the time-frequency representation of the signal – it is not possible to know what spectral component exists at what instant of time. We can only know the time interval Δt in which the certain band of frequencies $\Delta \omega$ exist

$$\Delta t \cdot \Delta \omega = const \tag{5.71}$$

It means that if we can more exactly determine the time Δt then the frequency bandwidth $\Delta \omega$ is less exactly determined (the improvement of the uncertainty of determination of Δt is performed on the expense of the uncertainty of determination of $\Delta \omega$ and *vice versa*). In the STFT, if the time window is narrow we obtain good time resolution, but poor frequency resolution. And if the time window is wide we obtain poor time resolution and good frequency resolution. Because the STFT uses the time window function of constant width, the problem is to choose the correct width of the window and it is not possible to obtain the solution enabling the analysis of low frequency and high frequency signals with the same uncertainty.

This dilemma is solved in a much better way in the *wavelet transform*, where the width of the time window is scalable, as illustrated in Fig. 5.73. The dimensions of the window can be selected by an adaptive method to obtain the best fitting to the analyzed signal. Instead of frequency in Fourier Transform the wavelet transform utilizes the *scale*, which describes scaling of the signal in time – compression or dilatation of the signal.

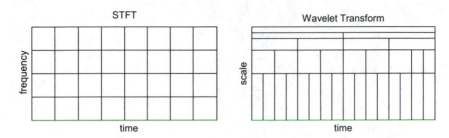

Figure 5.73. The dimensions of the time and frequency windows in the STFT and Wavelet transform

Another important difference between Fourier transform and the wavelet transform is that the signal is not composed from the sine and cosine functions, but from a finite set of well defined, limited in time *wavelets*. The wavelets are scaled in the frequency domain (compressed or dilated) and shifted in time. There are various types of wavelets: Haar, Morlet, Daubechies, Coiflet, Symlet, etc.

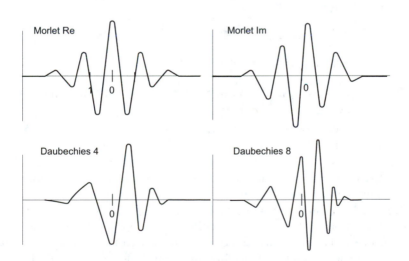

Figure 5.74. The examples of typical wavelets (approximated pictures)

The continuous wavelet transform can be expressed by the following relation

$$W(\tau,s) = \frac{1}{\sqrt{|s|}} \int_{-\infty}^{\infty} x(t)\psi^*\left(\frac{t-\tau}{s}\right)dt \qquad (5.72)$$

where s is the scale factor, τ - is the time shift, and ψ is the *mother wavelet* function.

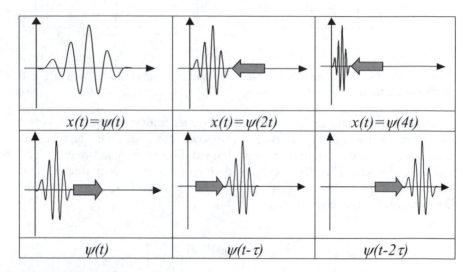

| $x(t)=\psi(t)$ | $x(t)=\psi(2t)$ | $x(t)=\psi(4t)$ |
| $\psi(t)$ | $\psi(t-\tau)$ | $\psi(t-2\tau)$ |

Figure 5.75. The scaling and shifting of wavelet (Rak 2004)

The discrete wavelet transform can be expressed by the following relation

$$W(k,s) = 2^{s/2} \sum_{n} x(n)\psi\left(2^s n - k\right) \qquad (5.73)$$

In a discrete wavelet transform the signal is sampled in selected points of the t/s plane, where the scale is selected as 2^{-s} and the wavelet shift is selected as $2^{-s}k$. This kind of sampling is called *dyadic sampling*.

As in the case of the Fourier transform, also for the discrete wavelet transform there have been developed numerical algorithms enabling us to perform efficient and fast realization of the transform. One of them is developed by Mallat (Mallat 1989, Mallat 1999) algorithm known as the

Mallat pyramid, *subband coding* or sometimes as *Fast Wavelet Transform FWT.*

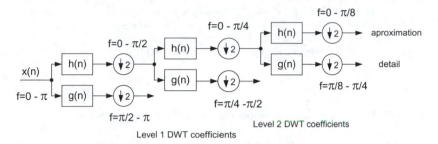

Figure 5.76. The principle of wavelet filter decomposition (Polikar 2002]

The *subband multiresolution coding* decomposition of the signal is divided into two parts: *coarse approximation* and *detail information*. In Fig. 5.76 the frequency expressed in radians $f = 0 - \pi$ is divided into two parts $0 - \pi/2$ and $\pi/2 - \pi$. This decomposition is realized by passing the signal through two filters – low-pass for approximation and high-pass for detail

$$y_{high}(k) = \sum_n x(n) \cdot g(2k - n)$$

$$y_{low}(k) = \sum_x x(n) \cdot h(2k - n)$$

(5.74)

The impulse responses $g(n)$ and $h(n)$ of the filters correspond to the kind of wavelets used for analysis, for example the filter coefficients for Daubechies wavelet of 4-th order are as follows

$$h(n)_{D4} = \left\{ \frac{1+\sqrt{3}}{4\sqrt{2}}, \frac{3+\sqrt{3}}{4\sqrt{2}}, \frac{3-\sqrt{3}}{4\sqrt{2}}, \frac{1-\sqrt{3}}{4\sqrt{2}} \right\}$$

The procedure of filtering is repeated many times after subsampling the signal by 2 (decimation) and next the details are determined, as shown in Fig. 5.76.

The wavelet transform should be reversible – it should be possible to recover the time varying signal from its wavelet representation. To perform such an operation it is necessary to reconstruct the signal in the reverse direction as demonstrated in Fig. 5.77.

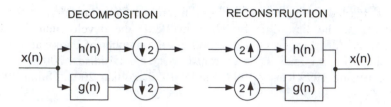

Figure. 5.77. The wavelet decomposition and reconstruction of signals

The reconstruction of the signal can be performed according to the following equation (Polikar 2002)

$$x(n) = \sum_{k=-\infty}^{\infty}\left[y_{high}(k)\cdot g(-n+2k)\right]+\left[y_{low}(k)\cdot h(-n+2k)\right] \qquad (5.75)$$

Fig. 5.78 presents the example of the results of wavelet analysis with one approximation wavelet and three detail wavelets.

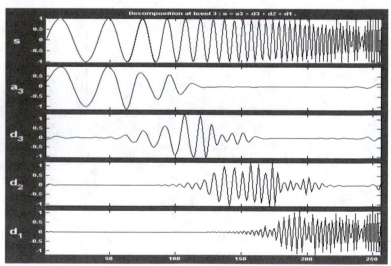

Figure 5.78. The example of the wavelet decomposition of the signal: time varying signal (s), approximation wavelet (a_3) and detail wavelets (d_1, d_2, d_3) (after Rak 2004)

Although the wavelet transform was introduced some time ago (this term was used for the first time by Haar in 1909) the development of this transform is still in progress. Therefore, a large number of books and other publications on this subject is currently available (Addison 2002, Burrus 1998, Goswami 1999,Keiser 1994, Mallat 1999, Polikar 2002, Quian 2002, Valens 2004, Walker 1999, Walnut 2001).

5.3.4. Digital filters

<u>5.3.4.a. Digital filters – an introduction</u>
In comparison to the analogue filters described earlier (built from the RC elements and the amplifiers) the digital filters exist mainly as computer programs (thus they are some kind of virtual instruments) – although, there are digital filters available in the form of integrated circuits. The digital filters can be easily modified by the software, including also the possibility of the alteration of the parameters during the filter operation, in special kinds of filters, called adaptive filters. The performance of the filter does not depend on the quality of RC elements, but some hardware factors can limit their implementation (organization of the memory, speed of the processor etc).

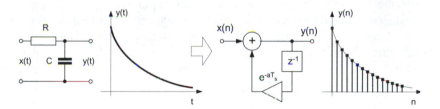

Figure 5.79. The analogue RC low-pass filter and its digital equivalent

The digital filters (as the analogue ones) should exhibit relatively flat characteristics in the passband, the transition band should be as narrow as possible, the filter should be linear (without the phase distortions), and the step response in time should be fast and without overshoot. Additionally, it is demanded to deliver optimal design (relatively simple), taking into account the necessary time and number of numerical operations. A correctly designed digital filter has a performance of filtering much better than its analogue counterpart.

In digital filters the function of RC elements is realized by the delayed element of the transmittance z^{-1} (Fig. 5.79). Examples of two typical forms of digital filters are presented in Fig. 5.80.

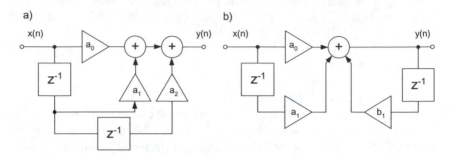

Figure 5.80. Two main structures of the digital filters: the FIR filter (a) and the IIR filter (b)

In the filter presented in Fig. 5.80a the response depends only on the input signal – this filter is without the feedback. This filter is called the *Finite Response Filter FIR* or *non-recursive filter*. In the filter presented in Fig. 5.80b the response depends not only on the input signal, but also on the output signal due to the feedback. Therefore, such filters are called *recursive filters*, or *running back filters*. If the unit impulse is fed to the input of such filter, then theoretically at the output an infinitively long sequence of impulses appears due to the feedback. Therefore, recursive filters are also called *Infinite Response Filters IIR*[1].

The transmittance of a digital filter is described by the dependence

$$H(z) = \frac{Y(z)}{X(z)} = \frac{\sum\limits_{k=0}^{N-1} a(k) z^{-k}}{1 + \sum\limits_{k=1}^{M} b(k) z^{-k}} \qquad (5.75)$$

---

[1] The term "infinite response" is slightly misleading because in real IIR filter structures the impulse response reduces to zero in a finite time.

The samples with coefficients $b(k)$ are the feedback samples, while the samples with coefficients $a(k)$ are the input samples. In the case of FIR filter the coefficients $b(k) = 0$ and the transmittance is expressed by

$$H(z) = \sum_{k=0}^{N-1} a(k)z^{-k} = a_0 + a_1 z^{-1} + a_2 z^{-2} + .. \tag{5.76a}$$

or

$$y(n) = \sum_{k=0}^{N-1} a(k)x(n-k) = a_0 x(n) + a_1 x(n-1) + a_2 x(n-2) + .. \tag{5.76b}$$

The expression (5.76) is the same as the expression describing the convolution. Therefore the digital filter can be described by

$$y(n) = h(n) * x(n) \tag{5.77}$$

and according to the Eq. (5.48e)

$$Y(k) = H(k) \cdot X(k) \tag{5.78}$$

The characteristic of the filter is explicitly described by its *impulse response h(n)* called also the *filter kernel*. Therefore the sequence of coefficients $h(n)$ is also called the *FIR filter coefficients*.

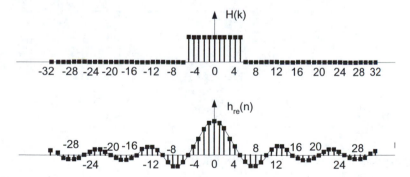

Figure 5.81 The inverse Fourier transform of the rectangular sequence of impulses in the frequency domain

It was shown in Section 5.3.2 if the input signal $x(t)$ is represented by the rectangular time window, then the output signal (after DFT) is represented by the relation $X(k) \approx sinx/x$. We can invert this relation and ask which time window of the input impulses $x(t)$ guarantees that the output impulse sequence is represented by the "rectangular window in the frequency domain". If the output impulses are in the form of a rectangular window we obtain the ideal rectangular transfer characteristic of the filter. To answer this question it is necessary to perform an inverse Fourier transform of the sequence of impulses multiplied by the rectangle window in the frequency domain, as shown in Fig. 5.81.

As could be expected if the sequence of the impulses in the time domain is described by the $sinx/x$ function then the response in the frequency domain is the same as the frequency response of the ideal filter. Thus we can see that the window in time function is crucial for the performances of the digital filter. It would be the best case if such a window was similar to the function $sinx/x$.

5.3.4.b. The non-recursive FIR digital filters

Two examples of non-recursive filters are presented in Fig. 5.82. These filters consist of $N+1$ multipliers and N adders, where N is the order of the filter.

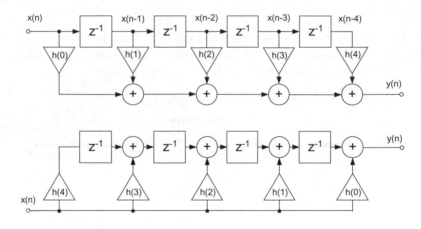

Figure 5.82. Two examples of the FIR filters

The higher the order of the digital filter, the better the frequency characteristic (close to the rectangular one). But even a large number of coefficients of the impulse response does not guarantee that the response is rectangular with the flat part in the pass-band. There are always oscillations in passband and stopband called *Gibbs phenomenon* (Fig. 5.83).

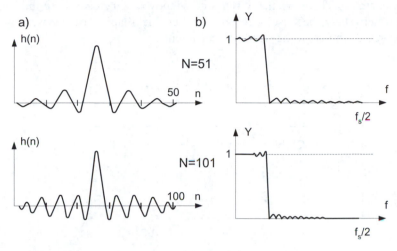

Figure 5.83. The comparison of two filters of the 51 and 101 order: the impulse response (a) and the frequency characteristic (Lai 2004)

Figure 5.83 presents the comparison of the frequency characteristics of the digital filers of 51- order and 101 - order. We can see that the increase of the order of the filter improves the steepness of the frequency characteristic in the transition band, but does not remove the Gibbs oscillations. The way to decrease the ripples is to apply the appropriate window function (Fig.5.84).

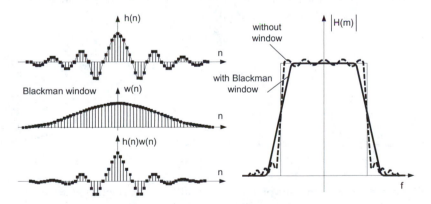

Figure 5.84. The impulse response of the filter with the Blackman window and the transfer frequency characteristic of the 63-order filter (Lyons 2004)

The application of the Blackman window decreases the ripples in the frequency characteristic, but at the expense of the steepness of this characteristic. There are many types of windows. Very useful are the Keiser and Chebyshew, because such windows enable shaping of the window by appropriate choice of the window coefficients.

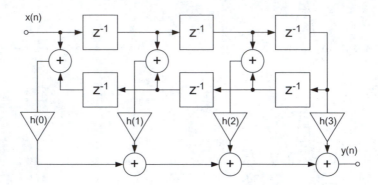

Figure 5.85. The FIR filter with decreased number of multipliers

It is possible to decrease of the number of elements in the filter structure taking into account the symmetry of coefficients $h(k) = h(N–k)$. The example of such a filter structure is presented in Fig. 5.85.

5.3.4.c. The recursive IIR digital filters
Figure 5.86 presents the comparison of the frequency characteristics of recursive and non-recursive digital filters.

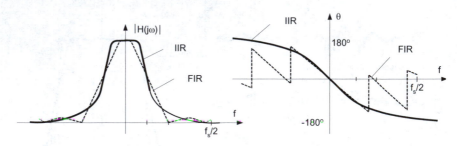

Figure 5.86. The comparison of the 5th order FIR and IIR filters (Lyons 2004)

The non-recursive filters are simpler in design and since they are without feedback there is no problem with the stability of the filter. The non-recursive filters have much better phase linearity than the recursive filters. But recursive filters, due to the feedback, enable us to obtain much better other performances, most of all the steepness of the filter frequency characteristic in the transition band.

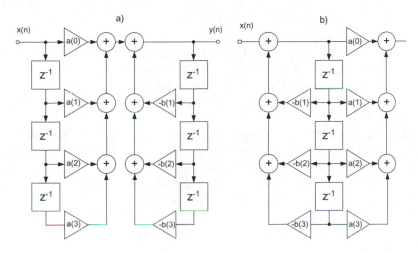

Figure 5.87. The typical structure of the recursive filter (a) and the structure with decreased number of elements (b)

Better performances of the recursive filters enable us to design the filters with a lower number of the multiplier elements and therefore such filters can be faster and less demanding for the processing power and memory requirements. The main drawback of recursive filters is the danger that they can be unstable. It is necessary to perform analysis of the stability conditions – among other the position of the poles in the z-plane.

Figure 5.87 presents the typical structure of the 6th order recursive filter. Also in this case it is possible to decrease the number of elements taking into account the symmetry conditions.

5.3.4.d. Realization of digital filters

We can realize digital equivalents of practically all described earlier most important analogue filters: Butterworh, Bessel, Cauer, Chebyshev, etc. There are developed various methods of design of digital filters (Parks 1987, Thede 2004, Winder 2002). Because procedures of design of analogue filters are very well developed, sometimes it is reasonable to design an appropriate analogue filter and then to convert it into digital one. There are tools enabling conversion between analogue and digital techniques, for example the bilinear transformation from the *s*-plane to the *z*-plane

$$s = \frac{2}{T}\left(\frac{1-z^{-1}}{1+z^{-1}}\right)$$ (5.79)

On the market there are available professional software for filter design as well as free programs on the Internet. Also, programming platforms, like LabVIEW, Agilent VEE Pro offer user-friendly tools for filter realization. Developed tools for filter design are also available in MatLab (Jackson 1995, Lutovac 2000).

Digital filters can be realized practically in all computers equipped with a data acquisition board. The *digital signal processors DSP* in comparison with usual microcontrollers are equipped with a special module MAC (multiplier/accumulator) suitable for digital filter realization. The manufacturers of DSP usually enclose suitable Application Notes enabling design of the digital filter – an example is Application Note No. SPRA669 published by Texas Instruments for digital signal processor TMS320C54x (SPRA669 2000).

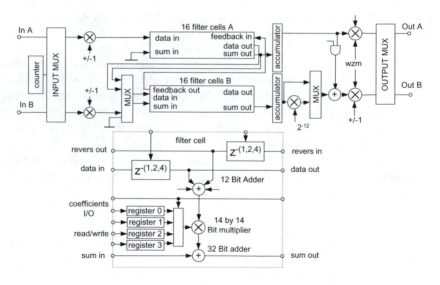

Figure 5.88. The functional block diagram of GC2011A digital filter chip of Texas Instruments

Figure 5.88 presents the digital filter form of the 3.3 V chip developed by Texas Instruments. This device enables realization of FIR, IIR and adaptive filters with 32-bit internal precision and 106 MSPS input rate.

5.3.4.e. Special types of digital filters

Digital filters due to their versatility are one of the most often used tools in digital signal processing. Every year brings new achievements in this area. Indeed, digital filters enable the realization of the special filters practically possible only in a digital version. For example currently many TV equipment manufacturers promote their products as the TV with digital comb filters.

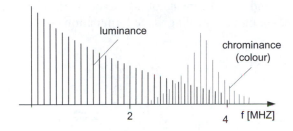

Figure 5.89. The spectral lines of the TV signal

The signals representing luminance (black and white) and chrominance (color) utilize the same frequency bandwidth and it is necessary to separate them correctly. Thus, it is required to design a filter with multiple pass-bands and multiple stop-bands. This function is realized by the *digital comb filter* and application of such filter improves significantly resolution and reduces mixing of a colors.

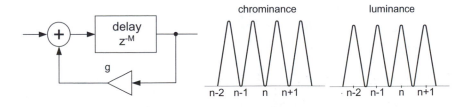

Figure 5.90. The principle of operation of the comb filter and examples of frequency responses of such filter used to separation of the signals in TV

The comb filter is realized by application of the delay element z^{-M} (or series z^{-1} elements). Fig. 5.90 presents the application of such a filter to separation of the luminance and chrominance signals in TV sets. The SONY company developed special chip digital comb filter model CXD2073s enabling high precision separation of signals.

Since digital filters are typically the counterparts of analogue RC filters, the special kind of digital filers called *wave digital filters – WDF* enables realization of practically all electrical networks, including LC filters. The principle of wave filters has been introduced in the late 1960s by Fettweis (Fettweis 1986, Lawson 1991). The Fettweis theorem enabled to digitalize electrical circuits composed of resistors, capacitors, inductors, transformers, gyrators, circulators and other elements of classic network theory. The wave digital filters are based on the traveling wave formulation.

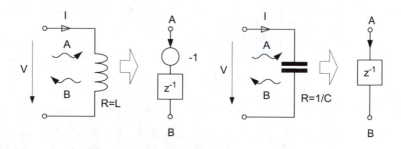

Figure 5.91. Functional equivalent of the on-port linear filter elements

The linear wave filter elements are derived from the electrical transmission line theory where the wave is reflected depending on the wave resistance R (Fig. 5.91)

$$A(s) = V(s) + R I(s)$$
$$B(s) = V(s) - R I(s) \qquad (5.80)$$

By substituting $V(s)=sL\ I(s)$ and after bilinear transform we obtain

$$B = -z^{-1}A \qquad (5.81)$$

in the case of an inductor and

$$B = z^{-1}A \qquad (5.82)$$

in the case of a capacitor. Thus the capacitance can be simulated by a unit delay while the inductor as the unit delay in cascade with an inverter, as presented in Fig. 5.91.

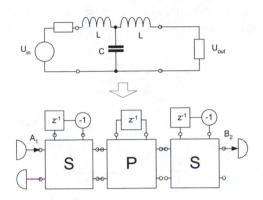

Figure 5.92. Analogue LC filter and its equivalent as the digital wave filter

The equivalent elements of wave filter are represented by delay element and invertors. These elements can be connected to the filter circuits by so-called *wave adaptors*, serial or parallel respectively. Fig. 5.92 presents an example of realization of a digital equivalent of a LC filter.

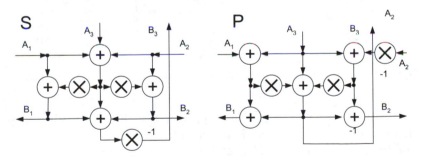

Figure 5.93. Two examples of wave adaptors: serial (S) and parallel (P)

The wave adaptors serial S or parallel P can connect several ports. Figure 5.93 presents the examples of realization of three-port wave adaptors.

5.3.4.f. Analogue versus digital filters

Figure 5.94 presents the comparison of analogue and digital filters. Steven Smith (Smith 2003) has compared two low-pass filters: analogue six pole Chebyshew filter with *1 kHz* cut-off frequency and digital filter with sampling frequency *10 kHz* and 128 samples in the time window. The analogue filter was constructed using 3 op-amps, 12 resistors and 6 capacitors.

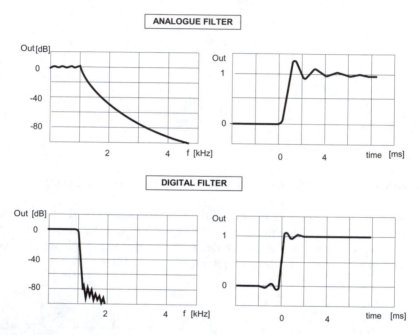

Figure 5.94 The comparison of the analogue and digital filter characteristics (Smith 2003)

The comparison of both filters, presented in Fig. 5.94 without doubt indicated that the digital filter is much better. Its characteristic is flat in the pass-band while for the analogue filter *6%* ripples are detected. In digital filter we can observe *100 dB* attenuation of the signal in the transition band between *1 kHz* and *2 kHz* while in analogue filter such effect is for about *4.5 kHz*. Also, the time response of digital filter is more linear and with smaller overshoot.

However, in this comparison Smith also pointed out several advantages of the analogue filters. The first advantage is the speed – to obtain fast digital filter large sampling frequency is required while an analogue filter can operate without problems up to about *1 MHz*.

Also the amplitude dynamics of the analogue filter can be better. In 12-bit ADC the quantization noise is *0.29 LSB* for *4095* bits which results in dynamics of about $14 \cdot 10^3$. In comparison, the dynamics of a typical op-amp is about $10 \cdot 10^6$. The frequency range of an op-amp is between *0.01 Hz* and *100 kHz* which is about seven decades. To obtain the same frequency range in a digital filter for the sampling frequency *200 kHz* it is necessary to process *20* million samples.

5.4. THE EXAMPLES OF APPLICATION OF DIGITAL SIGNAL PROCESSING IN MEASUREMENTS

5.4.1. The spectral analysis

The spectral analysis is a versatile tool because by using it we obtain almost complete information about all components of the analyzed signal. We know which harmonic components are present, which components are dominant, the frequency bandwidth, the proportion of magnitudes of these components, etc. The spectral analysis is crucial for acoustic measurements, testing of vibrations, in medicine and other technical areas.

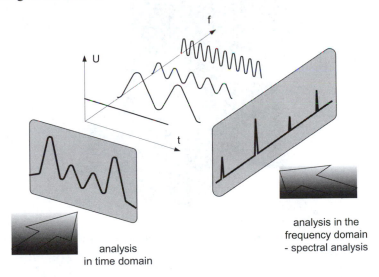

Figure 5.95. The two points of view at the signal: analysis in the time domain and analysis in the frequency domain (spectral analysis) (Agilent 150 2005)

As illustrated in Figure 5.95 the spectral analysis is the observation of the signal from another point of view. We are accustomed to performing the time domain analysis – for example by observation of the signal on the oscilloscope screen. Knowing this signal we can determine its *rms* value, frequency, magnitude. Both kinds of analysis (in time domain and in frequency domain) are complementary and reversible (if the circuit is linear). Thus having results of the frequency analysis we can perform the synthesis of the signal – and vice versa.

The significant step in the progress in frequency analysis was the development of FFT analysis. Currently, the spectral analysis is available in most digital oscilloscopes. On the market there are available special instruments called *spectrum analyzers* – usually these instruments are sophisticated and expensive, but also simple and portable instruments are offered. An example of portable spectrum analyzer is presented in Fig. 5.96. There are also available the spectrum analyzers in form of computer plug-in boards, or special software (professional and freeware in the Internet). And it is not difficult to design such spectrum analyzers by using standard tools of MatLab or LabVIEW.

Figure 5.96. The portable spectrum analyzer model R&S FSH3 of Rohde Schwarz (permission of Rohde Schwarz)

We can perform the spectral analysis off-line, but currently it is also possible to obtain FFT results on-line in the real-time systems. The off-line analysis is usually available in digital oscilloscopes – first we save the signal and next we analyze it. In the past typical spectral analysis required several second of time, today due to development of fast computer processors such operation requires time less than several ms.

Figure 5.97 presents the operation principle of the spectrum analyzer realized in the FFT technique. The analogue signal is converted into the digital one with sampling frequency f_s. Next the series of 2^n samples is formed. This series of samples is additionally shaped by the window function. As the result of analysis we obtain the spectral lines with period f_s/n and in the bandwidth $f_s/2$.

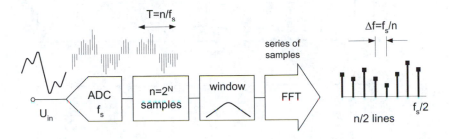

Figure 5.97. The FFT used as a tool for frequency analysis

Figure 5.98 presents a typical example of the spectral analysis. In the best case the harmonics are represented by single lines – this is possible if we perform synchronous analysis.

It is necessary to discuss the vertical and horizontal scales of the spectrum graph. In the case of continuous Fourier analysis this problem is simple – we obtain spectral lines exactly representing the frequency of harmonics and the length of the lines represents the magnitude of harmonics. In the case of discrete Fourier transform we can consider this method as the $m=N/2$ parallel connected band filters which represent the frequency f_m

$$f_m = m\frac{N}{f_s} \tag{5.83}$$

For example, if we analyze the signal with $f_s = 3200\ Hz$ and $N = 128$ samples as a result of the analysis we obtain 64 lines distanced by $25\ Hz$. If there is a harmonic of multiple of $25\ Hz$ in our signal then we obtain the spectral lines exactly representing such harmonics. But for instance the harmonics of $60\ Hz$ cannot be represented by a single line but rather by several adjacent lines.

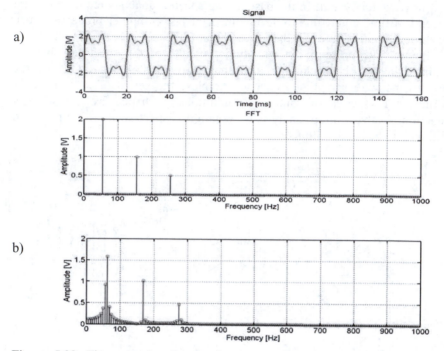

Figure 5.98. The example of spectral representation of the distorted signal for synchronous (a) and asynchronous (b) FFT analysis (Matlab simulation for $f_s = 2$ kHz, $N = 320$, $A_1 = 2V$, $A_2 = 1V$, $A_3 = 0.5V$, $f_1 = 50$ Hz (case a), $f_1 = 54$ Hz (case b))

Of course we can improve the resolution of the analysis by increasing the number of samples. In our example discussed above for *1024* samples the distance between lines would be *3.125 Hz*, but also in this case there would be no line representing exactly *60 Hz*. The increase of the number of samples causes increase of time necessary to analyze and the on-line analysis is very difficult to perform. Moreover, also the screen of typical spectrum analyzer exhibit limitation of resolution and usually the maximum number of lines is about *200 – 300*.

Sometimes the horizontal line is scaled not in frequency units but in *radian/sec*. In such case the lines are distanced by $2\pi f_s / N$ and the range of analysis is πf_s. After assumption that the $f_s = 1$ we obtain so called normalized spectral characteristic with the range equal to π and distance between lines $2\pi / N$.

The length (or rather height) of the spectral line should correspond to the magnitude of the harmonics A_m . In discrete Fourier transform the length B_m of the line for the real input signal is

$$B_m = \frac{N}{2} A_m \tag{5.84}$$

The length of the spectral line is related to the magnitude of the harmonic, but only in the case when the analysis is synchronous. If the analysis is asynchronous then the spectral line is composed of several lines due to the leakage effect described earlier, and even the line nearest to the analyzed frequency does not represent the magnitude of such components. Fortunately in most cases we are not interested in precise determination of the value of spectral component but rather in the relation between these components. Usually the height (length) of the spectral line is described as a logarithmic relation to the reference value U_o

$$dB = 10\log\frac{P}{P_o} = 20\log\frac{U}{U_o} \tag{5.85}$$

Figure 5.99 presents the spectrum analyzers employing filters. In this circuit each filter is tuned to one frequency. In order to obtain sufficient resolution of analysis it would be necessary to use a huge number of filters. For that reason such an analyzer is used for visualization of limited number of selected harmonics– for example in HiFi sound equalizers.

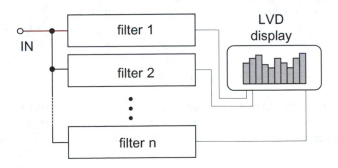

Figure 5.99. Multiple filter spectrum analyzer

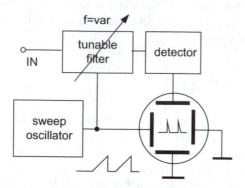

Figure 5.100. The spectrum analyzer with tuned filter

A large number of filters can be substituted by one or several tuned filters, as presented in Fig. 5.100. But such solution significantly increases the time of analysis. Instead of a filter it is more convenient to use a tuned oscillator as shown in Fig. 5.101.

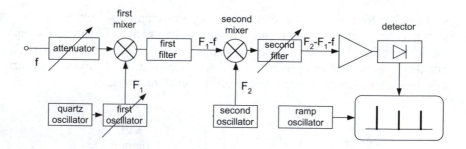

Figure 5.101. The superheterodyne spectrum analyzer

The spectrum analyzer presented in Fig. 5.101 utilizes the old[1] superheterodyne principle. In superheterodyne the mixing of two signals results in the signals on the output: $f, F, F-f, F+f, \ldots$ Using the filter we can

---

[1] The superheterodyne was commonly used for radio frequency signals from lower LF to intermediate frequency IF in all radio receivers – recently this technique is substituted by the PLL digital technique.

obtain the signal of frequency *F-f*. This coarse signal can be further processed to obtain better resolution in next step of mixing.

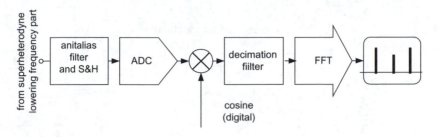

Figure 5.102. The digital spectral analyzer type PSA (Agilent 150 2004)

The main advantage of the superheterodyne analyzers is the possibility of processing signals with very large bandwidth – even up to hundred of GHz which is practically beyond the range of digital FFT technique. The superheterodyne analyzer is an analogue device. Recently, there are available the spectrum analyzers combining advantages of both techniques – the first step is the heterodyne device lowering the analyzed bandwidth to the level convenient for digital technique (for example *50 kHz*). Next, the FFT part is used to perform the spectrum analysis. An example of such an analyzer is presented in Fig. 5.102.

Digital spectral analyzer presented in Fig. 5.102 is a part of *Performance Spectrum Analyzer – PSA* developed by Agilent (Agilent 243 2000). To avoid great number of expensive anti-alias filters for various frequency bandwidths there is only one anti-alias filter connected at the input, and the subdivision to the various ranges is realized by digital mixing and decimation filter that reduces the sampling rate at its output to the rate needed for the frequency span.

Another model of all-digital spectrum analyzer of PSA series of Agilent is presented in Fig. 5.103. In this instrument it is possible to perform the analysis using two techniques. In the first one, traditional FFT technique can be used. In the second one, the 160 narrow step resolution bandwidth digital swept filters can be used. Because both techniques exhibit different advantages it is possible to select the best technique individually for analyzed case (also in automatic mode). The *FFT analysis* is faster – for example for analysis of pulsed RF signal with dynamic *94 dB* it requires *0.17 s*. The *swept analysis* requires for such case *3.1 s* while the dynamic range is *116 dB*. The compromise between dynamic range and speed is application of several FFT. This mode is called *multiple – FFT*. For presented above example the 5 FFTs/span requires *0.37 s* for analysis with dynamic range *104 dB*.

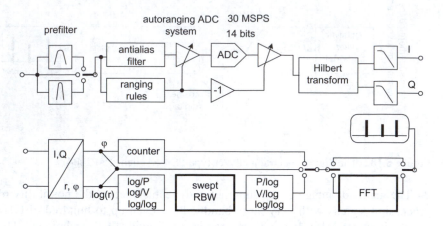

Figure 5.103. The all-digital spectrum analyzer PSA of Agilent (Agilent 150 2005)

In the spectrum analyzer presented in Fig. 5.103 the Hilbert transform is used to decompose the signal into *In-phase I* and *quadrature Q* components. Next, these signals are converted into the signals representing real part *r* and phase *φ*.

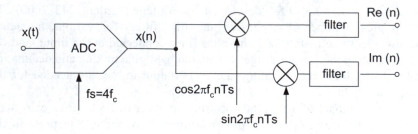

Figure 5.104. The quadrature sampling of the vector signal

Figure 5.104 presents another technique of vector signal analysis called quadrature sampling (Agilent 150-15 2004). The analyzed signal is digitally mixed with cosine and sine signals. As a result we obtain both components of the vector signal. It is possible to obtain the same result by mixing the analogue signal with sine signal and cosine signal (for example by shifting the sine signal of 90°) and next by converting both signals into digital ones.

a)   b)   c)

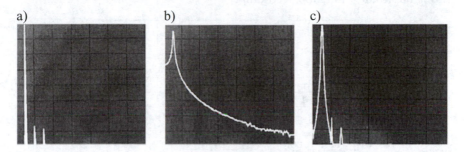

Figure 5.105. The example of synchronous spectral analysis (a), asynchronous one without window (b) and asynchronous one with Hanning window (c) (Agilent 150 2005) (permission of Agilent)

In spectral analysis important can be an appropriate choice of the window function. Figure 5.105 presents the comparison of the results of the analysis of the same signal. In the synchronous case the results are excellent. In the case of asynchronous analysis the first line due to leakage practically covers the other harmonics. But after application of the Hanning window the results of asynchronous analysis are significantly improved.

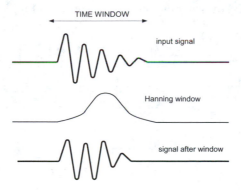

Figure 5.106. The distortion of the short time signal after application of the window function (Agilent 150 2004)

In the case of short transition signals the window function can disturb the result of the analysis as demonstrated in Fig. 5.106. To avoid such situations

Agilent included in the spectrum analyzers the function "auto-window" which automatically detects the start and end points of a short signal and fitting the window function accordingly.

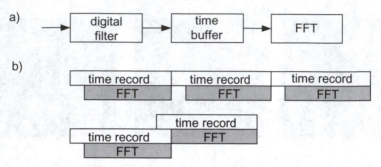

Figure 5.107. The time buffer for real time FFT analysis

Only parallel connecting filters (Fig.5.99) enable us to perform the spectral analysis in real time. The digital FFT usually requires certain time for analysis of the packet of samples and during this time incoming samples can be unavailable for analysis. To avoid losing the data between the ADC and FFT devices a time buffer can be inserted (Fig. 5.107a). The time buffer should be dynamically fitted to the time of analysis, which depends on the frequency span. Sometimes, for very short signals the time buffer can be substituted by several overlapping buffers, as shown in Fig. 5.107b.

Figure 5.108. The musical notation as the example of the time-frequency signal analysis

The real time FFT analysis (or described earlier Short Time Fourier transform) enables to demonstrate both components of the signal – frequency spectral and time analysis. It resembles of popular musical notation as it is demonstrated in Fig. 5.108.

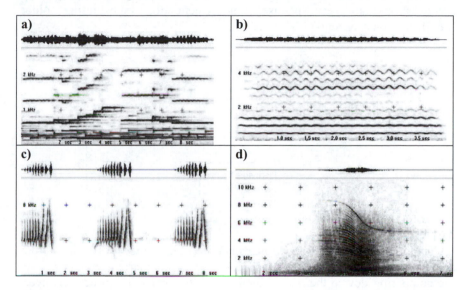

Figure 5.109. The examples of the spectrograms of various sound effects: flute (a), singing voice (b), song of bird (c) and flying jet (d) (Horne 2005) (permission of Richard Horne)

The time-frequency presentation of the signal is called a spectrogram. Four examples of spectrograms representing various sound effects are presented in Fig. 5.109.

5.4.2. Digital signal synthesis

Digital signal processing used for analysis of the signals can be also used for the synthesis of the signal. The most important development of this technique is visible in high quality sound processing. Most of the audio components are equipped with *digital sound processor DSP* enabling sophisticated processing of the sound (for example supplementing the sound with artificial reverberation). On the market there are available various sound synthetizers. Most personal computers are equipped with a sound board or a chip enabling the software processing and creating of the sound.

There are also available integrated circuits designed for signal synthesis and numerically controlled signal generation. Recently, the most popular are three systems: *PLL – phase locked loop*, *DDS – direct digital synthesis* and *AWG – arbitrary wave generators*.

The PLL system is used for very exact control of frequency of the generated signal. It is used in radio communication systems (including radio broadcasting) as very precise tuned generator (substituting older superheterodyne systems).

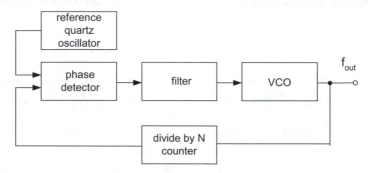

Figure 5.110. The principle of PLL frequency synthesis

The principle of the PLL synthesis is presented in Fig. 5.110. The output signal f_{out} is compared with precise frequency of the quartz oscillator f_{ref}. As the comparing device the *phase frequency detector PFD* is used to convert the difference of frequencies to the voltage. This voltage is used to control the *voltage controlled oscillator VCO*. The value of the output frequency is set by the change of parameters of the frequency divider. The precise setting of the output frequency is guaranteed by the feedback.

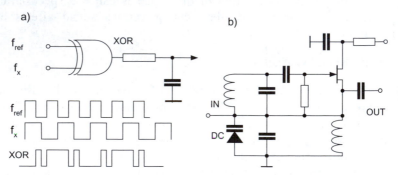

Figure 5.111. The example of the PFD device (a) and the voltage controlled oscillator (b)

As the simplest phase-frequency detector a *XOR type gate* can be used – in such a gate the output signal is equal to zero when both inputs are the same (Fig. 5.111a). In PLL systems also more sophisticated PFD circuits are used, so called PFD type II modules, in which the output signal is proportional to the phase shift between slopes of two impulses. Usually, an oscillator with the *varicap type diode* is used as the voltage controlled oscillator (Fig. 111b). A varicap diode (indicated as DC in Fig. 5.111b)) is a diode of the capacity depending on the voltage.

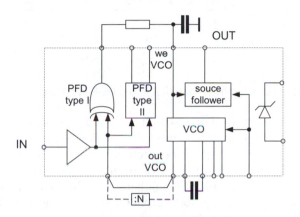

Figure 5.112. The example of IC PLL device – model CD4046 of Fairchild

There are available integrated circuits of VCO devices as well as the whole PLL systems. As an example in Fig. 5.112 is presented the PLL circuit type CD4046 of Fairchild. This device can work as the PLL oscillator and as a voltage controlled oscillator.

The PLL synthesis is performed in the frequency domain. Due to development of fast processors it is possible to perform the signal synthesis in the time domain (in real time) which is realized by the DDS system. The direct digital synthesis enables us to generate the signal with synthesis of the frequency as well of the wave shape. The principle of operation of DDS system is presented in Fig. 5.113.

The DDS system generates the sine wave with the frequency depending on the clock frequency and binary number – *tuning word M* at the input. To generate a fixed frequency sine wave phase increment, which is determined by tuning word, is added to the phase accumulator with each clock cycle.

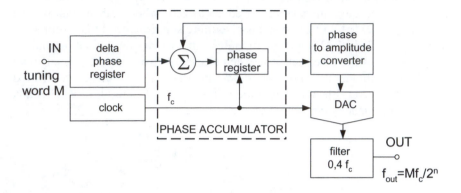

Figure 5.113. The principle of operation of the DDS synthesis

The phase accumulator acts as a phase wheel presented in Fig. 5.114. The sine-wave oscillation can be considered as the vector rotation around the phase circle and each point of the circle corresponds to equivalent point of the wave. One revolution of the vector around the phase wheel means the full sine wave cycle. In the n-bit accumulator the wheel can be divided into 2^n points. Thus for 32-bit register we obtain the resolution of phase equal to *4 294 967 296* points around the wheel.

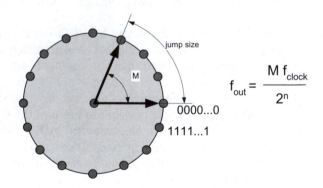

Figure 5.114. The phase circle of the DDS system

In the phase accumulator with every pulse of the clock generator the pointer in the wheel is moved by the binary coded input word *M*. The output frequency is determined by the M word

$$f_{out} = \frac{M \times f_{clock}}{2^n} \tag{5.86}$$

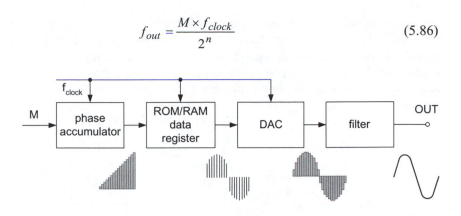

Figure 5.115. The signal synthesis in the DDS system

The amplitude of the output signal is formed by the phase to amplitude converter related to the data registered in the RAM memory (usually this can be the sine wave – amplitude/sine converter algorithm).

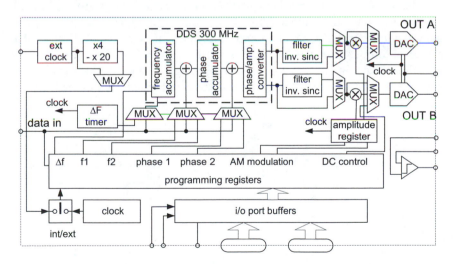

Figure 5.116. Functional block diagram of the 300 MSPS Quadrature DDS circuit – model AD9854 of Analog Devices

Figure 5.116 presents the example of the DDS system – model AD9854 of Analog Devices. The 12-bit 300 MHz processor enables the synthesis of the output signal with frequency resolution *1 µHz*. At the output there are two 12-bit digital-to-analogue converters – one of them is controlled by the signal shifted by *90°* which enables us to obtain two components of the output signal. The DDS can be used as a programmable generator. The phase and frequency of a signal can be adjusted digitally by stored data or the real time systems.

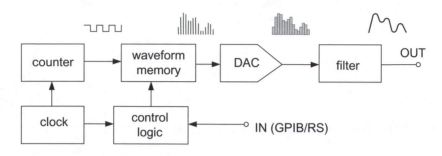

Figure 5.117. The principle of operation of the arbitrary wave generator

The *arbitrary wave generator AWG* can be used as the source of signals of precise adjusted frequency, phase and amplitude waveform. This instrument is currently one of the most frequently used signal generators. The principle of operation of AWG is presented in Fig. 5.117. The user stores the waveform data in the memory and according to the clock signal desired signal is then reconstructed sample by sample and transferred to the DAC.

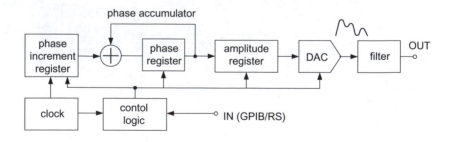

Figure 5.118. The DDS system used as the arbitrary wave generator

As mentioned earlier the DDS system can also operate as the waveform generator. An example of the arbitrary wave generator utilizing the direct digital synthesis DDS is presented in Fig. 5.118.

The arbitrary wave signal is programmable and can be realized from device memory as well as from an external device or computer via a suitable interface. In such way this instrument can work in computer measuring system as the signal source.

5.4.3. Improvement of the signal quality and the signal recovery

Analogue methods described earlier as well as digital ones are used to improve the quality of the signal. This improvement means first of all rejection of the noises and interferences. Modern techniques are capable of reconstructing the signal covered by much larger noises. This is called *signal recovery*. Signal recovery methods are very useful in biomedical measurements where the signal (for example the encephalography signal used to monitor the brain activity) is accompanied by much larger interference signals. Especially, digital methods are very efficient in *digital image processing*.

The obvious method of signal improvement is to record the signal many times and to *calculate the average value*. This method is frequently used in digital oscilloscopes. Fig. 5.119a presents the single waveform captured from the oscilloscope screen and the same waveform after averaging from 128 records (Fig. 5.119b).

a)   b)

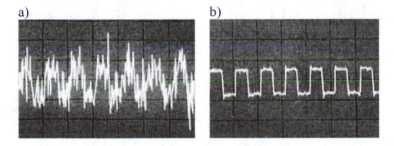

Figure 5.119. The noisy signal registered on the oscilloscope screen (a) and the same signal after averaging from 128 records (b) (Agilent 150 2004) (permission of Agilent)

To perform the averaging all registered pictures should be synchronized in time – for example they all should start with crossing the zero value line. This type of averaging is called linear or *coherent averaging*. The signal should be larger than the noises and the trigger system should start the sampling for the same phase. It is also possible to use the external trigger source to start the averaging in a precisely determined moment.

Because the standard deviation σ of the average value depends on the number of observations n (see Eq. 2.23) the improvement of the SNR for coherent averaging is

$$\frac{SNR_{av}}{SNR} = \frac{A / \sigma_{ave}}{A / \sigma} = \sqrt{n} \tag{5.87}$$

$$\frac{SNR_{av}}{SNR}[dB] = 20\log\left(\sqrt{n}\right) = 10\log N \tag{5.88}$$

It is possible also to average the measuring results in the frequency domain. Fig. 5.120 presents the averaging of the spectrum analysis. In such cases it is possible to perform *vector averaging*, where additionally real and imaginary components are averaged and then the modulus is calculated.

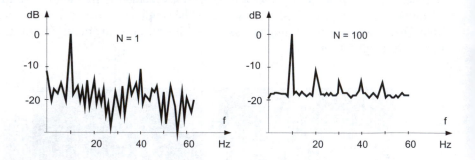

Figure 5.120. An example of the averaging in the frequency domain (Lyons 2004)

If it is not possible to perform coherent averaging we can use *rms averaging* (or non-coherent averaging). In this case we calculate average value of several *rms* values of the signal. In this case the profit of averaging is not as large as in the case of coherent averaging but we obtain the improvement of the *SNR* ratio according to the relationship

$$\frac{SNR_{av}}{SNR}[dB] = 10\log\sqrt{n} \tag{5.89}$$

It is also possible to perform *exponential averaging*, in which to the current sample is added the previous sample with selected weight α

$$y(n) = \alpha x(n) + (1 - \alpha)y(n - 1) \qquad (5.90)$$

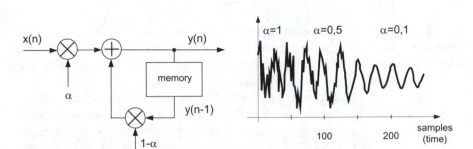

Figure 5.121. The exponential averaging method

The weight factor α can be set in the range $0-1$, the lower the coefficient α the better is the filtering, but at expense of the increase of the time constant (as it is demonstrated in Fig. 5.121). For exponential averaging the improvement of the SNR depends on α

$$\frac{SNR_{exp}}{SNR} = 10\log\frac{\alpha}{2-\alpha} \qquad (5.91)$$

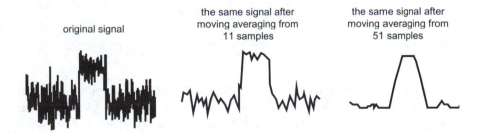

original signal   the same signal after moving averaging from 11 samples   the same signal after moving averaging from 51 samples

Figure 5.122. The rejection of noises by application of moving average filter (Smith 2003)

One of the most frequently used averaging digital filter is the moving average filter. This filter calculates the average value from the sample and M neighboring samples – it averages the samples in the window around the sample. Thus it realizes the following operation

$$y(i) = \frac{1}{M} \sum_{j=0}^{M-1} x(i+j) \qquad (5.92)$$

The value M can be selected that M previous samples or M next samples are used, but the best results are for the samples around the processed sample ($M/2$ previous samples and $M/2$ next samples). Figure 5.122 presents the examples of moving averaging results.

The averaging filters are very useful for the improvement of the images affected by noises. The moving average filter calculates the average value of the pixel and neighboring pixels, for example from the 3×3 area as demonstrated in Fig. 5.123a. The area used for averaging can be of various values and shapes (not necessary the square one).

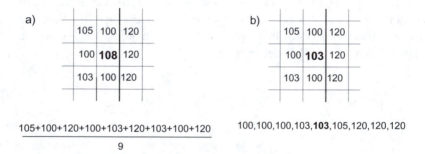

Figure 5.123. The central value of the 3×3 area calculated as the average value (a) and median value (b)

Sometimes better results can be obtained with the *median filter*. The median value is a value calculated in such a way, that 50% of samples are larger than processed sample and 50% of samples are smaller. Practically, the median value is calculated as follows: the samples are ordered from smallest one to the largest, and next the central value is selected (Fig. 5.123b).

The median value is more efficient in noise rejection because extreme values of the samples are rejected (in the case of average filter this samples influence the processed value). Moreover, the median filters in better way improve the sharp edges in the picture, while the average filters smooth the edges. Figure 5.124 presents two examples of the picture improvement after application of the appropriate filter.

a)　　　　　　　b)　　　　　　　c)

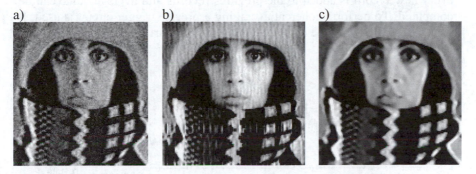

Figure 5.124. The improvement of the image quality of the noisy picture (a) by applying of the Wiener filter[1] (b) and median filter (c) (Young 1998) (permission of Ian T. Young)

To average the signal two measuring instruments can be used: digital signal averager or Boxcar averager. *Digital signal averager* calculates numerically the average value from the several recorded samples and coherent averaging is guaranteed an by appropriate triggering system.

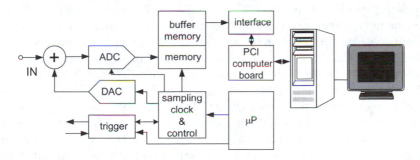

Figure 5.125. System block diagram of Eclipse Digital Signal Averager of Signal Recovery (Signal Recovery 2005)

Figure 5.125 presents the block diagram of the Eclipse Digital Signal Averager. At the input is used *2 GSPS* ADC circuit (four times *500 MSPS*). DAC device at the input enables precise control of offset. Due to very fast operation it is possible to transfer the data to the bank memory where the processing is performed practically in real time (averaged records are streamed to hard disk at *4 MB/s*). The data acquisition can be repeated -

---

[1] Described later – Section 5.5

every new record is added to the previous records and average value can be computed. To perform coherent averaging a precise trigger circuit is used. It is also possible to use the output trigger signal to start the experiment.

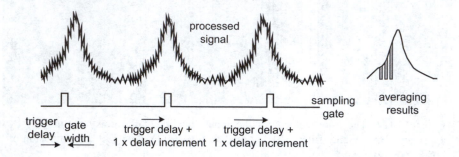

Figure 5.126. Boxcar average operation – waveform recovery mode (Signal Recovery TN 1005 2005)

The *boxcar averager* (called also as *boxcar integrator* or *gated integrator*) calculates the average value of the signal during the time of the sampling gate (Signal Recovery TN 1005 2005). When the sampling gate is open the input signal is connected to the integrating amplifier. As an example we can consider the Boxcar Averager model 4100 of Signal Recovery. This device can operate in two modes: the static gate mode and the waveform recovery mode. In the static gate mode the device determines the average value of the signal during the gate time, which is fixed. This way it is possible to determine the average value of the small part of the signal – for example peak value. In waveform recovery mode (Fig. 5.126) the gate is swept with constant increment and this way it is possible to determine the waveform of the processed signal (similarly to sample/hold operation of osciloscopes).

Application of the correlation and convolution function can be efficient for the signal recovery. Remember that correlation and convolution are calculated in similar way, but with inverse direction of k sample:

correlation $$r_{12}(k) = \frac{1}{N} \sum_{n=0}^{N-1} x(n)h(n+k)$$ (5.93)

convolution $$x(n) * h(n) = \sum_{k=-\infty}^{\infty} x(k)h(n-k)$$ (5.94)

The *algorithm of correlation* can be used to recover the signal if we look for the similar signal (correlated) to the other known signal. A typical application is searching for the radar signal covered by noises (Fig. 5.127). Because the reflected (received) signal is correlated to the transmitted signal it is possible to find it even if the noises are larger than the useful signal.

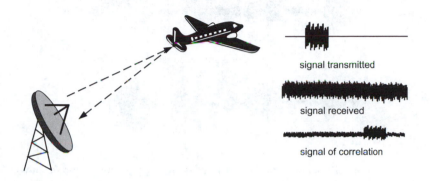

Figure 5.127. The example of the correlation algorithm used for the recovery of the radar signal

Also very useful for signal processing can be *convolution and deconvolution algorithm*. As we discussed earlier the output signal of the filter *y(n)* can be determined if we know the impulse response (filter coefficients) *h(n)*

$$y(n) = x(n) * h(n) \qquad (5.95)$$

We can invert the problem assuming that a virtual filter with impulse response $h_{noise}(n)$ is responsible for the noises and interferences. For example if we analyze the noisy pause between musical parts of recording we can determine such filter parameters. Next we can try to remove the noise from the musical record performing the deconvolution (asking what is the input signal *x(n)* if we know the output signal *y(n)* and impulse response).

An interesting example of the application of the deconvolution algorithm is presented in Fig. 5.128. In the first step, the impulse response of the filter representing the original signal was determined for a musical instrument (from the Baroque time). Next, this impulse response was used for reconstruction of the old, disturbed and damaged record. As a result pure sound was obtained (without noises and distortion); moreover, this reconstruction was not only technical one, but it also recovered the musical character of the Baroque period.

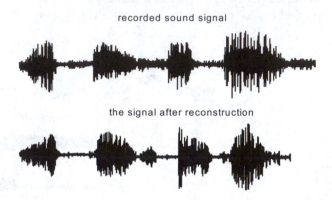

Figure 5.128. The sound signal reconstructed after application of the deconvolution algorithm (Czyzewski 1998)

The deconvolution function is widely used for improvement or reconstruction of images. There were developed various filter impulse responses, called *mask* or *kernel*. Such masks can improve various features of a picture: the edges (smoothing or enhancement), contrast, color, etc.

The calculation of the convolution in the time domain is rather difficult and time consuming. Even more difficult is the calculation of the deconvolution in the time domain. Both operations are much easier to perform in the frequency domain taking into account the relationship

$$y(n) = x(n) * h(n) \qquad Y(f) = X(f)H(f) \tag{5.96}$$

To perform the convolution operation in the frequency domain it is necessary:

- to calculate the Fourier transform of input signal and impulse response $X(f)$, $H(f)$
- to multiply both transform (both components real and imaginary),
- return to the time domain by using the inverse Fourier transform.

Figure 5.129 illustrates the realization of the convolution operation in time and frequency domain. The multiplication of the transform can be performed according to the relationships

$$Y(f) = H(f) \times X(f)$$
$$\operatorname{Re} Y(f) = \operatorname{Re} X(f) \operatorname{Re} H(f) - \operatorname{Im} X(f) \operatorname{Im} H(f) \quad\quad (5.97)$$
$$\operatorname{Im} Y(f) = \operatorname{Im} X(f) \operatorname{Re} H(f) + \operatorname{Re} X(f) \operatorname{Im} H(f)$$

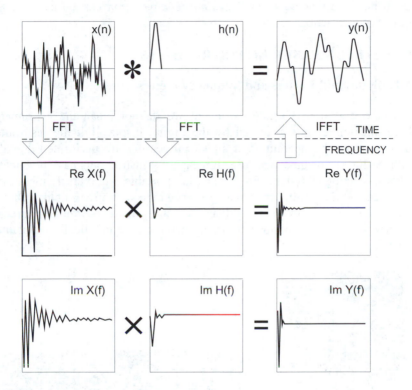

Figure 5.129. Realization of the convolution operation in the time domain and the frequency domain (Smith 2003)

The deconvolution can be realized in the frequency domain as the division operation, according to the relationships

$$H(f) = Y(f) : X(f)$$
$$\operatorname{Re} H(f) = \frac{\operatorname{Re} Y(f) \operatorname{Re} X(f) + \operatorname{Im} Y(f) \operatorname{Im} X(f)}{\left[\operatorname{Re} X(f) \right]^2 + \left[\operatorname{Im} X(f) \right]^2} \quad\quad (5.98)$$
$$\operatorname{Im} H(f) = \frac{\operatorname{Im} Y(f) \operatorname{Re} X(f) - \operatorname{Re} Y(f) \operatorname{Im} X(f)}{\left[\operatorname{Re} X(f) \right]^2 + \left[\operatorname{Im} X(f) \right]^2}$$

The deconvolution algorithm is very useful for image processing, for example when the picture is blurred. Usually, to perform the deconvolution the impulse response is used (called in the case of image processing *point spread function PSF*). When the PSF is unknown or poorly determined then special iterative techniques called *blind deconvolution* can be used for the picture reconstruction.

5.5. DIGITAL MEASURING INSTRUMENTS

5.5.1. Digital multimeters and frequency meters

As the digital measuring instrument we do not mean the instruments where the pointer is substituted by the digital display but the instrument where most of the operations of signal processing are performed digitally (Rathore 2004). In some areas digital measuring instruments have practically replaced analogue ones. For example portable, universal measuring instruments (Fig. 5.130) are available everywhere with prices comparable to analogue ones but with performances much better. Similarly precise voltmeters, ammeters, ohmmeters (multimeters) have practically supplanted the analogue instruments.

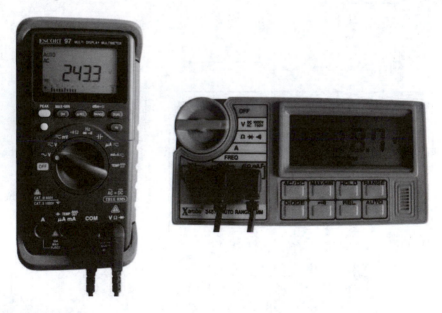

Figure 5.130. Two examples of portable digital measuring devices

Fig. 5.131 presents the block diagram of a typical digital instrument. The input circuits contain the conditioning circuits: voltage dividers or amplifiers for measuring the voltage in various ranges, shunt resistors for current measurement, supply source for resistance measurement, DC/AC converters. The example of a typical input circuit of a digital multimeter is presented in Fig. 5.132.

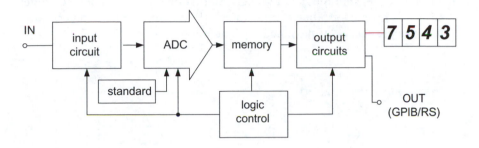

Figure 5.131. The functional block diagram of typical digital measuring instrument

Usually in the case of DC measurements the input circuit is separated by the capacitor. Therefore AC and DC measurements are performed separately and the resultant value is determined according to the formula (2.36). On certain instruments it is indicated (usually as *"AC+DC"*) that it is possible to measure both components of the signal. To measure the AC values in the input circuit is inserted AC/DC converter. If this converter calculates the *rms* value of the AC input signal often such instrument is indicated as *"True rms"*.

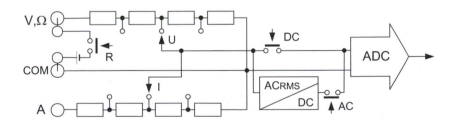

Figure 5.132. The input circuit of the typical multimeter

The logic circuit controls all functions of the instrument: automatic change of the input ranges (in certain instruments), triggering of the measuring cycle, control of ADC, saving the data into memory, etc. Sometimes the instrument saves a certain number of last measurement results and these data can be transmitted through available interface. More expensive measuring instruments are equipped with GPIB interface, cheaper ones with serial interface RS232 or USB interface. The interface enables us not only to transmit the data to other external devices (including computer) but also allows controlling the instrument by the external computer system (for example change of the ranges or functions).

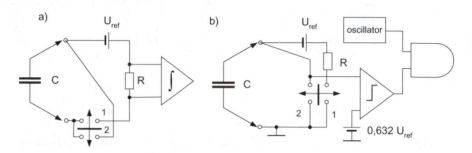

Figure 5.133. The input circuits for capacity measurements

Typically digital multimeters enable the measurements of voltage, current, resistance, often also frequency. Sometimes the measurement of capacity is possible as well. Fig. 5.133 presents two examples of the input circuits designed for capacity measurement. In both circuits the measuring process is divided into two cycles: in the first one (switch in position 1) the measured capacitor is charged from the voltage source U_{ref} connected by the resistance R. In the second cycle (switch in position 2) the capacitor is discharged.

In the first circuit (Fig. 5.133a) measured capacity is proportional to the charge Q and

$$C = \frac{Q}{U_{ref}} = \frac{\int_0^T i\,dt}{U_{ref}} \tag{5.99}$$

Thus in the output of the integrating circuit the signal is proportional to the measured capacity.

In the second circuit (Fig. 5.133b) the measured voltage across the capacitor increases exponentially

$$U_c = U_{ref}\left(1 - e^{-t/RC}\right) \qquad (5.100)$$

After a time equal to the time constant $\tau = RC$ this voltage reaches value $0.632\ U_{ref}$. If the voltage U_c is connected to the one input of the comparator and to the second input is connected to voltage $0.632\ U_{ref}$ then the comparator closes the gate after the time equal to the time constant. Thus we obtain conversion of the measured capacity into the time period. If to the gate is connected to a reference oscillator the digital counter directly indicates the measured capacity.

The integrating device is most often used as the analogue-to-digital converter. Sometimes it is substituted by the delta-sigma converter, seldom by the SAR converter. The advantage of the integrating converter is that this device is insensitive to the interference of frequency corresponding with integrating time (most often it is 50 Hz interference).

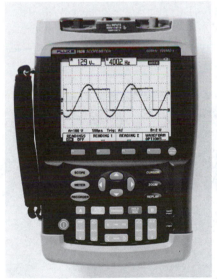 

Figure 5.134. The examples of the ScopeMeter instruments of Fluke (Fluke 2005) (permission of the Fluke Corporation)

On the market there are also available extended measuring devices combining in one instrument functions of a digital multimeter and an oscilloscope. The example of such devices (called sometimes *ScopeMeters*) are presented in Fig. 5.134

Currently, there are available various types of digital multimeters and the portable multimeters sometimes exhibit performances comparable to the parameters of laboratory ones. Table 5.5 presents a comparison of typical performances of digital multimeters.

Table 5.5. Performances of typical digital multimeters

type	portable	lab instrument	precise
model	110	34401A	2002
manufacturer	Fluke	Agilent	Keithley
Number of digits	3¾ (6000)	6½ digits	7½ or 8½ digits
measure	U, I, R, C, f	U, I, R, f	U, I, R, f, T
DC uncertainty	0.7%	0.0035+0.0005%	0.0006+0,00008%
R_{in} in DC voltage 200mV measurement	10 MΩ	>10 GΩ	>10 GΩ
AC uncertainty	1%	0.06+0.03%	0.02+0.01%
bandwidth	50 – 500 Hz	10 Hz – 300 kHz	1 Hz – 2 MHz
Speed of readings	40/s	1000/s DC, 50/s AC	2000/s 4½ digits
memory	-	512 readings	30 000 readings
interface	-	RS232C , HPiB	GPIB

Figure 5.135. The laboratory multimeter type HP 34401 of Agilent (Agilent 2005)

Figure 5.135 presents one of the most popular digital multimeters, HP 34001 of Agilent. This instrument enables measurements of voltages, currents (both DC and AC), resistances and frequencies with uncertainty of about 0.004% DC and 0.1% AC. In the case of measurements of AC values the instrument operates as a DC voltmeter with *AC rms/ DC* converter. It is also possible to determine a maximal (peak) value of the input signal.

An another operating principle is used in the time/frequency measurements (electronic counters). Fig. 5.136 presents the block diagrams of the frequency meters (Fig. 5.136a) and period meters (Fig. 5.136b). In the case of frequency meter the triggering input circuit converts the measured signal to the rectangle waveform and starts counting. The signal of the standard quartz oscillator after frequency division opens the gate for a precisely set period of time. The number of pulses counted in this time period is a direct the measure of frequency.

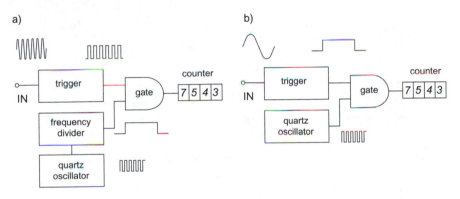

Figure 5.136. Digital measurement of the frequency (a) and period (b)

For the low frequency signal the gate should be open for a long time period to obtain sufficient resolution (sufficient number of counts). Therefore, for low frequency signals it is better to substitute the frequency measurement by the period measurement. The input signals of the gate are reversed. The measured signal opens the gate for the period or multiple of periods. The pulses of standard oscillator are counted as the measure of the period. Typical frequency meters enable also to measure width of the pulse, phase shift and time.

On the market there are available rather sophisticated and expensive digital impedance meters (or capacity or inductance meters). Using an AC voltmeter it is possible to determine the value of impedance using *three-*

voltmeter method (Fig. 5.137). According to the phasor diagram presented in Fig.5.137 we can write following relations

$$\cos\varphi = \frac{U_1^2 - U_2^2 - U_3^2}{2U_2 U_3} \qquad (5.101)$$

$$Z_x = \frac{U_3}{I} = \frac{U_3}{U_2} R \qquad (5.102)$$

$$R_x = Z_x \cos\varphi \qquad (5.103)$$

$$X_x = \sqrt{Z_x^2 - R_x^2} \qquad (5.104)$$

Instead of three voltmeters it is sufficient to use one voltmeter connected to the three signal sources U_1, U_2 and U_3.

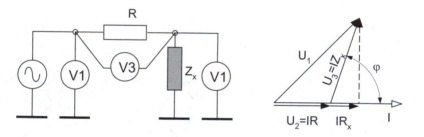

Figure 5.137. The impedance measurement by three-voltmeter method

5.5.2. Digital oscilloscopes

Figure 5.138 presents the block diagram of a typical digital oscilloscope. We can notice in this diagram similar functions as are present in the analogue instruments: triggering circuit and horizontal/vertical position logic. Although there are significant differences of the operation principle of digital and analogue oscilloscope, the manufactures take into account certain tradition and equip the digital instrument with very similar functions as are in the analogue instruments.

The main differences between the digital and analogue oscilloscope are as follows: after conversion of the input signal to digital one all operations (signal processing) are performed digitally, in digital oscilloscopes usually there is storage function enabling simple recording and reproduction of signals, due to the digital processing oscilloscopes are equipped with various additional functions, such as FFT analysis, averaging function, integration of the signal, measurement of the value and frequency of the signal, etc (Banks 1997, Hickman 1997, Kularatna 2003).

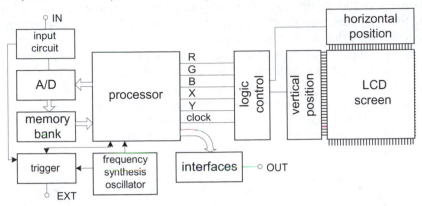

Figure 5.138 Block diagram of a typical digital oscilloscope

In the analogue instruments the picture on the screen was stopped artificially – by appropriate fitting of the synchronization frequency to the tested signal frequency. In the case of digital conversion the reproduction of the picture representing signal seems to be much simpler. After trigger signal the input voltage is sampled according to the clock impulses and as result we obtain series of samples (Fig. 5.139). This kind of sampling is called *sequential sampling*. We can save this sampling result in memory and then reproduce the signal "sample by sample" on the screen.

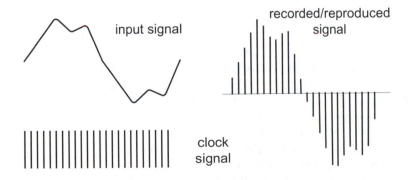

Figure 5.139. The signal before and after sequential sampling

It is important to fit the time axis to the clock impulses. Of course it is possible to reproduce the picture on the screen directly, bypassing the memory circuit. In the case of digital processing it is possible to reproduce pictures of many signals at the same time – the multi-channel operation is much simpler than in the case of analogue instruments.

In the input circuit apart from conventional attenuators there should be inserted a sample-and-hold circuit and an anti-alias filter. This filter is very important, because if the frequency of sampling is too small due to the aliasing effect the processor can fit to the sample an incorrect signal (as illustrated in Fig. 5.140). The observer will see the signal of lower frequency (continuous line) instead of correct signal (dashed line).

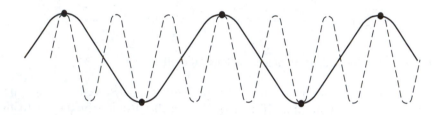

Figure 5.140 The aliasing effect in digital oscilloscopes

The digital technique solves many problems of the analogue case but at the same time some new troubles emerge. One of them is that analogue operation is true real time (small time delay is negligible). In the case of digital operation the sampling, saving and reproduction needs time, which is not always negligible. The solution is to use fast analogue-to-digital converters, with sampling frequency *1 GHz* or better.

Another problem is the frequency limitation of typical analogue-to-digital converters. Even if we use *1 GHz* converter the bandwidth is limited to about *500 MHz* (which is easy attainable by analogue instruments). One of the methods of increasing the number of samples is *multiple-point random sampling* in which the same signal is sampled several times with randomly shifted samples (Fig. 5.141a). In such technique very important is to reference the sampled wave to the trigger point.

In *sample-and-hold technique* (Fig. 5.141b) each sample is shifted by a small time increment. In this way each time other point of the signal is sampled. Knowing the time interval it is possible to reconstruct the signal, although we sampled it only with too small sample number per period (in our case one sample per period).

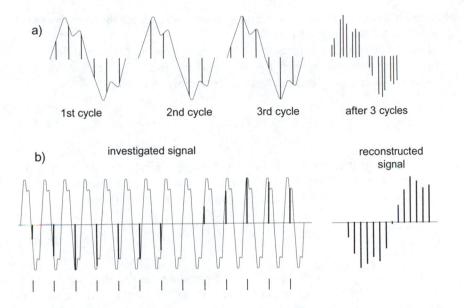

a)

1st cycle   2nd cycle   3rd cycle   after 3 cycles

b)   investigated signal   reconstructed signal

Figure 5.141. Sampling techniques: multiple-point random sampling (a) and sample-and-hold technique (b)

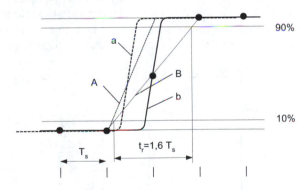

Figure 5.142. The error of rise time

Another problem of digital reconstruction results from the fact that we do not know what happened in the time between samples (in contrast to the analogue technique where signal is continuous). If the investigated pulse is shorter than the sample period we cannot say anything about its shape. Consider two cases of sampling of rise slope presented in Fig. 5.142. Depending on the pulse placement (with respect to the sampling pulses) we can obtain various results – for the pulse *a* we approximate the slope by the *A* line, while for the pulse *b* by the *B* line.

Placement of pulse *b* (exactly in the same time as the sampling pulse) results in the interpretation of the rise time as the *1.6 Tₛ*. This is the most disadvantageous case. Thus the shortest detectable rise time is

$$t_r = 1.6 \cdot T_s = \frac{1.6}{f_s} \tag{5.105}$$

We can say that the larger the sampling frequency the easier it is to reconstruct the signal because we have more samples. It is assumed that the *useful storage bandwidth USB* depends on the sampling frequency

$$USB = \frac{f_s}{C} \tag{5.106}$$

where *C = 2.5 – 25* is the coefficient depending on the method of wave interpolation. When we do not interpolate the wave (the shape is reconstructed from the sampled points) the coefficient *C = 25*. When the points are connected by line (linear interpolation) it is assumed that *C = 10* guarantee correct reconstruction. When we interpolate the sampling results by the sine wave, *C = 2.5* is sufficient. But we can use the sine interpolation only when we are sure that the investigated signal is a sine wave (for the rectangle pulses the error of sine wave interpolation can be significant).

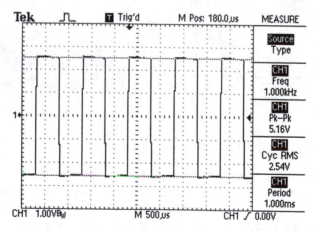

Figure 5.143. The example of the screen of digital oscilloscope – model TDS 220 of Tektronix (Tektronix 2005)

Modern digital oscilloscopes can be used as the instrument for observations of the signals. But due to the enhanced storage capability and signal processing possibility this instrument can be used as a recorder or also as measuring and analyzing device (for example for FFT analysis). Figure 5.143 presents the example of the screen of digital oscilloscope. It is possible to obtain also a multicolor picture of investigated signals. Fig. 5.144 present one of the most popular digital oscilloscopes – oscilloscope of Tektronix.

Figure 5.144. The example of the digital oscilloscope – oscilloscope of Tektronix (Tektronix 2005)

5.5.3. Digital measurement of power and energy

In Chapter 3 the induction *watt-hour meter* (electrical energy meter) was described. In the near future probably the meter readers should be substituted by the computer system. To realize such a system it is necessary to replace the induction watt-hour meters by digital ones. Therefore, the methods of digital power measurements are still being developed. The energy meters are similar to the power meters – they are only supplemented by the voltage/frequency converters (the output pulses are counted as the energy).

There are two strategies of digital power measurement. The first one uses power to voltage converters and next there are connected typical digital devices. In the other method all-digital conversion is used. Figure 5.145 presents the power-to-voltage converter using the multiplier device.

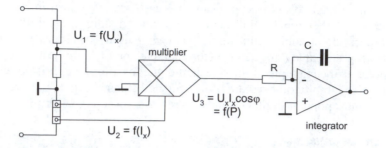

Figure 5.145. The power to voltage converter using the multiplier device

Another frequently used power-to-voltage converters utilize the *Time Division Multiplier TDM* principle. The example of such a converter is presented in Fig. 5.146. The sum of the first voltage U_1 and the reference voltage U_r is connected to the input of the integrator. The integrator output voltage increases linearly by the time period T_1 to the moment when it is equal to the value of the ramp voltage U_t connected to the second input of comparator. In this moment the switches are changed over and at the input of integrator the difference of voltages U_1 and U_r is connected. The output voltage of integrator decreases by the time period T_2 and

$$(U_1 + U_r)T_1 = -(U_1 - U_r)T_2 \tag{5.108}$$

The width of impulse depends on the U_1 voltage

$$T_1 - T_2 = -\frac{U_1}{U_r}(T_1 + T_2) = -\frac{U_1}{U_r}T_g \tag{5.109}$$

Time T_g is the period of the ramp U_t voltage. The output signal of the comparator switch also connects to the second integrator voltage U_2 or $-U_2$ respectively. Thus

$$U_{out} = \int_0^{T1} u_2(t)dt - \int_{T1}^{T1+T2} u_2(t)dt = U_2(T_1 - T_2) = \frac{T_g}{U_r}U_1U_2 \tag{5.110}$$

The output voltage is the waveform of impulses with the width of the impulse proportional to the first voltage U_1 and the amplitude proportional to the second voltage U_2. According to the relationship (5.110) the output voltage is proportional to the result of multiplication of both voltages $U_1 \cdot U_2$.

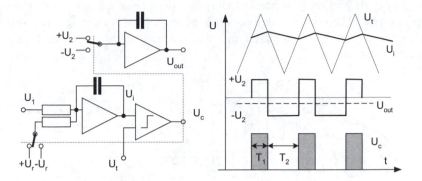

Figure 5.146. The TDM multiplier

Fig. 5.147 presents the IC all-digital power and energy converter. Two input signals after analog-to-digital conversion (delta-sigma *16-bit 450 kHz*) are digitally multiplied and at the output there is signal proportional to the power. The frequency output is adapted for electromechanical counters and 2-phase stepper motors. The uncertainty of conversion is *0.1%* for power and *0.1 °* for phase.

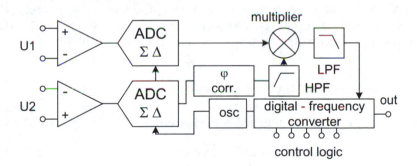

Figure 5.147. Energy metering IC – model ADE7757A of Analog Devices (Analog Devices 2005)

Fig. 5.148 demonstrates the enhanced version of IC energy and power converter. This circuit is supplemented by two *rms* converters and an additional multiplier. Due to this add-on elements it is possible to determine also apparent power and energy (*VA* power), as well as *rms* values of current and voltage. It is also possible to determine the period (frequency) of the voltage signal. The circuit is equipped with two outputs – the frequency one for energy counters and a serial one for communication with computer measuring systems.

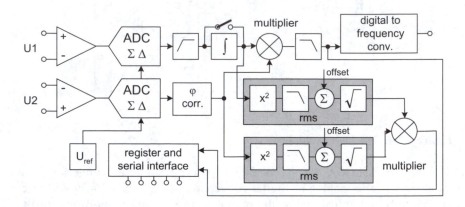

Figure 5.148. The active and apparent energy metering IC – model AD7763 of Analog Devices (Analog Devices 2005)

An interesting extra element is the integrator in one of the channels. This integrator enables the converter to cooperate with inductive sensor of current (for example Rogowski coil) where the output signal is proportional to dI/dt.

5.5. INTELLIGENT DATA ANALYSIS

5.5.1. The artificial intelligence in measurements

The methods, algorithms and instruments described in previous chapters usually operated in precisely determined conditions. But often the measuring instruments work in varying conditions where parameters of the measuring devices should be changed according to the unexpected deviation of the situation. In such cases we should use *adaptive measuring methods*.

The adaptive methods are used when the analyzed phenomenon is ambiguously described or we do not have sufficient information about the investigated object, when the conditions are varying in an unexpected way (for example external interferences and noises), when we do not know exactly which algorithm of operation is the best. Thus the adaptive methods require non-routine operations and selective choice of the best solution. For such operations we have to use intelligent methods (or artificial intelligence) (Berthold 2003, Widrow 1985, Zaknich 2005).

The term *artificial intelligence* appeared with the development of computer technology. Even the phrase *computational intelligence* is in use. The main question is: can the computers learn? This activity was earlier reserved only for the human brain. However, modern instruments supported by the informatics are able not only to analyze the incoming data, but they can also predict the result of an action or adapt to unexpected situations.

The adaptive methods often imitate the natural adaptive methods which are used with success from thousands of years by nature, most of all by biology. By imitation of the human brain activity the *artificial neural networks ANN* were developed. By the imitation of the human way of thinking the *fuzzy logic* methods were developed. By observation of the evolution process the *genetic algorithms* were developed.

5.5.2. The adaptive filters

Digital filters are programmable and therefore they are easily modified in on-line mode. The filters, which are able to learn, can change their parameters under the influence of varying conditions. Such filters are called *adaptive filters* (Bellanger 2001, Diniz 2002, Forhang-Boroujeny 1999, Haykin 2001, Sayed 2003). The most frequently used method of adaptation is the *least mean square LMS* method (Haykin 2003). In the learning process the output signal $y(n)$ is subtracted from the reference signal $u(n)$ and the difference $e(n)$ (error) is used to adapt the setting of the weight filter coefficients h. We calculate the estimated value E of LMS error as the sum of the squares of these differences and next we look for the minimum of this sum (derivative equal to zero).

$$E\left(\sum_{i=0}^{k} e^2(n)\right) = E\left(\sum_{i=0}^{k} [u(n) - y(n)]^2\right) \quad (5.111a)$$

$$\frac{\partial E|e^2(n)|}{\partial h} = 0 \quad (5.111b)$$

The algorithm modifies the filter coefficients until the error is minimal. As an example of application of the adaptive filter we can consider the noise rejection method. The signal of the noises is subtracted from the output signal and this difference can be used to modify the parameters of the filter.

The classic application of such filter is the recovery of the signal from the noisy background – the so called *cocktail party problem,* when we try to select the voice of the interlocutor from the background (other conversations, sound of orchestra, glass clinks, etc). Consider the similar situation when we attempt to select the voice from the noise of a car engine (Fig. 5.149). From the detected by microphone signal (voice and noise) we subtract the signal of noise (detected by other microphone – distanced from the voice source and close to the car engine). The signal of error e_n influences the filter coefficients h until the interferences caused by the engine are rejected.

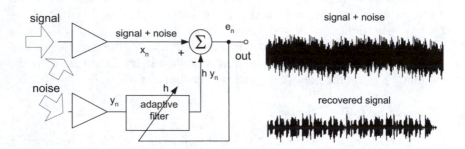

Figure 5.149. The adaptive filter used for noise rejection

The conditions for efficient performance of the adaptive filtration of the noises in the example described above are as follows: the noise in signal and reference noise should be correlated and signal and reference noise should be non-correlated.

One of the most frequently used adaptive filters is the Wiener filter. The typical structure of the Wiener filter is presented in Fig. 5.150. The output signal of the investigated system $y(n)$ is sent to the input of the filter of the M-order and at the output of the filter is

$$\hat{x}(n) = \sum_{k=0}^{M-1} h_k y(n-k) \qquad (5.112)$$

The error signal is

$$e(n) = x(n) - \hat{x}(n) = x(n) - \mathbf{h}\mathbf{y} \qquad (5.113)$$

where $x(n)$ is desired non-disturbed signal, $\mathbf{h}$ is a matrix of filter coefficients and $\mathbf{y}$ is a matrix input signals of the filter $y = [y(n), y(n-1),...,y(n-M-1)]$.

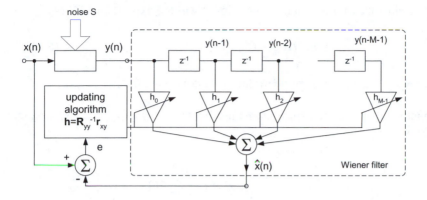

Figure 5.150. The example of the Wiener filter (Vaseghi 2000)

The condition of optimization of the filter is minimal value of the sum of square of the errors e. From equation (5.111) we obtain the conditions of the optimal Wiener filter (Wiener-Hopf conditions) (Vaseghi 2000)

$$\mathbf{R}_{yy}\mathbf{h} = \mathbf{r}_{yx} \tag{5.114}$$

where $\mathbf{R}_{yy}$ is the autocorrelation matrix of the $y(n)$ signal and $\mathbf{r}_{xy}$ is correlation matrix between the input signal $x(n)$ and the input signal $y(n)$.

For the optimization of the filter it is convenient to use the relationships in the frequency domain. The expression (5.114) can be rewritten as

$$\sum_{n}\sum_{k=0}^{M-1} h_k r_{yy}(n-k)e^{-j\omega n} = \sum_{n} r_{yx}(n)e^{-j\omega n} \tag{5.115}$$

After Fourier transform the filter coefficients are

$$W(f) = \frac{P_{xy}(f)}{P_{yy}(f)} \tag{5.116}$$

where P_{xy} is the cross-power spectra of $Y(f)$ and $X(f)$ and $P_{yy}(f)$ is the power spectra of $Y(f)$.
We can also obtain the expression (5.116) as the minimum of the square of the dependence $F(f)=X(f)-H(f)Y(f)$.

If to the signal x(n) is added the noise signal s(n) the expression (5.114) is

$$\mathbf{h} = (\mathbf{R}_{xx} + \mathbf{R}_{ss})^{-1}\mathbf{r}_{xy} \qquad (5.117)$$

where $\mathbf{R}_{ss}$ is the matrix of autocorrelation of noise signal.

Respectively in the frequency domain (for $Y(f)=X(f)+S(f)$) the expression (5.16) is

$$W(f) = \frac{P_{xx}(f)}{P_{xx}(f) + P_{ss}(f)} \qquad (5.118)$$

where P_{xx} and P_{ss} are the power spectra of signals $X(f)$ and $S(f)$.

Dividing the numerator and denominator by power spectra $P_{ss}(f)$ and substituting $SNR(f) = P_{xx}(f)/P_{ss}(f)$ the relationship (5.118) is often presented as

$$W(f) = \frac{SNR(f)}{SNR(f) + 1} \qquad (5.119)$$

To determine the filter coefficients it is necessary to know the power spectra of both signals (or *SNR* ratio). This is not always possible and therefore it is assumed a character of the noise, such as the white noise. Sometimes the power spectra of noise can be determined experimentally, for example in the silent fragment of the restored musical noisy record.

We can use the adaptive Kalman filter to determine the estimated value of the distorted signal (Grewal 2001, Haykin 2001, Ristic 2004, Zarchan 2000). *The Kalman filter* is the optimal recursive data processing algorithm utilizing prior knowledge about the investigated system to produce the estimate of variable in such a way that the error is minimized statistically. And it is very important, taking into account the subject of our book, that the Kalman algorithm includes also the measurement uncertainty.

The Kalman theory is based on the state-space approach in which the state equations are used to predict the signal or state parameters based on the observation of previous samples. Thus it is a real learning system because it works according to the recursive adaptation formula (Voseghi 2000):

next parameter estimate = previous parameter estimate + update

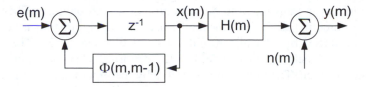

Figure 5.151. The model of signal in the Kalman filter

For example the state equations of the noisy signals can be written in the form (Voseghi 2000)

$$x(m) = \boldsymbol{\Phi}(m, m-1)\, x(m-1) + e(m) \qquad (5.120)$$

$$y(m) = \boldsymbol{H}(m)\, x(m) + n(m) \qquad (5.121)$$

where $x(m)$ is a P-dimensional signal or state parameter, $y(m)$ is M-dimensional noise and distorted observation vector, $n(m)$ is M-dimensional noise process, $\boldsymbol{\Phi}(m)$ is $P{\times}P$ dimensional state transition matrix related to the states of process at times m-1 and m, $e(m)$ is P-dimensional uncorrelated input excitation vector of the state equation, $H(m)$ is the $M{\times}P$ channel distortion matrix.

The Kalman algorithm requires statistical description of noises and knowledge about the uncertainty of the measurement. Due to rather sophisticated theory it is practically impossible to describe it in the limited area of this book. Therefore readers are requested to benefit from the large number of publications on this subject (Haykin 2001, Grewal 2001, Maybeck 1979, Voseghi 2000).

5.5.3. Artificial neural networks

Artificial neural networks initiated new possibilities in digital signal processing. These networks imitate the human brain – parallel transmission of data by very large number of junctions, learning by training, learning through exposure to external stimuli, storing the knowledge by the change of the weight of connection between neurons, resistance to disturbances.

Biological neurons are organized in networks. The incoming neural fiber (the dendrite) receives electrical signals from other neurons through synapses (via biochemical process). If incoming signal reaches the threshold then the action potential is transmitted to other neurons through outgoing fiber (the axon). Presented in Fig. 5.152 is the mathematical model imitating activity of neuron introduced by McCulloch and Pitts (McCulloch 1943). The input circuits $x_1,..., x_n$ operate in this model like the dendrites . The output circuit y_i works like the axon. The threshold is imitated by the input w_o. As the nucleus operate two elements: an adder that sums the input signals, each with appropriate weight w_{ij}, and activation function $f(u_i)$.

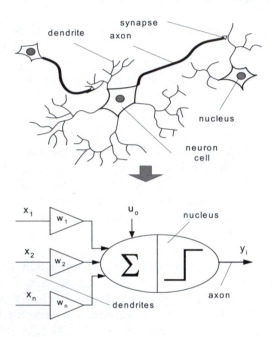

Figure 5.152 The biological neuron and its mathematical model

The output signal of the neuron model is

$$y_i = f(u_i) = f\left(\sum_{j=1}^{N} w_{ij} x_j(t) + w_{i0}\right) \tag{5.122}$$

In the McCulloch-Pitts model the activation function is binary

$$f(u) = \begin{cases} 1 & if \quad u \geq 0 \\ 0 & if \quad u < 0 \end{cases} \tag{5.123}$$

The real biological neurons are not binary – they work rather as analogue devices because the output signal depends on the value of excitation. Therefore more universal is the activation function in form of the sigmoid function $f(u) = tanh(\beta u)$ (Fig.5.153). If the coefficient β is equal to *1* the activation function is nearly linear. If the coefficient β is larger than *10* the activation function is nearly binary.

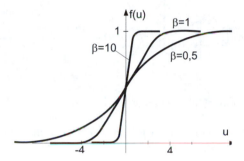

Figure 5.153. Various forms of the activation function *f(u*

Picture 5.154 presents the elementary model of a neuron network called *perceptron*, which is able to learn. The method of learning (training) is somewhat similar to the adaptation model of filters (described in the previous chapter). The output signals y_i are compared with the reference (training) signal d_i. The difference (error) δ modifies the weights w_i as the feedback.

The process of training is continued until the error function E reaches minimum, where E is

$$E = \frac{1}{2} \sum_{k=1}^{p} \left(y_i^{(k)} - d_i^{(k)} \right)^2 \tag{5.124}$$

The single perceptron exhibits rather small power of computation – the strength of neuron network results from the great number of neurons. The neural network can be a very efficient tool of signal processing when in such network huge a number of neurons cooperate (it is assumed that the human brain consists of about 10^{15} synapses – for comparison contemporary computer use only about 10^8 of transistors).

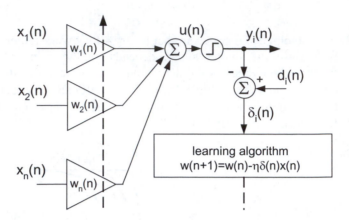

Figure 5.154 The model of a perceptron

Fig. 5.155 presents the most frequently used non-recursive network – the *multilayer perceptron*.

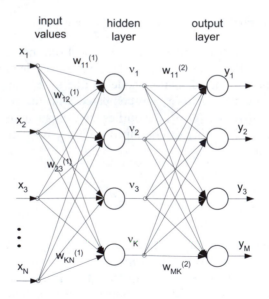

Figure 5.155. The multilayer perceptron

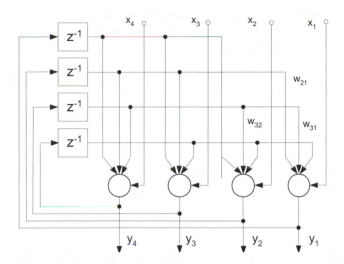

Figure 5.156. The recursive Hopfield net

Fig. 5.156 presents the *Hopfield net* – recursive network with feedback. The non-recursive networks are used for classification and optical character recognition. The recursive networks are used for optimization and prognosis.

It is necessary to minimize of the error function by "teaching" the network – by using the training procedure. The error function can be represented by the error plane with several local minima. At these local minima the process of learning could be stopped. Thus, it is necessary to organize the learning in a way, which allows reaching the absolute minimum. The learning efficiency depends on the speed of learning. If this speed is too large we can obtain a final result faster, but there is a risk of missing the absolute minimum. There were developed various algorithms of learning.

The problem of learning is rather complex in the case of the multilayer network. It is the problem of the error estimation in the hidden layer, because we only know the errors at the output layer (due to comparison with the training values). We do not know the reference values for the intermediate layers. Therefore for learning of multilayer network the *back propagation algorithm* (Fig. 5.157) is used.

In the back propagation method we determine the values of errors in the output layer and next the direction of data transmission is reversed. From the output layer the signal proportional to the errors is sent back and taking into account the activation function we can determine the errors of this layer. This way by the subsequent back error propagation towards the input terminals we determine errors of all layers and update the weights, thus achieving the "learning" of the neural network.

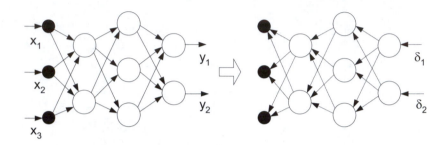

Figure 5.157. Forward propagation of the input values and back propagation of errors to update the weight values

Consider the optimization gradient descent algorithm, when the change of the weights is computed using the *delta rule*. According to this rule the weight change is

$$\Delta w_{ik} = -\eta \frac{\partial E}{\partial w_{ik}} \tag{5.125}$$

where η is the learning rate parameter (describing the speed of learning).

Taking into account equations (5.122) and (5.124) we can write

$$\frac{\partial E}{\partial w_{ik}} = \frac{\partial \frac{1}{2} \sum_{k=1}^{p} (y_i - d_i)}{\partial w_{ik}} = $$

$$= (y_i - d_i) \frac{\partial f\left(\sum_{k=1}^{n} w_{ik} x_k\right)}{\partial w_{ik}} = (y_i - d_i) f'(u_i) x_k = \delta_i x_k \tag{5.126}$$

where the delta error is

$$\delta_i = (y_i - d_i)f'(u_i) \qquad (5.127)$$

and the increment of the weight of the output layer is

$$\Delta w_{ik} = -\eta \delta_i x_k \qquad (5.128)$$

For the hidden (before the last layer) the delta error is

$$\delta_i^m = f'(u_i^m) \sum_{l=1}^{n_{m+1}} \delta_l^{m+1} w_{lj}^{m+1} \qquad (5.129)$$

thus we can compute the weight increment in the m layer knowing the error of the next $m+1$ layer. For computation of the weight increments in the hidden layer by back propagation it is necessary to calculate the derivative of the activation function $\partial f(u) / \partial u$.

The algorithm of the back propagation consists of the following steps (Silipo 2003):

a) initializing the weight w_{ik} values as the random values;
b) applying the training values x_q to the input layer;
c) propagation x_q values forward to the output layer (Eq. 5.122);
d) calculation of the error E of the output layer (Eq. 5.124);
e) calculation of the delta of the output layer (Eq. 5.125-5.127);
f) back propagation of the delta to the previous layer and computation of the delta of current layer (Eq. 5.129)
g) calculation of the weight increments
h) modification of the weight;
i) repetition of the same process for other training values.

Fig. 5.158 presents the example of application of the neural network to optical character recognition. The letters can be represented by a set of pixels (or generally the pictures). The neural network is first trained to link various sets of pixels with certain letters and then depending on the "intelligence" of the network the system is able to recognize more and less blurred or illegible letters (for example by writing the text by hand). Using this method also other elements can be analyzed – sounds for the speech recognition, defects in material evaluation, etc (Bishop 1995).

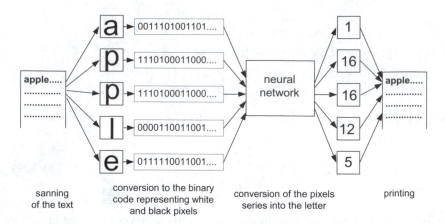

sanning of the text / conversion to the binary code representing white and black pixels / conversion of the pixels series into the letter / printing

Figure 5.158. Application of the neural network to the optical character recognition – OCR (Neuro Solution 2005)

Described neural networks require assistance of trainer (teacher, supervisor) who delivers input pattern and reference output signal (answer). Such method of learning is easily understandable. But it is surprising that there are neural networks that can learn themselves without the teacher. These networks are called *self-organizing* or *unsupervised networks*. As the basis of such networks the *Hebb's theory* (Hebb 1949) is utilized – it has been observed that the faster or more frequently stimulated neurons are privileged.

If the neural network is excited by the input patterns of some kind of organizations (for example the same letter but written with various font types) and we repeat the learning many times the neurons which meet the black pixels become privileged. There are various approaches of unsupervised learning – the most important are *competitive learning* and *Hebbian learning*. One of the competitive strategies is the *winner-take-all*. It means that the neuron that is fitting to the input pattern in the best way changes its weight. In the Hebbian strategy the most frequently excited neuron gives the strongest corresponding answer.

Fig. 5.159 presents the example of the most widely used self-organizing network called the Kohonen network (called also *self organizing feature maps – SOFM*) (Kohonen 1991, Kohonen 2000). In Kohonen map neurons are organized as a lattice (for example 2D lattice) where all inputs are connected to all neurons with appropriate weights. The self-organizing Kohonen algorithm consists of following steps: initially all weights have random values, next, to the inputs the learning pattern is sent, the winner neuron is found, which is most similar to the input pattern (the smallest error), the weight of the winning neuron (or eventually its neighbours) is modified, the process is repeated with new input pattern.

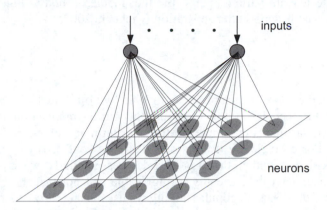

Figure 5.159. The example of the Kohonen network

Fig. 5.160 presents the example of a self-organizing network used in *blind signals separation*. Blind separation is able to separate mixed signals, although the network did not learn the mixed signals or the rule of mixing. The required condition is that the signals are statistically independent. One of the methods of separation is *independent component analysis ICA* .

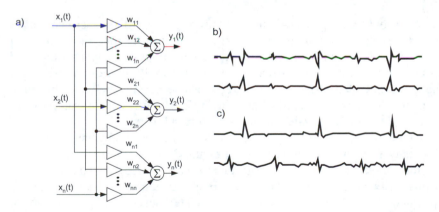

Figure 5.160. The example of the neural networks used for signals separation (a), mixed EKG signals of mother and fetus (b) and these signals after separation (c) (Cichocki 2002)

Fig. 5.160 demonstrates the separation of two signals of an electrocardiogram – EKG of mother and EKG of the fetus. Usually the signal coming from the mother covers the fetus signal as shown in Fig. 1.160b – the signals detected in various parts of the mother's body. Fig. 5.160c demonstrates both signals after separation (Cichocki 2002).

5.5.4. Fuzzy Logic

The term *fuzzy logic* was introduced in 1965 by Lofti Zadeh (Zadeh 1965) who stated that people describe various phenomena more notionally than digitally (we say that the people prefer to determine values as *linguistic variables*). For example we usually say that it is cold, frosty, warm, hot, mild, very cold and seldom that it is *22.2* degrees. Similarly, we say that the person is nice, agreeable, friendly, disagreeable and not that the person is agreeable in *70%*. Thus, in popular talking we use inexact, fuzzy values.

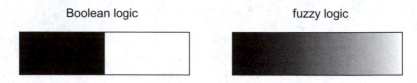

Figure 5.161. The same value described in Boolean logic and fuzzy logic

We should distinguish voluntary fuzzy logic from the described earlier uncertainty resulting from not precise measuring instruments. As the basis of fuzzy logic we assume that often it is not necessary to know exactly analyzed value or this value is very difficult to assess precisely. It does not mean that the value described using fuzzy logic is coarse in comparison with digital one. Just the opposite – as illustrated in Fig. 5.161 fuzzy logic enables us to determine the analyzed value in another, more subtle way.

An important date in fuzzy logic development was 1985 when the Hitashi Company simulations demonstrated the superiority of fuzzy logic systems for the railway control. Two years later such a system was introduced in Sendai with great success. The new control system of subway in Sendai appeared as very flexible, soft and without disturbances. Today fuzzy control is widely applied in Japan[1], also in popular appliances, as in washing machines or rice cookers.

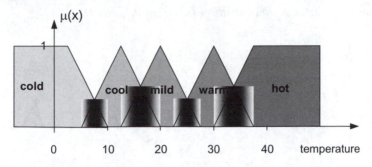

Figure 5.162. The example of description of temperature using fuzzy logic

The fuzzy set A is described by the membership function μ_A

$$A = \{x, \mu_A(x)\} \tag{5.130}$$

The membership function can be represented by various shapes: triangular, trapezoidal, Gaussian etc. Temperature in Fig. 5.162 is described by a triangular or trapezoidal membership function. Parameter $\mu_A=1$ means that this value is a complete member of this fuzzy set, for example temperature $0°$ is in *100%* cold, but *16°* is *50%* cool and *50%* mild (*0.5* membership). Triangular membership function can be described as

$$\mu(x) = \frac{\Delta x - |x|}{\Delta x} \tag{5.131}$$

We can perform the logical operation on fuzzy numbers (Fig. 5.163), for example *fusion* described by the membership function

$$\mu_{A \cup B}(x) = \mu_A(x) \vee \mu_B(x) = \max(\mu_A(x), \mu_B(x)) \tag{5.132}$$

or *intersection*

---

[1] The fuzzy logic is especially appreciated in Asian countries. It is sometimes ideologically justified as the opposition of the ambiguous Buddhism philosophy (*the world is filled with contradictions and almost everything contains some of its opposite*) and European Aristotle binary logic expressed in the work "*The law of the excluded middle*".

$$\mu_{A \cap B}(x) = \mu_A(x) \wedge \mu_B(x) = \min(\mu_A(x), \mu_B(x)) \qquad (5.133)$$

We can also perform *concentration* $(\mu(x))^2$ or *dilation* $(\mu(x))^{1/2}$.

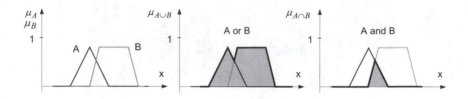

Figure 5.163. The logical operations on the fuzzy numbers

In the Boolean logic we use the expression: *IF x=A THEN y=B*. The fuzzy logic is expressed by the deduction logical rules *Modus Ponens* in the form

$$IF...AND...THEN$$

or

$$R:\ IF\ x_1\ IS\ A_1\ AND...AND\ x_n\ IS\ A_n\ THEN\ y\ IS\ B$$

For example we can form the following *fuzzy rule set*:

R1: IF age IS young AND car power IS high THEN risk IS high
R2: IF age IS middle AND car power IS medium THEN risk IS low
.....
RN: IF age IS old AND car power IS high THEN risk IS high

The fuzzy rule set can be used for the *fuzzy logic inference*. Consider the following example: we intend to control car speed (or better to control of "foot on accelerator pedal") as the dependence of the speed and distance from the car before us. We can construct the following fuzzy rule set:

R1: IF distance IS small AND speed IS medium THEN accelerator is not changed
R2: IF distance IS medium AND speed IS medium THEN accelerator can be pressed
R3: IF distance IS medium AND speed IS high THEN accelerator should be slow

Of course we can deduce much more rules.

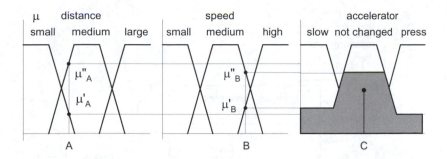

Figure 5.164. Inference about the speed change taking into account the distance A and speed B

The design of the resultant membership function of the accelerator is presented in Fig. 5.164. We can assume the distance *A* as small with coefficient μ'_A or medium with coefficient μ''_A. Similarly we can assume the speed as high with coefficient μ'_B or medium with coefficient μ''_B. We construct the resultant membership function as the AND rule –by choice of the smallest possible coefficients. As result we obtain the area (indicated as grey) as the sum of three possibilities: slow, not changed or press. We can determine the resultant "defuzzified" decision as the gravity center *C* of the grey area. The presented example is very simple – we can add several other variables, such as visibility, fogginess, age of driver etc.

Fig. 5.165 presents the application of fuzzy logic to the protection of a power transformer (Rosolowski 2002). After short-circuiting the protection device should disconnect the transformer. As the criterion we can use the value of the differential current $I_d(k)$. However, many other incidents can cause an effect similar to short-circuiting (increase of I_d) and there are many various kinds of short-circuiting with various levels of danger.

It would not be reasonable and expensive if after every such incident the transformer was switched off. Because the whole object is very complex, precisely describing all cases (and the answers related to this cases) would be difficult and could result in too frequent disconnection of the transformer (for trivial reasons). Using fuzzy logic we can describe many rules including rules related to various experiments. This way we can realize the protection system as more flexible and more subtle fitting to various incidents.

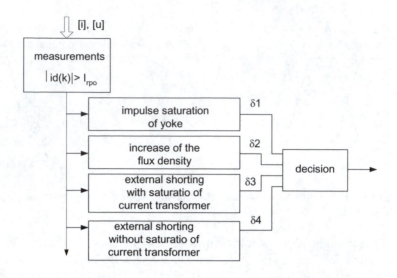

Figure 5.165. The application of fuzzy logic to the classification of shorting in the multi-criterion protection of a transformer (Rosolowski 2002)

Fuzzy logic and its engineering applications are described widely in many publications (Harris 2002, Kartepoulos 1995, Patyra 1996, Ross 2004, Tanaka 2003)

References

Addison P.S., Addison N. 2002, *The Illustrated Wavelet Transform Handbook*, IOP Publ.
AD 2005 Analog Devices, www.analog.com
Agilent 150 2000 *Spectrum Analysis Basics*, Agilent Application Note No.150
Agilent 243 2000 *The Fundamentals of Signals Analysis*, Agilent Application Note No.243
Agilent 150-15 2004 *Vector Signal Analysis Basisc*, Agilent Application Note No.150-15
Agilent 2005, Agilent technologies, www.agilent.com
Antoniou A. 2005 *Digital Signal Processing*, McGraw-Hill
Banks A.J. 1997 *Oscilloscope Guide*, Promt
Bellanger M. 2001 *Adaptive Digital Filters*, Marcel Dekker

Berhold M. and Hand D.J. (Ed) 2003 *Intelligent Data Analysis*, Springer

Bishop C.M. 1995 *Neural Network for Pattern Recognition*, Oxford Univ. Press

Bracewell R.N. 1999 *The Fourier Transforms and its Applications*, McGraw-Hill Science

Brighan E.O. 1988 *Fourier Transforms and its Application*, Prentice Hall

Burrus C.S., Gopinath R.A and Guo H. 1998 *Introduction to Wavelets and Wavelet Transform. A primer*, Prentice Hall

Candy J.C., Temes G.C. 1991 *Oversampling Delta Sigma Data Conversion*, Wiley & Sons

Cichocki A., Amari S.I. 2002 *Adaptive Blind Signal and Image Processing*, Wiley & Sons

Coley J.W., Tukey J.W. 1965 *An Algorithm for the Machine Calculation of Complex Fourier Series*, Math. Computation, 19, pp. 297-301

Czyzewski A. 1998 *Digital Sound* (in Polish), Exit

Deziel J.P. 2000 *Applied Digital Signal Processing*, Prentice Hall

Diniz P.S.R. 2002 *Adaptive Filtering*, Springer

Fettwes A. 1986, *Wave Digital Filters: Theory and Practice*, Proc IEEE, 74, pp. 270-327

Fluke 2005 Fluke Electronics, www.fluke.com

Forhang-Boroujeny B. 1999 *Adaptive Filters – Theory and Applications*, Wiley & Sons

Geerts V., Steyaert M., Sansen W.M.C. 2002 *Design of Multibit Delta-Sigma A/D Converters*, Springer

Goeschele D.F 1994 *Analog to Digital and Digital to Analog Conversion Techniques*, Wiley & Sons

Goswami J.C. and Chan A.K. 1999 *Fundamentals of Wavelets*, Wiley & Sons

Grewal M.S., Andrews A.P. 2001 *Kalman Filters – Theory and Practice using Matlab*, Wiley & Sons

Harris J. 2002 *An Introduction to Fuzzy Logic Applications*, Kluwer

Haykin S. 2001 *Adaptive Filters Theory*, Prentice Hall

Haykin S. 2001 *Kalman Filtering and Neural Networks*, Wiley & Sons

Haykin S. 2003 *Least Mean Square Adaptive Filters*, Wiley & Sons

Hebb D.O. 1949 *The Organization of Behaviour: a Neuropsychological Theory*, Wiley & Sons

Hickman I. 1997 *Digital Storage Oscilloscopes*, Newnes

Horne R. 2005, *Audio Spectrum Analysis*, www.visualizationsoftware.com

Jackson L.B. 1995 *Digital Filter and Signal Processing with Matlab Exercises*, Springer

James J.F. 2002 *A Students Guide to Fourier Transform*, Cambridge Univ. Press

Jespers P.G. 2001 *Integrated Conversion*, Oxford University Press

Kartepoulos S. 1995 *Understanding Neural Networks and Fuzzy Logic*, IEEE

Keiser G. 1994 *A Friendly Guide to Wavelets*, Birkhauser Boston

Khahn A. 2005 *Digital Signal Processing Fundamentals*, Charles River

Kohonen T. 2000 *Self-organizing Maps*, Springer

Kohonen T., Makisara K., Simula O., Kangas J. 1991 *Artificial Neural Networks*, Elsevier

Kularatna N. 2003 *Digital and Analogue Instrumentation*, IEEE Publ.

Lai E. 2004 *Practical Digital Signal Processing*, Newnes

Lawson S., Mirzai A. 1991 *Wave Digital Filters*, Prentice Hall

Lutovac M.P., Tosic D.V., Evans B.L. 2000 *Filter Design for Signal Processing using MATLAB and Mathematica*, Prentice Hall

Lyons R.G. 2004 *Understanding Digital Signal Processing*, Prentice Hall

Mallat C.G. 1989 *A Theory for Multiresolution Signal Decomposition: the Wavelet Representation*, IEEE Trans. on Pattern Analysis and Machine Intelligence, 11, n.7, pp.674-693

Mallat C.G., Mallat S. 1999 *A Wavelet Tour of Signal Processing*, Academic Press

Maxim 644 2000 *ADC and DAC Glossary*, Maxim Application Note No. 644

Maxim 2102 2003 *Migrating from Integrating ADC Architectures to Sigma Delta*, Maxim Application Note No. 2102

Maxim 2005 Maxim – Dallas Semiconductor, www.maxim-ic.com

Maybeck P.S 1979 *Stochastic Models, Estimation and Control*, Academic Press

McClellan J.H., Schafer R.W., Yoder M.A. 1998 *DSP First – a Multimedia Approach,* Prentice Hall

McCulloch W.S and Pitts W. 1943 *A Logical Calculus of the Idea Imminent in Nervous Activity*, Bulletin of Mathematical Biophysics, pp.115-133

Mitra S.K. 2002 *Digital Signal Processing*, McGraw Hill

NeuroSolution 2005 *What is the Neural Network*, www.neurosolution.com

Norsworthy S.R., Temes G.C. 1996 *Delta Sigma Data Converters,* IEEE Computer Society

Oppenheim D.V., Schaffer S.W. Buck J.R. 1999 *Discrete-time Signal Processing*, Prentice Hall

Qian Shie 2002 *Introduction to Time-frequency and Wavelet Transform*, Prentice Hall

Parks T.W. 1987 *Digital Filter Design*, Wiley & Sons

Patyra M.J., Mlynek D.J. 1996 *Fuzzy Logic – Implementation and Application*, Wiley & Sons

Polikar R. 2002 *The Wavelet Tutorial* , http://users.rowan.edu/~polikar

Proakis J.G Monolakis D. 1995 *Digital Signal Processing – Principles, Algorithms, Applications*, Prentice Hall

Rak R. 2004 *The Virtual Measuring Instruments* (in Polish), WPW

Rathore T.S. 2004 *Digital Measurements Techniques*, CRC Press

Ristic B., Arulamparan S., Gordon N. 2004 *Beyond the Kalman Filter*, Artech House

Rorabaugh C.B. 2005 *Complete Digital Signal Processing*, McGraw-Hill

Rosolowski E. 2002 *Digital Processing of Signal in Power Electric Control* (in Polish), Exit

Ross T.J. 2004 *Fuzzy Logic with Engineering Application*, Wiley & Sons

Sayed A.H. 2003 *Fundamentals of Adaptive Filtering*, Wiley & Sons

Schreier R. Temes G. 2004 *Understanding Delta Sigma Converters*, Wiley & Sons

Signal Recovery 2005 www.signalrecovery.com

Signal Recovery TN1005 2005 *What is a Boxcar Averager*, Signal Recoverey Technical Note No. TN 1005

Silipo R. 2003 *Neural Network*, Chapter 8 in Intelligent Data Analysis, Springer

Smith S.W. 2003 *Digital Signal Processing*, Newnes

Sneddon I.N. 1995 *Fourier Transform*, Dover Publ

SPRA669 2000 *TMS320C54x digital filters*, Texas Instruments

Stranneby D. 2001 *Digital Signal Processing*, Newnes

Tanaka K. 2003 *An Introduction to Fuzzy Logic for Practical Applications*, Springer

Tektronix 2005 Tektronix, www.tektronix.com

Thede L. 2004 *Practical Analog and Digital Filter Design*, Artech

TI 2005 Texas Instruments, www.ti.com

Tran Tien Lang 1987 *Electronics of Measuring Systems*, Wiley & Sons

Valens C. 2004, *A Really Friendly Guide to Wavelets*, http//perso.wanadoo.fr/polyvalens

Vaseghi S.V 2000 *Advanced Digital Signal Processing and Noise Reduction*, Wiley & Sons

Van de Plesche 2003 *CMOS Integrated Analog to Digital and Digital to Analog Converters,* Kluwer Academic Publishing

Walker J.S. 1999 *A Primer on Wavelets and their Scientific Applications*, CRC Press

Walnut D.F. 2001 *An Introduction to Wavelet Analysis*, Birkhauser Boston

Widrow B. Stearns S. 1985 *Adaptive Signal Processing*, Prentice Hall

Winder S. 2002 *Analog and Digital Filter Design*, Newnes

Wolfson Microelectronics 2000 *Wolfson Microelectronics Audio DAC Architecture* (white paper)
Young I.T., Gerbrands J.J., van Vliet L.J. 1998 *Fundamentals of Image Processing,* Delft University of Technology
Zadeh L. 1965 *Fuzzy Sets*, Information and Control, pp.407-428
Zaknich A. 2005 *Principles of Adaptive Filters and Self-Learning Systems*, Springer
Zarchan P., Musolf H. 2000 *Fundamentals of Kalman Filtering*, AIAA

6

Computer Measuring Systems

6.1. INTRODUCTION

A typical sequence of operations during the measurement experiment is as follows:

a) proposition of physical and mathematical model of observable fact or object,

b) conversion of the measured values into electrical signals,

c) conditioning of signals (linearization, amplification, filtering, etc.),

d) acquisition of the data (multiplexing, conversion to digital form),

e) processing of the data signals, assessment of uncertainty,

f) visualization of the results or transmission the data via the network.

Today, practically whole experiment can be performed without the activity of the researcher. Prior the measurement it is only necessary to prepare the model of investigated phenomena and to plan the necessary operations – the routine operations can be realized later under control of a computer system (Nawrocki 2005, Park 2003a, Putten 2003).

Consider the following example: we plan to investigate the properties of various magnetic yokes. The computer system can perform automatically (without the intervention of the researcher) the following operations:

a) connecting the investigated yoke to the excitation source and to the measuring equipment (the data acquisition board in computer or a measuring instruments connected through an interface to the computer),

b) performing the investigation according to the earlier planned algorithm – set desired values of excitation (flux density or magnetic field strength) and collect output data (voltages induced in the secondary coil, current in primary coil),

c) in more exact investigations – ensuring the excitation conditions, as in controlling the waveform of flux density to be sinusoidal,

d) processing obtained data – for example integrating the coil signal to determine the flux density B because $E = -nA\, dB/dt$ (n – number of turns, A - cross section area of the yoke), computing parameters as specific power loss or permeability, performing spectral analysis of the obtained data,

e) determining the characteristics of the investigated yoke, as hysteresis, magnetizing curve, dependence of the losses on flux density,

f) saving the obtained data and eventually sending these data through local net or Internet.

We can analyze the obtained results the next day (*off-line*) or control the whole process of measurement by Internet (*on-line*).

The measuring system can be composed of measuring devices connected through interfaces to a computer or almost all operations can be performed by the data acquisition board in the computer. Also, we can use a simple microcontroller instead of a computer. By *computer measuring system* we usually mean the set of tools, methods and operations (software and hardware) designed for realization of operations necessary to perform measurements: setting of excitations, collecting measured data, processing data, saving or transmitting data. The word *computer* in such a system is rather important. An example of a typical measuring system is presented in Fig. 6.1. Apart from the computer (which is the "brain" of the system) connections and interfaces operating as the nervous net of the system are also important.

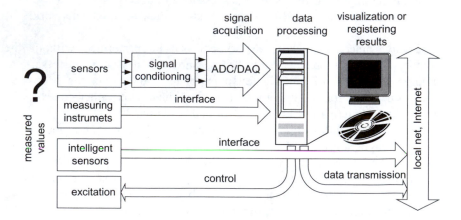

Figure 6.1. An example of the computer measuring system

The computer system can be designed and realized by a professional company – such systems are often used in the manufacturing process in factories. The user of such a system should accept, understand, know and use the documentation. However, nowadays the potential user (sometimes non-specialist in measurements) is also able to design and construct the measuring system. Most components of the system are prepared as standard devices, with well defined interfaces and software often in "plug and play" technology. The user should connect together all elements and design an appropriate program/software. Also in this area there is "user friendly" software enabling to design an appropriate program.

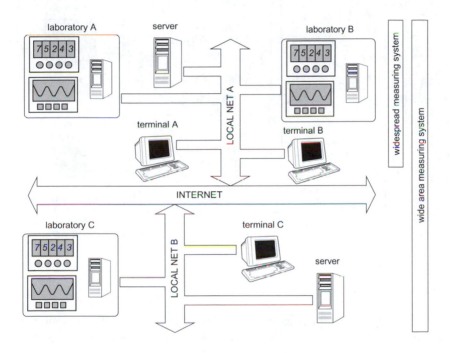

Figure 6.2 The widespread and wide area measuring systems

Due to the development of the data transmission systems the measurements are not limited to the laboratory area. Various systems (or various elements of the system) can be connected by the local net in the *widespread measuring system* (Fig. 6.2). Also, it is possible to connect systems (or elements of the system) via the global nets (e.g. Internet) as *wide area measuring system WAMS* enabling co-operation of various system positioned at the distances of thousands of kilometers. The factory on one continent can be controlled and managed from headquarters on the other continent.

Figure 6.3 presents the basic components of a typical measuring system. The sensors usually operate as the primary elements. Signals from the sensors are processed by the input elements of the system: conditioners (to match the signals to the system), data loggers and transmitters (for collecting the input signals), measuring instruments. Especially important is the data acquisition board DAQ because often its performance determines the system accuracy and quality. The conditioned and acquisitioned signals are transmitted within the system by appropriate interfaces. As presented in Fig. 6.3 there is rather large number of various interface systems.

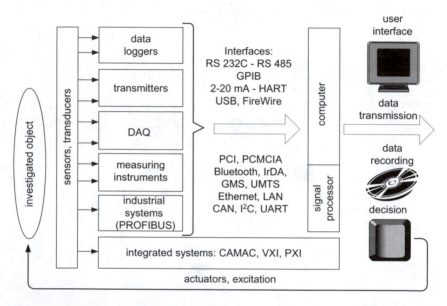

Figure 6.3. The main components of the measuring system

The *computer* operates as the central unit. Apart from the computer for the management of the system a *microcontroller* or specially designed for data processing type of microcontroller (*digital signal processor DSP*) can be used. By applying the DSP device it is possible to design a complete miniature measuring instrument or measuring system (even the measuring system in single chip – *measuring microsystem*).

The *real-time systems* form a special class of measuring system (Burns 2001, Cooling 2002, Liu 2000). A typical user of an ordinary computer system can believe that such a system is very fast. But for certain applications, for example for control of fast processes, a typical computer system can be too slow, because it is required to process the signals practically immediately after measurement to make the decision with minimal delay, called *latency*.

In popular operating system as MS Windows the user has limited possibility to control the timing in data flow. The real time systems use special kinds of instruments (hardware) with very fast processors and special kinds of operating system. The operating systems such as MS Windows with not enough fast input/output circuits, memory managing system or signal processing are not suitable for real time systems. Therefore, special operating systems for real time data processing, as QNX or iRMX were developed.

Apart from measuring systems designed for research work there are also special kinds of measuring/control systems used mainly for control of technological processes. Such systems are called *Fieldbus systems*, and the most known of such system used in Europe is the *ProfiBus* system.

6.2. INPUT CIRCUITS OF THE MEASURING SYSTEMS

6.2.1. Circuits for data conditioning and acquisition

Many sensors convert the measured value into the change of resistance (or impedance). The first step in signal conditioning is to convert this change into an electrical signal – voltage or current, convenient for further amplification and transmitting. The second important aim of signal conditioning circuits is to correct potential nonlinearity or temperature errors.

Since there are a lot of various sensors, often with various output signals, it is necessary to introduce certain systems of signal conditioning. Sometimes it is enough to amplify the signal to the level acceptable by most of analogue-to-digital converters, for example ± 5V or ± 10V. But often the current output signal is preferred.

One of the most popular standards of the output signal is the DC current in a range of *4 – 20 mA*. The advantage of such a signal is that it is independent of the changes of the resistance of connections (for example caused by the change of the temperature). Often, it is assumed that the change of output resistance in a range of *0 – 2kΩ* should not cause additional errors larger than the total error of the transducer. Under such assumption the data transmission at distances up to *2 km* is possible.

More advanced sensors are equipped with a standard digital interface, for example serial interface RS232C (or lately USB). Sometimes, wireless connections can be used as interface connections: infrared, mobile phone signal or Bluetooth.

The development of microtechnology enables us to introduce electronic circuits to the sensor. Such sensors are often called *smart sensors* or *intelligent sensors*. Such sensors can include conditioning circuit, linearization system, error correction system, serial interface or even network interface. To attain such performances sometimes the microcontroller circuit is included in the sensor. The term intelligent or smart is often abused for marketing purposes. Currently the IEEE 1451 standard describes the smart type transducers.

Another concept of signal conditioning is based on the assumption that the sensors cannot be expensive and it is simple and better to design versatile input circuits adapted to cooperate with various sensors. These circuits are called the transmitters – *two-wire transmitters* with current DC output or *digital transmitters*.

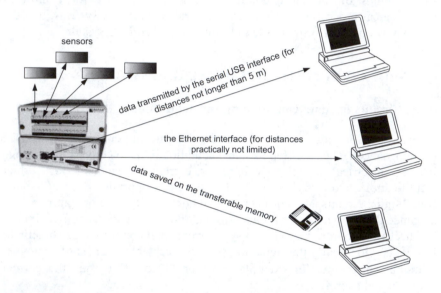

Figure 6.4. The methods of communication of the input conditioning device and the computer - as an example of data-logger DI-710 of Dataq Instruments (Dataq 2005)

Another strategy of cooperation with sensors is to register the data from the sensors, for example into solid state transferable memory elements or internal memory. Such devices called *data loggers* can be connected to the computer for transferring the data from time to time or can operate without a computer (*stand-alone data loggers*). Figure 6.4 presents various modes of data transferring from the data logger input circuit.

6.2.2. The sensors with built-in interface – intelligent sensors

In the intelligent sensors more and more often the communication interface is embedded, for example the serial interface. Figure 6.5 presents the intelligent type temperature sensor with a built-in 1-wire interface.

Figure 6.5 The temperature sensor *iButton* of the Dallas Semiconductor (Dallas Semiconductor 2005) (permission of Dallas Semiconductor)

The sensor called *iButton* developed by Dallas Semiconductor includes a semiconductor temperature sensor (silicon based) and with dimensions of the pile 17 mm in diameter, 6 mm high and 4 g weight converts temperature to an 8-bit digital signal. This digital data can be stored in internal memory (2 048 readings) or can be transmitted to the external devices using the 1-wire interface (and serial interfaces RS 232C or USB with special converter).

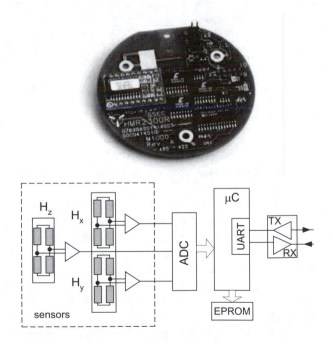

Figure 6.6. The three-axis magnetic field sensor *Smart Digital magnetometer* HMR2300 of Honeywell (Honeywell 2005) (permission of Honeywell)

Figure 6.6 presents another intelligent type sensor – the magnetoresistive MR thin film magnetic field sensor – a compass developed by the Honeywell Company. Integrated with three MR sensors (for three axes) microcontroller circuit controls the measurement operation, enables us to correct the temperature errors and additionally performs basic numerical calculations, such as components of magnetic field, direction of measured field, declination and inclination angle with resolution $0.02°$.

6.2.3 Analogue and digital transmitters

The *data transmitters* operate as intermediate circuits between sensors and computer (or data acquisition circuit). These circuits use various form of the data transmission – most often serial interface connection RS232C or USB, Ethernet connection or current loop $4 – 20\ mA$. This last case is called *two-wire transmitter*. In the case of serial interface it is easy to extend the output possibility to the modem or wireless connection.

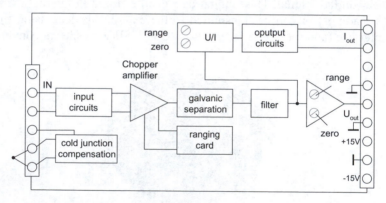

Figure 6.7. Functional block diagram of typical two-wire transmitter (Park 2003b)

Fig. 6.7 presents the typical transmitter/conditioner "two-wire" type. Because this circuit is designed for operate with thermocouple sensor it is equipped with additional temperature sensor for cold junction compensation. The sensor signal is amplified, isolated, filtered and converted to the $4 – 20\ mA$ output for data transmission. For co-operation with data acquisition board the voltage output $0 – 10\ V$ can be used. As an example of such transmitter we can point AD693 *Loop-powered 4 – 20 mA Sensor Transmitter* of Analog Devices (AD 2005).

Data transmitters are important part of measuring systems used for investigations of weather conditions. Fig. 6.8 presents the example of Multisensor Weather Transmitter developed by Vaisala Group (Vaisala 2005).

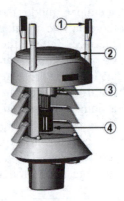

Figure 6.8. The Vaisala Weather Transmitter WXT510 (1 – wind transducers, 2 – precipitations sensor, 3 - pressure sensor, 4 – humidity and temperature sensors (Vaisala 2005) (permission of Vaisala Group)

Weather transmitter presented in Fig. 6.8 enables measurements of wind speed, wind direction, rainfall (including rain intensity and rain duration), barometric pressure, air temperature and humidity by means of originally developed sensors. The output signal is transmitted by serial connection (RS-232, RS-485, RS-422 or SDI-12 standards). Interesting extension of weather transmitter function is cooperation with GPS (Global Positioning System). It enables us not only precise determination of time and location of transmitters but also by measuring the delays in the GPS signals between several GPS satellites and several GPS met sites it is possible to estimate precipitable water vapor in the atmosphere (GPS meteorology).

6.2.4 Data loggers

Data loggers can operate in two modes – as *stand-alone data loggers* and *real-time data-loggers*. Fig. 6.9 presents a typical Data Logger Device.

Figure 6.9. Data-logger DT800 of DataTaker (Data Taker 2005) (permission of DataTaker)

In stand-alone mode the logger supplied by internal battery system collects the data into memory circuits and later transmits the data through typical interface. Often, the data are registered in memory and can be transmitted via memory card reader/writer. The functional diagram of a typical data logger is presented in Fig. 6.10.

In the real-time mode the data-logger operates like the data transmitter. The data logger presented in Fig. 6.10 uses the *UART – universal asynchronous receiver/transmitter* for connection between microcontroller and serial interface. The expansion connector enables us to connect several data-loggers in series. The input slots are four-terminal one – two terminals for input voltage, two terminals for supply of the resistive sensors.

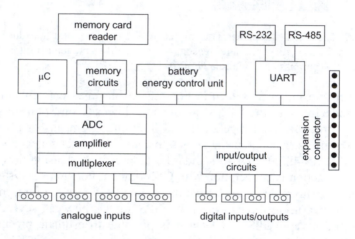

Figure 6.10 The functional block diagram of a typical data logger (Park 2003b)

Figure 6.11. Data-logger of DataTaker with extension module for receiving/sensing data by GSM interface in SMS format (DataTaker 2005) (permission of DataTaker)

Figures 6.9 and 6.11 present the data-loggers of DataTaker. These loggers can communicate with external unit via serial interface RS-232C, RS-485, USB or Ethernet. In the lack-of-network and even lack-of-electrical-power environment it is possible to communicate with the data logger by GSM telephony connection as SMS format (text messages).

6.2.5 IEEE P1451 standard – smart sensors

From 1985 work on new *standard IEEE-P1451* was started under the auspices of *IEEE Instrumentation and Measurement Society TC-9* and *National Institute of technology NIST* (IEEE 1451 2005). The main aim was to standardize the performances and requirements of smart type sensors in form of IEE P1451 standard titled "*A smart transducer interface for sensors and actuators*". Many important companies as Analog Devices (design of special microcontroller), National Instruments (design of special modules in LabVIEW), Honeywell, Agilent, and Signetics took part in this work, thus we can expect introduction of world-wide standardization.

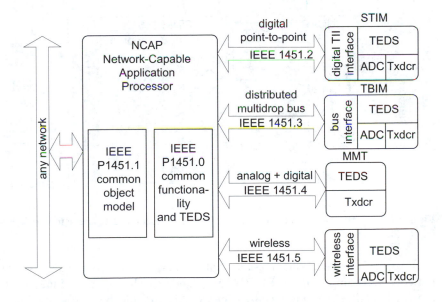

Figure 6.12. The main parts of IEEE 1451P standard (TXdcr – transducer of sensor or actuator)

Currently, the P1451 standard consists of four parts and it is planned to extend it by next part P14551.5 describing wireless communication with the sensors. In 2004, the work started on P14551.6 part of the standard concerning the communication between P1415 standard and CAN interface. Additionally, there is an IEEE 1588 standard for precise clock synchronization protocol for networked measurement and control systems.

Part 1451.1 describes the *NCAP – Network Capable Application Processor*. In this part there are defined hardware and communication protocols enabling the communication of the sensor with various types of computer networks. Part 1451.2 describes the "plug and play" standard of the sensor (incorporation to each sensor the *TEDS – Transducer Electronic Data Sheet* module) and module of STIM type. The *STIM – Smart Transducer Interface Module* is designed for communication between sensor and the NCAP module.

The 1451.3 standard is developed for a distributed multidrop transducer bus or network (in this case it is used TBIM – transducer bus interface module. The 1451.4 standard defines an interface specification for analogue transducers (including *4 – 20 mA* interface). Fig. 6.12 presents the basic components of the IEEE P1415 standard.

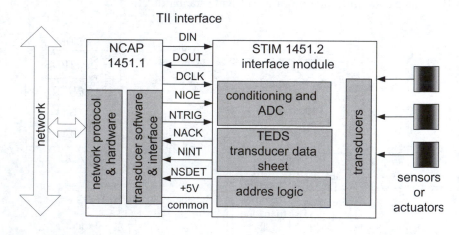

Figure 6.13. Network connection of the transducers according to the standard IEEE P1415.2

Fig. 6.13 presents the communication between sensors and external network according to the IEEE P1415.2 standard. The sensors connected to the SIM module are interfaced to the NCAD module by a 10-wire *TII – transducer independent interface*. This interface utilizes the following lines:

DIN (Data In) – address and data from NCAP to STIM,
DOUT (Data Out) – data from STIM to the NCAP,
DCLK (Data Clock) – clock line for synchronization DIN and DOUT,
NIOE (Output Enable) – signals that the NCAP is driving the data line,

NTRIG (Trigger) – for triggering the STIM (the signal of read or write request),

NACK (Acknowledge) – STIM acknowledge the trigger or data request,

NINT (Interrupt) – STIM send a request to the NCAP,

NSDET (STIM Detect) – used by NCAP to detect an attached STIM,

COMMON – signal ground.

The NCAP module performs network communication. It can be used as a node controller in a multiple-node network and can communicate with other NCAP nodes via the client/server protocol or publish/subscribe protocol. It contains a controller and the interface to the broader network that can support other nodes. For a DeviceNet control network can be used an 8-bit microprocessor, for Ethernet based network a 32-bit microprocessor. From TEDS the NCAP "knows" the data format and number of channels.

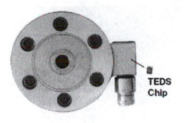

Basic TEDS	Manufacturer	Sensotec
	Model number	41
	Serial number	462992
	Version letter	53e
Standard and extended TEDS	Calibration date	April 22, 2002
	Measurement	200kN
	Response time	0.0005
	Sensitivity	1.998 mV/V
	Bridge impedance	350 Ohms
	Excitation nominal	10 V
User area	Sensor location	23 right dyno
	Calibration due	April 21, 2003
Templates	Special calibration	12.3+0.175
	Wiring code	15

Figure 6.14. The sensor of Sensotec Honeywell with TEDS chip and an example of the TEDS data sheet (Armson 2003)

The STIM module realizes the following functions: receiving the data across the interface, learning from the TEDS the data structure assembling the data, conversion of the unit according to specification of the TEDS, application programming interface provides access to the TEDS module, sensor readings, actuator control, triggering and interrupt request.

Plug and play standard means that the user can plug the sensor into the signal conditioning circuit and the system is automatically configured and ready to take the measurements. Therefore, important role plays the TEDS module which contains the most important information about the sensor. This module can be inserted as the EPROM directly in the sensor, in the cable part or virtually – delivered from the Internet. The standard data contains the main parameters, but in extended form the TEDS can contain more information, for example information about the linearization of the sensor.

Several manufacturers designed special STIM modules (for example TMI 1451.2-SA of TMI CogniSense), as well as network NCAP modules (for example TMI 1451.2-Na). With such modules it is possible to design wide area distributed networks with sensors or actuators connected via Internet, which can be controlled by master computer using typical Java Applet, as demonstrated in Fig. 6.15.

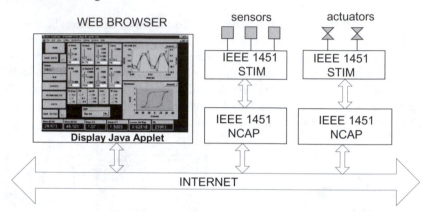

Figure 6.15. Wide area distributed network composed by application of IEEE P1451 components

The wide area computer system similar to presented in Fig. 6.14 was tested by NIST as the *DMC distributed measurement and control* system (Schneeman 1999).

6.3. DATA ACQUISITION CIRCUITS - DAQ

6.3.1. Plug-in data acquisition boards

The data acquisition devices are often available in a form of the *plug-in acquisition board* connected to a computer by PCI interface or PCMCIA interface.

Usually, the resolution (number of bits) and the speed (frequency of sampling) are considered and compared as the main DAQ performances. The most frequently used data acquisition boards exhibit the resolution 12-bit or 16-bit and sampling frequency *100 – 250 kSPS*. Such parameters are sufficient for many applications.

The prices of the DAQ increase significantly with the improvement of the considered parameters. For example, a typical high-quality DAQ of 16 bits and *200 kSPS* would cost about several hundred dollars while the sample-and-hold DAQ 16 bits and *800 kSPS* costs ten times more. The hardware is as important as software – it is important to have the drivers for typical program platforms, such as LabVIEW. Unfortunately, usually the DAQ operate without problems mainly with the software developed by the DAQ manufacturer.

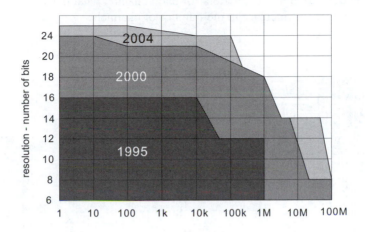

Figure 6.16. The progress in manufacturing of the DAQ (National Instruments Catalogue)

Fig. 6.16 presents the progress in development of the acquisition boards. More expensive cards offer better parameters and the most expensive are the DAQ with large sampling frequency and high resolution (for example 18 bit with *1 MSPS*). As can be seen from Fig. 6.15 the high resolution DAQ (24 bit and more) are usually used for DC and small frequency signals, while for high frequency 12/14 - bit DAQ are the best.

Fig. 6.17 presents the functional diagram of a typical plug-in data acquisition board (the photo of such device is presented in Fig. 1.6). Depending on the price, the DAQ can offer more or less enhanced options. More expensive boards have additional sample-and-hold circuits for all inputs, programmable change of gain (in cheaper boards the gain can be set by micro-switches). Typically there are 8 symmetrical (differential) or 16 unsymmetrical (single ended) inputs, but there are boards even with up to 64 inputs. Better boards are equipped with digital-to-analogue converters and a larger amount of memory to save the sampled data.

Usually, an unipolar or bipolar voltage signal in a range $\pm 5V$ or $\pm 10 V$ is used as the input signal. Often, in the board there is a programmable amplifier PGA with the gains *10, 100, 500 V/V*. The gain can be set to different values for various channels. It should be taken into account that the increase of the gain can result in decrease of sampling frequency (for example in popular KPCI-3108 DAQ of Keithley the throughput is *100 kSPS* for gain *1 – 10 V/V* but it decreases to *20 kSPS* for the gain *500 V/V*). Similarly, if we use more input signals the sampling frequency is divided between these inputs. Therefore most enhanced DAQs instead of multiplexer use analogue to digital converters for each channel what does not deteriorate the speed with the increase in the number of channels.

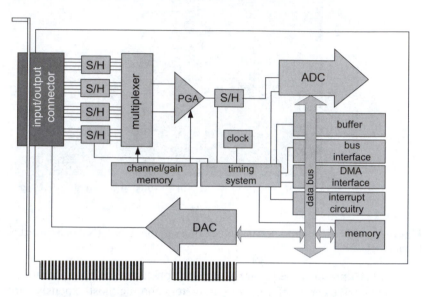

Figure 6.17. The functional diagram of a typical DAQ board

If we use several input signals each signal is sampled after some time necessary to sample the previous signal. Thus between first and the last signal there could be a time shift even of a few μs. This time delay can cause significant phase error, for example when we multiply the first and the last signal to determine the power $IUcos\varphi$. To eliminate this error it is necessary to use a sample-and-hold device, which is rather expensive and sometimes it is better to use two ADC operating at the same time, as it is for example realized in power/energy converters (see Fig. 5.148).

In default data conversion of multiple channels the *paced mode* of sampling is used, where the rate at which the single channel is sampled results from the pacer clock rate divided by number of channels. Instead of paced mode also the *burst mode* can be used (also termed as *pseudo sample-and-hold mode*). In the burst mode the signal is sampled with the maximum possible sampling of the board. Therefore the time delay between the samples is minimized.

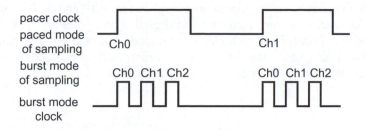

Figure 6.18. Paced mode and burst mode of sampling

The burst mode requires fast transfer of the data to the computer. As in paced default mode the data transfer can be performed in the interrupt mode, while in the burst sampling the DMA data transfer is required.

6.3.2. External data acquisition board

Recently, instead of plug-in DAQ external boards are also being developed. Such boards can be connected to the computer via USB or FireWire interface, which can be valuable in the case of laptops. Figures 6.19 and 6.20 present such an external board of Data Translation.

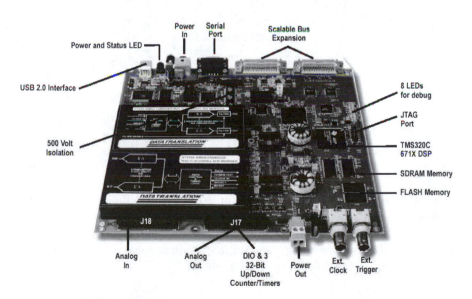

Figure 6.19. External DAQ board model DT9840 of Data Translation (Data Translation 2005)

The data acquisition system presented in Fig. 6.19 and 6.20 is expanded with enhanced memory elements and digital signal processor TMS320C of Texas Instrument. Thus such DAQ can operate practically in real time mode without a computer – the data can be saved in internal FLASH memory and next transferred to the computer via USB interface.

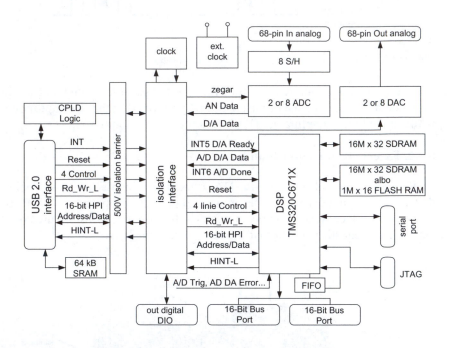

Figure 6.20. The functional block diagram of DT9840 data acquisition system

The presented DAQ system is manufactured with 24-bit delta-sigma ADC or 16-bit SAR ADC. Although both converters can operate with maximal sampling rate *100 kHz* for fast signals the SAR converter is recommended with a delay time error not larger than *10 μs*. The delta-sigma converter exhibits a time delay of about *400 μs*.

The embedded digital signal processor TMS320C671x – 150 MHz can be used not only for controlling the data acquisition process, but also for the data analysis, for example for spectral analysis.

6.4. DATA COMMUNICATION IN COMPUTER MEASURING SYSTEMS

6.4.1. Interfaces, buses and connectors

It is necessary to organize the communication between elements of the computer measuring system, often communication in both directions (Park 2003a). By the communication we mean not only the transferring of the pure data, but also sending messages, commands, instructions and synchronization signals. Fortunately, most of communication systems are standardized and such standardized systems are called *interface* (Axelson 1999, Axelson 2005, Caristi 1989, Gook 2004).

By *interface system* we usually mean the standardized equipment and programs designed for communication between two (or more) independent computer devices. The interface consists of hardware (wires, plugs, sockets) and software (languages and codes) called sometimes protocol. The *protocol* is a common set of rules governing the exchange of the data between the transmitter and the receiver in the communication network.

The elements of the system are connected with cables. A special type of connection is the interface bus. The *interface bus* is a connection subsystem that transfers the data between computer components. The bus consists of wires – it can be even a single wire – in such case we talk about a *bus line*. The data represented by *digital word* composed from bits can be transferred bit after bit through *serial interface* or through *parallel interface*, in which all the bits are transferred at the same time using several lines. Depending on the purpose we use various buses as data bus, address bus, etc.

Theoretically, for the serial data transfer it is sufficient to have only two wires (even one wire if we do not take into account the ground common wire). But usually in serial interfaces several wires are used with independent wires for data and commands (20 wires in RS interface) or data and supply voltage (4 wires in USB interface). The parallel interfaces usually use more lines because apart from the command lines the data words require several lines (8 bits in GPIB interface and even 64 bits in PXI system). The parallel interface is usually faster, but modern serial interfaces USB and FireWire enable the data transmission with speed comparable to the parallel interfaces.

The RS-232C is widely used as a serial interface. It is advantageous that almost all computers are equipped with such an interface. But RS-232C is an old-fashioned system and currently is often substituted by the USB interface. Among parallel interfaces for measurement data transmission more often the GPIB interface is used.

6.4.2. Serial interfaces: RS-232C and RS-485

The *RS-232C interface* is still most commonly used interface although it was designed in 1962 as RS-232 interface (*RS – Recommended Standard*). This interface was slightly modified in 1969 and is known as RS-232C. The RS-232C is USA standard and also is described in international standards (as V24 and V28) (Axelson 1999).

Because it has been designed a long time ago its performance is rather poor – speed of transmission not faster than *20 kbps (bps – bits per second)*, connection not longer than *15 m*, poor immunity to interferences. But it is still in use because it is traditional equipment of computers – as so-called COM serial port. Moreover, its relatively poor performances are sufficient in many applications. In comparison with the GPIB parallel interface it is useful that the connection cable is simpler and it is not necessary to install special card and drivers to the computer.

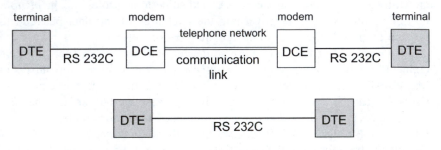

Figure 6.21. Connection of two data terminals using RS-232C interface

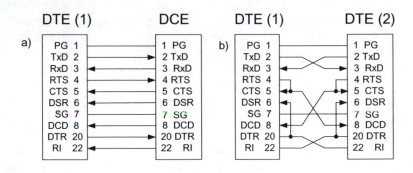

Figure 6.22. The standard RS-232C connection (a) and the example of null-modem connection (b)

Fig. 6.21 presents the typical structure of data transfer via RS-232C interface. Because the RS-232C was designed for telecommunication purposes typically two data terminals DTE (*DTE – Data Terminal Equipment*) are connected through modems DCE (*DCE – Data Communication Equipment*). It is also possible to connect directly both DTE data terminals – for example the measuring instrument and the computer. In such cases it is necessary to slightly modify the cable in order to obtain a so called null-modem cable (Fig. 6.22).

The RS-232C bus uses 20 lines in the full version: 4 lines of data, 11 lines for control, 3 lines for synchronization and 2 lines for ground. Not all lines are used – normally only 6–7 lines are used and in the simplest case it is possible to use only two lines for data and one for ground (Fig. 6.23).

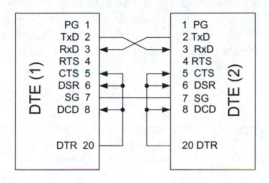

Figure 6.23. The simplified 3-wire connection via RS-232C interface

Table 6.1. The main lines in RS-232C bus interface

Pin No DB-9	Pin No DB-25	Code	Description
-	1	PG	Protective Ground
5	7	SG	Signal Ground
3	2	TxD	Transmitted Data
2	3	RxD	Received Data
7	4	RTS	Request to Send
8	5	CTS	Clear to Send
6	6	DSR	Data Set Ready
4	20	DTR	Data Terminal Ready
1	8	DCD	Data Carrier Detected
9	22	RI	Ring Indicator
	15	DB	Transmitter signal timing
	17	DD	Receiver signal timing
	24	DA	Transmitter signal timing

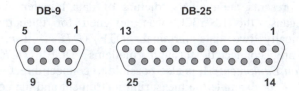

The RS-232C interface uses D-9 or D-25 pin connectors. The conversion between 9-pin connector and 25-pin connector can be easy realized using special adapter. The description of the main lines of RS-232C interface is presented in Table 6.1.

The amplitude of the signals can be varied in a wide range - low level corresponds with the voltage *-15 – - 3V* and high level corresponds with the voltage *3 – 15 V*. In the data transmission high level is related to the logical *1*, low level is related to *0*. In the control lines this relation is inversed. Fig. 6.24 presents the typical word of RS-232C interface.

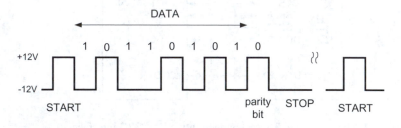

Figure 6.24. The serial data unit in asynchronous transmission

There are two types of serial communication – synchronous and asynchronous. In the *asynchronous* system the message is send in a form of well defined series of impulses called *SDU – serial data unit*, where first bit means the start of the transmission, next 7 or 8 bits are the data bits, next bit is used to detect the error of transmission (parity bit) and last bit (or two bits) means the end of transmission. An asynchronous frame format is presented in Fig. 6.24.

The parity bit (that may be even or odd) informs about total number of data bits – logic *1* of parity bit means that the total number must be an even number. Statistically such a method of error detection has only a 50% chance of detecting an error. That is why more sophisticated error detection systems (for example *CRC – cyclic redundancy check* with 99.99% of chance of error detection) substitute the parity method.

The common data transmission rates used by RS-232C are: *110, 300, 600, 1200, 2400, 4800, 9600* and *19200 bps*. Data rate is usually measured in *bps (bits per second)*. The term *baud rate* is also used and means how many times the physical signal changes per second. The cable length depends on the baud rate and for *19 200* it is limited to *50 m* (for baud rate *4800* the maximum cable length is *100 m*, and for *300* is *800 m*).

Usually, before asynchronous connection both transmitter and receiver should be adjusted in the following format: *BAUD=* (for example 9600), *PARITY=* (e- even, o – odd, n- none), *DATA=* (7 or 8 bits), *STOP=* (1 or 2 bits).

In the *synchronous* mode the receiver and the transmitter synchronize initially clock pulses, which enables the receiver to maintain the synchronization throughout large message, typically *4500* bytes (*36 000* bits). Usually there is no special clock signal line and the synchronization is derived from the data signal. The data bits and accompanying communication bits are sent in the form of a continuous stream called a frame.

FLAG	ADDRESS	CONTROL	DATA	FCS	FLAG

Figure 6.25. The example of the synchronous system frame format (HDLC protocol)

There are several various transmission protocols. One of them, called *HDLC – high level data link control,* is presented in Fig. 6.25. The frame starts with flag bits, next are two fields: address field and control field. After data transmission there is a field reserved for *FCS – Frame Check Sequence* used for the error detection.

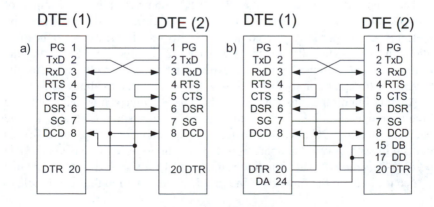

Figure 6.26. Two examples of device connection via RS-232C interface: asynchronous (a) and synchronous (b)

Although synchronous system is more efficient and enables to send more information at high data rate it is mainly used in telecommunication. For data transmission most often a much simpler asynchronous system is used. Figure 6.26 presents two examples of device connections via RS-232C interface.

The data transmission can be realized as a *simplex* one (sending messages in one direction only), *full-duplex* (data sending in both directions simultaneously) or *half-duplex* (data sensing in both directions but only in one direction at a time). The half-duplex mode is most often used for data transmission.

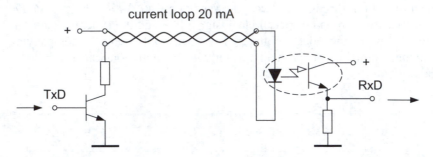

Figure 6.27. The application of current loop in RS-232C interface

One of the drawbacks of RS-232C interface is relatively short distance – about *15 m*[1]. The current loop method can be used in order to increase that distance. The current loop (application of the voltage/current converter) enables us to transmit the signal at a distance of about *2 000 m*. Fig. 6.27 presents the circuit with the current loop *20 mA*. In this loop the optical galvanic separation is used.

The main drawbacks of the RS-232C interface (short distance of transmission data, connection only two devices, and influence by interferences) is partially overcome in the RS-485 interface introduced in 1983. The significant difference between RS-232C and RS-485 is that the transmission is achieved with the differential lines, which increases the distance of transmission to about *1200 m* with a data rate to about *10 Mbps*.

Another important enhancement of the RS-485 interface is that due to the three-state operation it is possible to connect to the system *32* drivers and *32* receivers (in RS-232C only two devices can be connected). The three-state is realized in such a way that apart from logic *0* and logic *1* there is also a state of "*high-impedance*" when the line is disabled. This disabled state is initiated by a control pin on the line driver integrated circuit. Although it is possible to

---

[1] Although at the transmission rate *19 800* bauds it is possible to send the signal to about *50 m*, the standard RS-232C limits this distance to *15 m* as guaranteed length.

connect *32* devices only one device can be active (transmitting) at any given time.

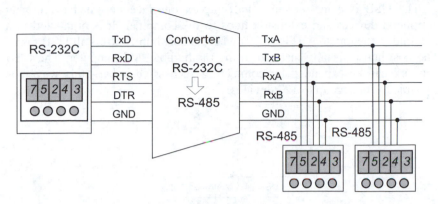

Figure 6.28. The measuring system with the RS-485 interface

The RS-485 interface is a significant improvement on the old RS-232C system. The drawback of the RS-485 is that it is necessary to install an additional card and driver in the computer. This obstacle can be omitted by addition of the RS 232/RS 485 adapter, as illustrated in Fig. 6.28.

6.4.3. Serial interfaces: USB and FireWire

The serial interface *USB* (*Universal Serial Bus*) was introduced in 1998 by Microsoft, Intel, Compaq and NEC (the enhanced version USB 2.0 in 2000) (Axelson 2005, USB 2005, Wooi Ming Tan 1999, Garney 1998). Primarily, the USB system was proposed as the solution of the problem of multi-cable labyrinth of periphery devices connected to the computer. The simplicity of the system and its efficiency results that today the USB system is widely used as a valuable interface system and many of the devices (including measuring instruments as oscilloscopes and data acquisition boards) are equipped with this interface. Currently, there are available parallel GPIB interfaces with USB communication port. Probably in the near future this system will substitute for the RS-232C interface.

The main advantage of the USB interface is high speed – *12 Mbps* in USB 1.0 and *480 Mbps* in USB 2.0. It is important that practically all computers are equipped with this interface. Another benefit of this system is also simple and inexpensive cable. The drawback is that the length of the cable is limited to *5 m* (although on the market there are devices enabling increase of the cable length to *25 m* with a special amplifier[1]).

The USB is a master/slave, half-duplex interface designed to connect peripheral devices and extension hubs to the computer. It is organized in a pyramid shape topology (Fig. 6.29) where a host hub controls all USB ports. The host hub is usually inserted in the motherboard of computer. The USB network can be expanded by connecting successive hubs – this way it is possible to connect up to 127 devices.

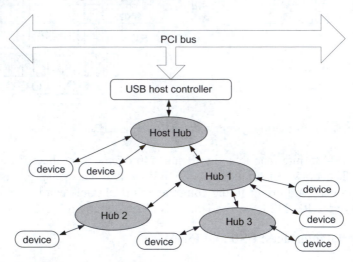

Figure 6.29. The topology of the USB interface

Every device after connection in "plug and play" mode obtains a description with information about required speed rate, data transfer mode, size of data packet, priority and address. The whole communication is organized and controlled by host hub that initializes and configures device port, enables or disables the port, recognizes if the device is connected, creates data packets and frames, sends the information to the bus, waits and recognizes response, tests the error.

In regular time intervals the host hub controller sends *token packets*. There are four types of token packets: start of frame, in packets, out packets and setup packet. The *start packet* informs the device that it is the beginning of *1 ms* of frame. The *packet in* transfers the data from the device to the computer whilst the *packet out* transfers the data from the computer to the device.

---

[1] Standard does not recommend such solution because *5 m* limit is due to timing and not level of the signal.

SYNC	PID	DATA	CRC	EOP
synchronisation	identification			end of packet
8 bits	2 x 4 bits	0...1023 bits	16 bits	4 bits

Figure 6.30. Example of the data packet in USB system

The data packet is formed as *1 ms* frame – more than one device can send a packet inside one frame. An example of the USB packet is presented in Fig. 6.30. It consists of a *synchronization* sequence, packet identification, data, CRC error test and EOP – end of packet.

The data transmission can be realized in four modes: interrupt transfer, isochronous transfer, control transfer and bulk transfer. The *interrupt transfer* is used for the device that does not need continuous and frequent data sending (for example a keyboard). Because it is not possible to initiate the data transfer by the device the host hub initiates poll of a device if such device needs periodic attention. This requirement of polling is established during setup of the device. For example keyboard is polled every 100th frame.

The *bulk transfer* is used for the devices that are not time dependent but need to transfer large packets of data (for example a printer). The *isochronous transfer* is time dependent (for example the data from an audio CD). The data are sent in 1023 bit packets in synchronous or asynchronous mode, but with synchronization of transmitter and receiver. There is no limit of data packets. The *control transfer* is used for configuration and the requested device must respond back.

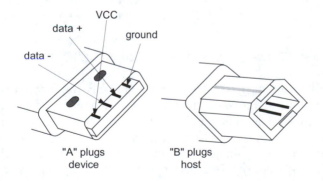

Figure 6.31. The USB connectors

The cable of a USB interface consists of two wires for data communication and two wires for power transmission. It is possible to supply the external device via the USB cable. There are two kind of connectors, type A and type B (Fig. 6.31). The power pins are slightly longer (*7.41 mm*) in comparison to the signal pins (*6.41 mm*), which means that the sequence of connection is established (supply voltage is connected first).

The *FireWire Interface* (known also as *IEEE-1394 standard*) was introduced by Apple as FireWire 400 (IEEE-1394a) in 1990 (FireWire 2005, Anderson 1998). It enabled data transmission with rate *400 Mbps* by the 6-wire cable with *4.5 m* length. It is possible to extend the total length to about *72 m* as daisy chain. In 2003 the enhanced version FireWire 800 (IEEE-1394b) was introduced with transmission speed *800 Mbps* through the 9-wire cable of total length to 100 m. The next version IEEE-1394c is planned as a *3.2 Gbps* interface with application of the optical connection.

A FireWire interface can connect up to 63 external devices in network structure. Like the USB, the FireWire supports "plug and play" technology and "hot-plug" mode (possibility of connection of the external device without restart of the computer). It enables users to supply external devices, with power consumption to *45 W*. Six-wire cable utilizes two pairs of wires for data transport and two wires for power (although in camcorder applications the connection is 4-wire without power supply) (Fig. 6.32).

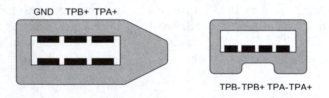

Figure 6.32. The IEEE-1394 six-pin and four-pin socket

Although there are many similarities between USB and FireWire systems several features make FireWire interface more interesting for measuring applications. FireWire accepts multiple hosts per bus (in USB system whole communication is controlled by host hub), which makes it possible that the data transfer can be initiated by the connected device. Also the communication between devices and without the computer (peer-to-peer connection) is permitted.

Currently, the FireWire interface enables serial data transmission with the highest speed and supports isochronous data transport (practically in real-time mode). This means that it is commonly used for very fast external devices, as for example with external hard disks. Starting in 1995 practically all digital camcorders use this interface – it is known also as *iLink* interface (name used by Sony).

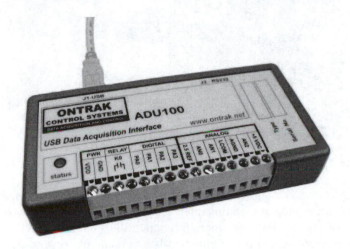

Figure 6.33. An example of the data acquisition interface model ADU100 of Ontrak (Ontrak 2005) (permission of Ontrak Control System Inc.)

The FireWire was introduced earlier than USB system. It is not as widely used as USB 2.0 although it enables faster data transfer, supplies larger power, works without host computer and works in most flexible network connection of devices. But because the laptop computers have become more and more popular and they are often equipped with a FireWire port, several manufacturers introduced data acquisition systems with both USB and FireWire interfaces. As the example we can point to the NI-DAQPad-6052E data acquisition FireWire system with 16-bit resolution and *333 kSPS* sampling rate 16 analogue inputs DAQ. There are available analogue-to-digital converters with USB output for example USB-9803_EC-I of SuperLogics with isolated 16-bit *100 kSPS* ADC (Superlogics 2005). Figure 6.32 presents the example of data acquisition interface designed for 16-bit resolution conversion of analogue or digital inputs to the USB format output.

6.4.4. Parallel GPIB interface (IEEE-488/IEC-625)

In 1965 Hewlett-Packard introduced a parallel interface designed specially for the measuring instruments. This interface called *HPIB – Hewlett Packard Interface Bus* was ten years later standardized as *IEEE-488* American standard or *IEC-625* international standard. It is also known as *GPIB – General Purpose Interface Bus*.

Figure 6.34. The connector of GPIB interface

An important advantage of GPIB is that almost all higher class measuring instruments are equipped with this interface. The GPIB enables us to connect up to 15 measuring devices (including controller). With a well defined transmission protocol and list of commands it is relatively easy to design the computer measuring systems operating with this interface. The speed of data transfer is limited to *1 MBps* (in special realization called High-Speed GPIB to *8 MBps*). The speed of data transfer depends on the cable length and although the length *20 m* is allowed, a *2 m* length is recommended.

The GPIB interface uses a special connector (Fig. 6.34) with stackable back-side slots that enables connection to other devices. The devices can be connected in star type network or chain (linear) type net (Fig. 6.35).

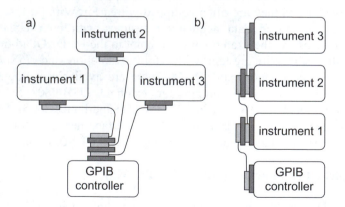

Figure 6.35. Connection of instruments in star net (a) and chain net (b)

There are several drawbacks of the GPIB interface. It is necessary to use relatively expensive and short 24-wire cable, it is necessary to install additional GPIB driver card in the computer. Moreover, it is slightly old fashioned and its performance in comparison with the modern serial FireWire interface is not as attractive as 30 years ago when it was designed.

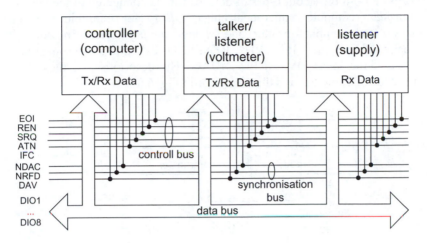

Figure. 6.36. The example of GPIB bus structure

Table 6.2. The main lines of GPIB interface

pin	name	description
1-4	DIO1-	Data
13-16	DIO8	
5	EOI	End Or Identify
6	DAV	Data Valid
7	NRFD	Not Ready For Data
8	NDAC	Not Data Acepted
9	IFC	InterFace Clear
10	SRQ	Service ReQuest
11	ATN	ATteNtion
12		Shield
17	REN	Remote ENable
18-23		Grounds

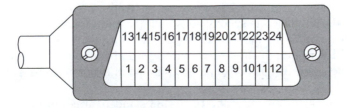

Figure 6.37. The pin assignment in GPIB connector

The GPIB bus consists of three lines of synchronization bus (*handshaking lines*), five lines of control bus (*general interface management*) and eight lines of data bus (plus several ground lines). The data bits are transferred as words of 7 – 8 bits. Four types of devices can be connected to the GPIB bus: *controller* (computer, microcontroller), *talker* (for example a counter), *listener* (for example a printer) and *talking/listening device*. The control bus consists of five lines: IFC, ATN, REN, SRQ, EOI and synchronization bus utilizes three lines: DV, NRFD, NDAC. The description of these lines is presented in Table 6.2 and the connector in Fig. 6.37.

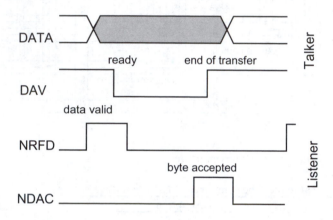

Figure 6.38. The GPIB handshaking sequence

Figure 6.38 presents the data transfer sequence of handshake signals. The talker waits until the listener changes the NRFD line to "high" which means that it is ready to receive the message. Then the talker sends byte of the data to DIO lines and next changes the level of NAV indicating that the valid data is available on the bus. After this signal the listener reads the data and changes the NRFD line indicating that no further data should be sent. When all listeners have read the data the NDAC is set to "high" which means that the last listener accepted data. After this, talker changes the DAV line and removes the data from the bus. The interface is ready to send next data after change of NDAC, which is synonymous with the end of data transfer. After another change of NRFD the listener is ready to receive new data.

The full description of commands, messages and communication protocol is beyond the scope of this book and is described in detail elsewhere (Caristi 1989).

In 1987 an enhancement of the IEEE-488 standard was introduced. This standard slightly improved the interface called now IEEE-488.1, as mainly hardware-oriented was supplemented by IEEE-488.2 standard. The IEEE-

488.2 standard is concerned with a more precise description of commands, data structure and transaction timeouts. Further extension of GPIB IEEE-488.2 was introduced in 1990 with the specific instrument command set called *SCPI – Standard Commands for Programmable Instruments*. SCPI is more programming language than interface and is described in the Section 6.4.13.

The extension of GPIB standard did not solve many problems such as relatively low speed of data transfer *1 MBps*, limited length of the cable and limited number of connected devices. In 1993 the National Instruments introduced improved protocol of synchronization HS488 – *High Speed Handshake Protocol* which enabled users to increase the transfer speed to about *8 MBps* (for the cable length *2 m*).

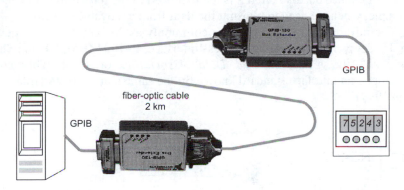

Figure 6.39. The increase of the transmission length by application of GPIB-140A extenders of National Instruments (National Instruments 2005)

The increase of the transmission length is possible after application of special extenders as GPIB-140A extenders of National Instruments (Fig. 6.39). Extenders enable users to transmit the data serially through the fiber-optic *2 km* cable with rate up to about *1 MBps*. Thus, the whole system operates as a serial-parallel hybrid interface.

Another extension of the performance of GPIB interface can be realized using GPIB expanders, which increase the number of connected devices. For example GPIB-120A expander of National Instruments enables to connect 29 devices to the GPIB interface (as compared to 15 in the original version).

In most of the computers a parallel interface called *Centronics* or *IEEE-1284* is still present. This interface can be used to connect the printer and is indicated as *LPT – Line Printer Terminal* (currently this mode of printer connection is often substituted by the connection through USB interface). Theoretically, Centronics could be used as a measuring system interface.

This interface after small hardware and software modifications can allow transmitting the data with speed up to *2MBps* (*as Enhanced Parallel Port mode EPP*). But an important advantage of GPIB interface in comparison with the IEEE-1284 standard is its well defined command set. In the case of IEEE-1284 it would be necessary to write practically a whole new program adopting this standard to measuring system requirements. Moreover, it is very probable that in the near future computers will be manufactured without the LPT parallel port.

6.4.5. Wireless interfaces: IrDA, Bluetooth and WUSB

The wireless connection of devices results not only in elimination of the chaos of cables, but also often it is the real need of the measuring system. For example, when we need to transmit the data from a movable element.

The infrared light system known commonly as the remote control device for TV set can be used as the wireless interface. In 1994 IBM, HP and Sharp organized the group *Infrared Data Association* or *IrDA*. This group developed an interface standard known by the same name – IrDA (Irda 2005, Rob 1999).

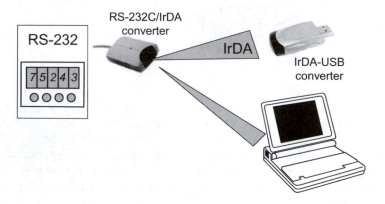

Figure 6.40 Data transmission using infrared interface

To complete a measuring system with an infrared connection it is possible to use converters, for example RS232C/IrDA converter and next IrDA/USB converter as it is illustrated in Fig. 6.39. Many laptop computers have as equipment the IrDa port.

The IrDA interface utilizes infrared radiation of the wavelength *875 nm*. The infrared LED can be used as the source of radiation. A typical length of infrared connection is up to one meter in presence of daylight and deflection (off-alignment) *± 15°*. Transmission speed of IrDA 1.0 ranges from *2 400* to

115 200 kbps, IrDA 1.1 extends this speed to *4 Mbps*. The pulses of LED light are modulated to the *3/16* of the width of the original duration of a bit which corresponds to *1.6 μs*. For high speed system IrDA this pulse width is enlarged to ¼ of original one bit pulse (Fig. 6.41) and for speed larger than *1.15 Mbps* system of modulation is more complex.

The IrDA interface is described by several protocols, for example IRPHY – infrared physical layer specification, IrLAP – infrared link access protocol, IrCOMM – infrared communication protocol, IrLAN – infrared local area network.

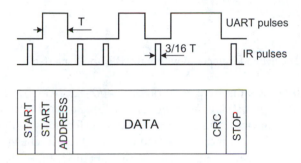

Figure 6.41. Modulation of the IR pulses and typical packet of IrDA interface

Recently, the infrared light data transfer is often substituted by *Bluetooth* radio interface (Held 2000, Morrow 2002, Muller 2000). Bluetooth connection overcomes one of the most inconvenient drawbacks of IrDa – the necessity of line-of-sight between connected devices. Bluetooth was introduced in 1998 by the Swedish company Ericsson[1]. In 1999 the Bluetooth Group was established: Sony, Ericsson, IBM, Intel, Toshiba and Nokia (Bluetooth 2005). Bluetooth interface enables to create the *personal area network*. In this network it is possible to transfer data with speed of *1Mbps* (Bluetooth 2.0 increases this speed to *2.1 Mbps*) at the distance of about *10 m*.

Up to eight devices can be mutually connected in the network – called a *piconet* (Fig.6.42). The device that initiates the piconet is the master, the other devices operate as slaves. At any time data can be transferred between the master and one slave. The master switches rapidly from slave to slave in a round-robin fashion. The piconets can be connected into larger network called *scatternet* (Fig. 6.41). One of devices acts as a bridge between the piconets playing a master role in one of them and slave in another.

---

[1] Harald Bluetooth was king of Denmark in the late 900s.

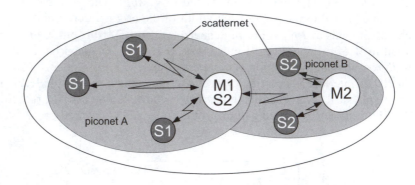

Figure 6.42. The network composed within the Bluetooth interface

Bluetooth utilizes the frequency range *2.400 – 2.4835 GHz*. The special technique called *frequency hopping spread-spectrum FHSS* enables co-operating many devices without additional interference. In this technique each of 79 devices use individual frequency, which is changed randomly 1 600 times every second. Thus each channel obtains a time slot *625 μs*. The master transmits in even time slots, slaves in odd time slots.

ACCESS CODE			HEADER						PAYLOAD
72 bits			54 bits						0 - 2745 bits
AC	synchronization word 64 bity	AC	AMA address	type	flow	ARQ	SEQN	HEC	DATA

Figure 6.43. The example of the data packet in Bluetooth interface

Fig. 6.43 presents the example of the data packet. The information about spread-spectrum frequency is included in synchronization word. Each member of the network obtains individual 32-bit address *BDA – Bluetooth Device Address*. There are two types of data transfer: SCO – synchronous connection oriented and ACL – asynchronous connectionless. The signal is relatively weak – about *1 mW* (for comparison – the mobile phone transmits a signal *3 W*) and therefore the distance is limited – to about *10 m*. This helps in avoiding the interferences between many networks and devices.

In 2005 an enhancement of USB interface was introduced – the *Wireless Universal Serial Bus WUSB*. This system combines the performance of USB 2.0 with wireless ultra band technology operating in the range *3.1 – 10.6 GHz*. Fig. 6.44 presents the topology of a WUSB network.

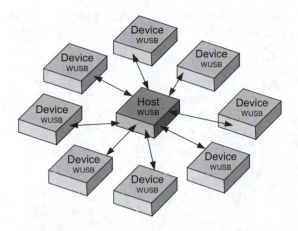

Figure 6.44 The topology of wireless USB

Similarly as in wired USB the whole system is controlled by one host WUSB device. Up to 127 devices can be connected to this host in a hub and spoke technology. The host initiates communication, and allots time slot and data bandwidth. The speed of data transfer is *480 Mbps* for *3 m* distance and *110 Mbps* for *10 m* distance.

6.4.6. Mobile telephony systems GSM and UMTS as a tool for data transfer

Mobile telephony can be competitive as compared to other interfaces after introduction of *UMTS – Universal Mobile Telecommunication System* with a speed of data transfer of *1,960 Mbps*. But also traditional *GSM Global System of Mobile Communications* with *115 kbps* can be an attractive tool of data transfer, especially in an environment without a computer network and even without supply of electrical energy (Bekkers 2001). To make possible the digital data transfer with GSM it is necessary to use a special kind of mobile phone and to obtain permission from a local GSM operator.

For data transmission the phone should have the input /output circuits for the data. In some phones it is RS-232C or IrDA interface but currently most popular is the Bluetooth interface. For the data transmission a special kind of mobile phone can be used, but also there are available palmtops and laptops with GSM interfaces as well GSM modules suitable to connect them to the computer.

The principle of operation of the GSM network is presented in Fig. 6.45. The mobile station consists of a terminal and smart cart called *SIM - subscriber identity module*. This mobile station links by radio with *base transceiver station* – BTS antenna. BTS serves a certain area called a cell (hence another name for mobile telephony is cellular telephony).

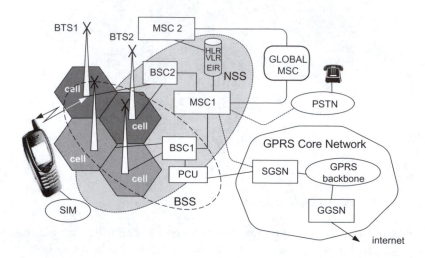

Figure 6.45. The structure of a GSM network

The base transceiver stations BTS are connected to the base station controller BSC and this network composed of several BTS and BSC units is called *BSS – base station subsystem*. These BSS networks are connected in larger network called *NSS – network subsystem*. The network subsystem consists of several BSS connected to *Mobile services Switching Center MSC*. The MSC unit cooperates with database VLR – Visitor Location Register, HLR – Home Location Register, EIR – Equipment Identity Register and others. MSC network can be connected to other networks as ISDN integrated services digital network, PSTN – public switched telephone network.

Part of the GSM system designed for the data packets transmission is organized as GPRS Core Network (GPRS – *General Packet Radio Services*). The data from BSC through PCU (packet communication unit) is connected to *SGSN – Serving GPRS Support Node* which is the input node to the GPRS Core Network. As the output node is a *GGSN – Gateway GPRS Support Node*.

In GSM it is possible to transfer the data using a following systems:
- *CSD (Circuit Switched Data)* – there is a time-slot allocated between phone and base station to transmit the data with the speed *9.6 kbps*. From base station this signal is distributed to MSC, where it can be transferred to the receiver.

- *HSCSD (High Speed Circuit Switched Data)* – development of CSD system where the speed of transmission is increased to *57.6 kbps* due to the allocation of several time slots at the same time (several channels)

- *GPRS (General packet Radio Service)* uses the *TDMA* (*Time Division Multiple Access*) technology, which enables sharing by several users the same frequency by dividing it into different time slots. GPRS is *packet switching*, in which multiple users can share the same transmission channel. Thus, although theoretically GPRS mode enables to transmit data with a speed up to *114 kbps*, practically the achievable speed is *30 – 70 kbps*. GPRS system enables us to connect the user to the Internet by *WAP – Wireless Application Protocol.*

- *EDGE (Enhanced Data rates for Global Evolution)* – called sometimes enhanced GPRS utilizes similar transmission technology as GPRS, but with a higher speed (*48 kbps* for one time slot and *384 kbps* for eight available time slots) is obtained due to introduction of special modulation technique and channel coding.

SMS-SUBMIT

type 1 B	No SMS 1 B	address 2-12 B	protocol 1 B	code 1 B	validity 1-7 B	length 1 B	data 0-140 B
PDU Type	MR	DA	PID	DCS	VP	UDL	UD

SMS-DELIVER

type 1 B	address 2-12 B	protocol 1 B	code 1 B	time 1-7 B	length 1 B	data 0-140 B
PDU Type	OA	PID	DCS	SCTS	UDL	UD

Figure 6.46. The packet format in SMS (text messaging)

Useful for data transfer is the SMS (text messaging), which can transfer information containing 140 characters. The format of the packet is presented in Fig. 6.46. In this packet the following data are sent: PDU – protocol data unit type, MR – message reference, DA – delivery address, PID – protocol identifier, DCS – data coding scheme, VP – validity period, UDL – user data length, UD – user data, OA – originator address, SCTS – service center time stamp.

Fig. 6.11 presents the data-logger with module for sending the messages as SMS via GMS network. The data-logger presented in Fig. 6.47 realizes similar operation. Data, for instance temperature, is determined every *30 s* and is saved into internal memory. The report is sent by e-mail once a day.

But it is also possible to obtain the actual data from the mobile phone. For example, if the tested parameter exceeds the assumed value, then an alarm signal is sent by SMS message.

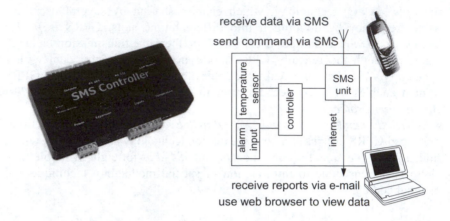

Figure 6.47. The SMS controller for data logging – model XE-8000 of Advanced Information (Advanced Information 2003) (permission of Advanced Information)

It is possible to design more advanced measuring systems with the GPRS or EDGE data transfer. In order to benefit from such high speed of communication we can use the WAP protocol to construct a measuring system with more sophisticated user interface.

For preparation of the pages in the WAP system can be used *WML (Wireless Markup Language) – equivalent of HTML*. Pages of WML are created as the deck consisting of a set of "cards". As example of sequence of cards in a GSM system for temperature measurements is presented in Fig. 6.48.

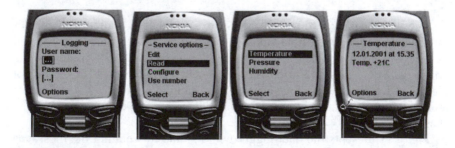

Figure 6.48. The measuring system using GSM communication and WAP technology (Winiecki 2003)

Further development of mobile phone communication offers "*3G – third generation*" telephony *UMTS – Universal Mobile Telecommunication System* (Castro 2001). This system applies *WCDMA (Wideband Code Division Multiple Access)* technology and higher *1 950/2 150 MHz* frequency enables data transfer with speed *1 960 kbps* for stationary terminals and *384 kbps* for mobile terminals. Moreover, it operates as a global satellite communication system, making large distance connection simpler and faster.

6.4.7. Radio data acquisition and transfer

When it is necessary to perform data acquisition continuously (for example for control of a moving robot) the GSM data transfer is inconvenient. The Bluetooth interface could be the solution, but the distance over which the data could be transferred is limited. Therefore radio acquisition and data transfer is for certain requirements very useful. The problem in radio modem application is that in many countries available radio frequencies are standardized and it is necessary to obtain a license to use a given frequency.

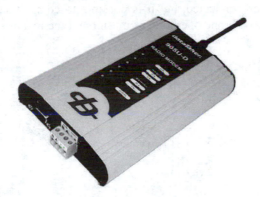

Figure 6.49. Radio modem model 905-U of Datataker (Datataker 2005) (permission of Datataker)

There are unlicensed frequency bands, available for all, called *ISM bands (Industrial, Scientific and Medical)*. For example, the frequency bandwidth *868-870 MHz* can be used with the power of the transmitter limited to *500 mW*. Very popular is the frequency bandwidth *433-435 MHz* commonly used in wireless electronic devices – for example home digital thermometers. Fig. 6.49 presents the example of a radio modem of frequency *869.4 MHz* and data transmission with speed *19 200 kbps* at the distance up to *5 km*. The data are transmitted by RS-232C or RS-485 interface.

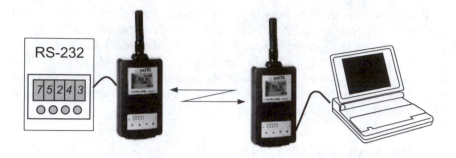

Figure 6.50. An example of measurement data wireless transmission (Satel 2005)

Fig. 6.50 presents an example of simple data wireless transmission using transmitter/receiver model Satteline 3As-869 of Satel. The data is delivered to the radio-modem by RS-232C or RS-485 interface. The transmitter working at the frequency *869 MHz* enables transmission of the data with a speed of *19.2 kbps* through the air in the half-duplex mode. The output power is programmable in a range of *10 – 500 mW*. In the presented example there are only two devices connected, but it is possible to connect several radio-modems in pico-net with one operating as master and the rest as slaves.

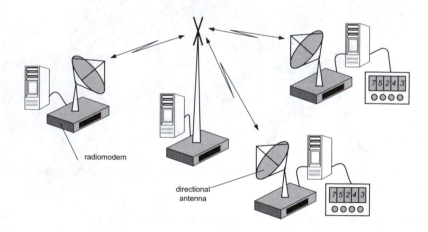

Figure 6.51. The data acquisition system with several radio-modems

The regulations limit the power of the transmitter, hence the distance of communication is not longer than *10 km*. The distance can be increased to about *100 km* for a transmitter power of about *10 W*. But usually for such communication a licence is required. Another way of increasing of the distance and constructing a small network is application of one main antenna and several directional antennas at the receivers. The stair topology of such a network is recommended (Fig. 6.51).

Figure 6.52. Transmitter/receiver BiM2/433 of Radiometrix: integrated circuit (dimensions 30×20×4 mm) and modem evaluation kit with RS-232 input/output (permission of Radiometrix)

There are also available transmitters and receivers in the form of integrated circuits - for example 433/868 MHz TDK 5100 receiver and TDA 5200 transmitter with VCO-PLL synthesizer of Infineon.

Fig. 6.52 presents the transmitter/receiver IC of Radiometrix (Radiometrix 2005). For power of transmitter of *10 mW* it is possible to transfer the data at a distance to *200 m* (*50 m* in buildings) with speed *64 kbps*. Fig. 6.53 presents the block diagram of this device. The transmitter operates as the SAW stabilized frequency modulated oscillator, while the receiver works as double conversion FM superheterodyne.

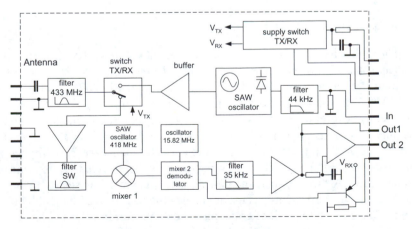

Figure 6.53. Block diagram of a BiM2 transceiver

6.4.8. Computer systems using Ethernet and Internet

The computers and computer associated devices can be connected in network – they can share a common communication line. Usually one of the computers can manage network resources as common software, memory, printers, network traffic. In this computer the *server software* is installed, which organizes the whole network – often this special computer is simply called a server. Traditionally computer networks are classified as *Local Area Network LAN* (limited to certain areas as building, company, factory, etc.) and *Wide Area Network WAN* covering large area or connecting computers within large distances, even thousands kilometers. One of the important applications of computer networks is data communication (Callaway 2003, Held 2002, Ilyas 2004, Reynders 2003, Shay 2003, Spurgeon 2000).

A special kind of LAN is *Ethernet* developed in 1980 by Xerox, DEC and Intel and later specified as standard in *Ethernet Blue Book 2* (Ethernet is sometimes called Blue Book Ethernet or *Ethernet 2*). In 1985 standard *IEEE 802.3* has been introduced, which describes the Ethernet type network. Both standards specify a physical layer (for example cabling, connectors) and a software layer (for example frame format and communication protocols). Although there are small differences between these two standards both are termed *Ethernet*. Ethernet is the most commonly used network. There are also other types of LAN networks, among others *Token Ring* (specified by IEEE 802.5 Standard) and *Fiber Distributed Data Interface FDDI*.

Preamble	Start	Destination address	Source address	Length	Data	CRC
8 Bytes	1 Byte	6 Bytes	6 Bytes	2 Bytes	46-1500 Bytes	4 Bytes

Figure 6.54. The format of Ethernet frame

The data transfer is in a form of well defined *frame*. The Ethernet uses *Carrier Sense Multiple Access Collision Detect CSMA/CD* protocol of communication between the nodes – specified in IEEE 802.3 standard. The CSMA/CD protocol describes the method of data transmission in the same channel by several devices. The device trying to send the frame tests whether the line is idle and if it is then the first frame is sent. If at the same time another device sends data a collision occurs. In such case both frames are discarded and after a random amount of time the devices try to send the frame again. Although such method of data transmission seems to be not reliable it is very efficient because it enables the use of one channel by many users. To diminish the risk of collision the data packet has limited length and access time. An example of the data frame used in Ethernet is presented in Fig. 6.54.

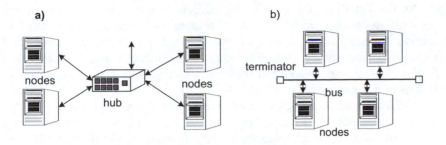

Figure 6.55. Typical network topology: star topology (a) and bus topology

The connection can be realized using coaxial cable (for example 10Base 2 standard or fiber optic cable 10Base F). One of the most popular Ethernet cabling standards is *10Base T* standard that uses twisted pairs of wires for communication with transmission speed up to *10 Mbps*. Each node of the network is connected to a central hub by two pairs of wires (for transmitting and receiving) with length up to about *100 m*. For connection the RJ-45 8-pin connector is used. Enhanced version of Ethernet is *Fast Ethernet* (or 100Base-T Ethernet), which supports the data transfer with speed up to 100 *Mbps*. Also faster 1 Gigabit Ethernet can be used.

Fig. 6.55 presents the computer connections in the network called network topology. To organize computers in most extended systems, special devices as *repeaters, hubs, bridges, switches and routers* are commonly used. Fig. 6.56 presents the example of Ethernet extended network.

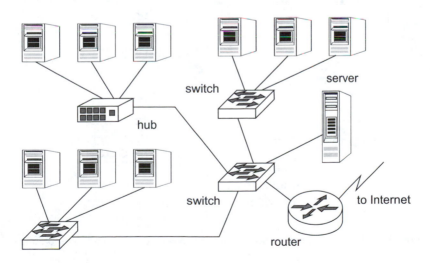

Figure 6.56. An example of the connection of devices in Ethernet

As the length of the connecting wire is limited the extension of the network is possible with the aid of bridges or repeaters. The *repeaters* receive signals from one part of the network and amplify, retime, and retransmit those signals to another network segment. Similarly operate the *bridge devices*, but in addition they can filter signals and reject incorrect frames or disturbances.

The *hub device* branches several connections to one wire and in this way decreases the number of wires. The *switch device* realizes a similar job, but a switch operates in more sophisticated way. It manages the signal traffic and is able to connect several devices simultaneously. Often the router operates as an input/output device in the Ethernet. A *Router* is an intelligent device that is able to determine the best route of the addressed signal. Usually the router is located at the gateway connecting two networks. Moreover, the router can translate various communication protocols (can connect various types of networks, for example Ethernet and Token Ring) to one common protocol of Internet.

The LAN type network is limited to an area of about 1 km². To connect larger areas or distances a special kind of network called *Wide Area Net WAN* has been developed. WAN uses special types of communication protocols (as TCP/IP, ATM, Frame Relay) and private communication lines or leased lines. Such network is rather expensive; hence it is used by large companies or organizations, as for example the electrical power industry.

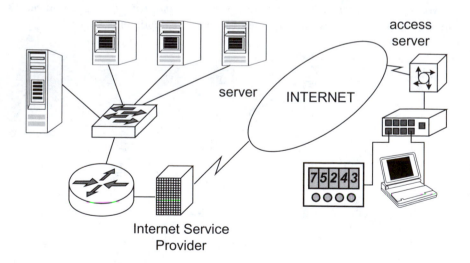

Figure 6.57. The application of the Internet in the wide area network (Cisco 2005)

A wide area network can be also organized in a cheaper way using the public network, for example Internet, as the connection line. The *Internet* is a system of connections of other networks, usually LANs. Today, the Internet is a system connecting most of the world networks[1]. To enable such inter-network connection a common communication protocol must be applied. As such a protocol TCP/IP is used.

The *TCP/IP (Transmission Control Protocol/Internet Protocol)* is a set of communication protocols describing a method of data transmission and standard of addressing (the main two are: TCP and IP). Fig. 6.57 presents the application of the Internet to communicate with the LAN type network (it can be a private LAN called an Intranet) and the remote user (for example measuring instrument). The gateway of network is usually associated with the router, which "knows" where to direct a given data packet.

The method of organization of wide area network by application of Internet is known as *Virtual Private Network VPN*. In virtual network sophisticated methods for data security and confidentiality and for organization of communication tunnels have been developed (Held 2004, Kosiur 2001).

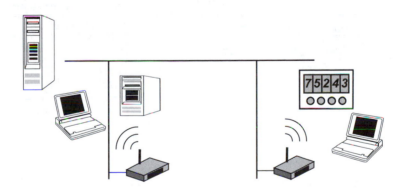

Figure 6.58. The wireless network connection - WLAN

The *Wireless Local Area Network WLAN* can be used as an alternative to the Bluetooth interface. WLAN is described by standard IEEE 802.11 and enables connection of devices with the speed *11, 22, 44* or *54 Mbps* and operates at frequency *2.4 GHz* (Callaway 2003, Roshan 2003). Currently, most of the laptop computers have built-in WLAN modules.

---

[1] Often Internet is considered as an example of WAN.

6.4.9. Dedicated interfaces: CAN, I²C, MicroLAN, SDI-12

Several companies designed bus interfaces for special purposes, for example the CANbus of Robert Bosch GmbH designed for the automotive industry, I²C of Philips designed for integrated circuits interfacing. Sometimes (as in the CANbus) the interface was accepted for other applications as a universal purpose interface.

The *CAN (Controlled Area Interface)* was developed by Bosch company as the solution of the problem of the huge number of sensors used in the automotive industry. The assumptions were as follows: interface should be reliable (enhanced error correction), working at real-time, the cabling and hardware should be not expensive and simple, in some crucial safety applications it should be sufficiently fast (speed of about *1 Mbps* of serial data transfer). Fig. 6.59 presents the application of CAN in car control.

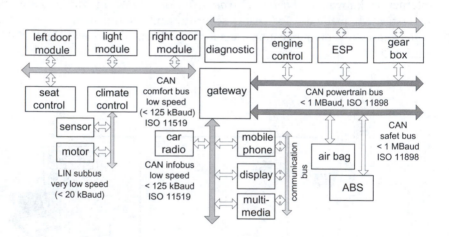

Figure 6.59. An example of the CAN interface in a modern car (Infineon 2005)

The CAN was introduced in 1990, and from 2003 is a standard ISO 11 898-1 (*1 Mbps* speed) and ISO 11 519 (for lower speed). It uses twisted pair wire connection indicated as CAN_H and CAN_L. The structure is open and additional devices can be easily connected. After connection, the device obtains an address, in which the priority is included – the higher the address number, the higher the priority.

The *Carrier Sense Multiple Access/Collision Avoidance CSMA/CA* is used as the data transmission protocol – similar to that used in the Ethernet, but slightly enhanced. In the case of collision the transmission is not interrupted, but the priority lies with the device with the higher number. Due to sophisticated error correction and short data frame (8 bytes) the CAN interface is assumed as very reliable and robust.

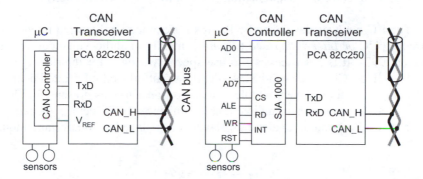

Figure 6.60. The connection of the device to the CAN bus: in the case when microcontroller has built-in CAN controller and for other microcontrollers

The message frame format is presented in Fig. 6.61. An arbitration field contains an identifier number and RTR (*Remote Transmission Request*) for discrimination between the data frame and the request for the data. In control field the bit IDR indicates that the data is standard or extended, the RBC bit is not used. The frame does not contain the address field – the information about the address is included in the message.

1	11	1	1	1	4	0..64	16	2	7
start	identifier	RTR	IDE	RBC	length code	data field	CRC	ACK	end of frame

arbitration   control

Figure 6.61. The message frame format of CAN

The CAN is a broadcast type bus. It means that all nodes obtain the message frame. If the message is relevant, it is processed; otherwise it is ignored. In 2000 an extended version of the CAN – *Time Trigerred Communication on CAN – TTCAN* was introduced. In this interface the messages can be sent with a defined timing. The TTCAN interface is compatible with CAN interface.

The CAN interface, although designed for automotive industry, is widely used in other areas due to its many advantages (simplicity and reliability), most of all in control and automation (Lawrenz 1997, Pfeiffer 2003).

The I²C bus was developed by Philips for connection between integrated circuits (*I2C or Inter-Integrated Circuit*) – mainly in audio-video equipment (Paret 1997). It is not necessary to use special wiring – for interfacing two paths, called SCL (clock) and SDA (data), are introduced on printed boards. Also the transmitter is very simple – just two transistors as presented in Fig. 6.62. The transmission is bidirectional with one master device (usually a microcontroller) and up to about 20 slaves. Each slave has to have an unique address.

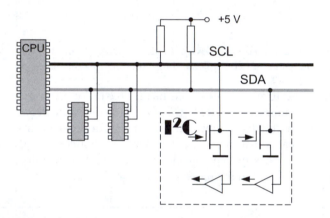

Figure 6.62. Connection of the devices to I²C interface

Figure 6.63 presents the frame format. The transmission is initiated by master, which sends the start bit, the address of a slave and information about whether it is *read* or *write* operation. If the master receives the acknowledgement bit the transmission of the data starts. After every byte of sent data the master waits for acknowledgement. The speed of transmission of about *125 kbps* is controlled by the clock pulses sent by the master.

M do S		S do M		M do S	S do M		M do S		
start	address slave	R/W	ACK	data	ACK	data	ACK	stop
1	7	1	1	1 byte	1	1 byte	1	1

Figure 6.63. The data frame format in I²C transmission

There are great number of integrated circuits compatible with the I²C standard, including microcontrollers. If the device is not compatible with the I²C standard it is possible to use an external controller – for example PCF8584 model of Philips.

Similar in structure to the I²C interface is the *MicroLAN* developed by Dallas Semiconductors and called the *1-wire interface* (practically there are two wires including the ground wire). The 1-wire interface enables us to connect several sensors, analogue-to-digital converters, memory elements, etc., in one measuring system. Usually to realize the interface it is necessary to use a special converter, for example RS232C/1-wire converter. Figure 6.64 presents the system with two sensors connected to the line.

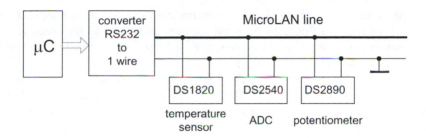

Figure 6.64. Connection of the devices to MicroLAN interface

In the mid 1980's a group of water monitoring instrumentation users introduced special interface called SDI-12 (*A Serial-Digital Interface Standard for Microprocessor-Based Sensors* (SDI 2005). Currently this standard is commonly used in environmental data acquisition – as an example we can point to the Vaisala Weather Transmitter system described in Section 6.2.3).

A SDI-12 standard is used to interface battery powered data recorders with microprocessor-based sensors. It is a set of rules that define how a sensor must communicate with a datalogger. The interface uses three wire cable – signal line, a ground line and *12 V* supply. It enables us to transmit along one cable several sensors. To apply SDI-12 interface the special RS-232C to SDI-12 converters can be used.

The microprocessor is used to convert the analogue signal into digital one, to perform the self calibration procedure and to compute the data in engineering units. Application of a such intelligent microprocessor-based sensors enable investigator to use one standardized recorder unit for various sensors. The SDI-12 interface enables data transmission with speed of a *1200 baud* using the cable up to *60 m*.

6.4.10. HART interface and the 4 – 20 mA standard

The analogue data transmission systems are still in use in the industry mainly due to their simplicity and low cost. Commonly, the current DC output signal *4 – 20 mA* is chosen. The *4 – 20 mA* standard (standardized by ANSI/ISA -50.1-1982 standard) exhibits following advantages:
– the same range of output signals for all sensors, transducers and actuators enables easy replacement of devices;
– the current output is immune to the changes of the load circuit (in certain range, typically *0 – 2 kΩ*), changes of the connecting wires resistance (caused for example by the ambient temperature variations),
– it is easy to transmit the data at a distance of several km (typically to about *3 km*).

The standard *4 – 20 mA* is convenient because zero value (corresponding to *4 mA*) of measured data differs from the zero value caused for example by the fault in the connection. Moreover, the range *0 – 4 mA* can be also used for supplying the power to the device.
On the market there are available various converters of *4 – 20 mA* signal into digital form, for example USB or RS 485 interface. It is also possible to inverse the conversion, for example with AD429 serial input 16-bit current output converter.

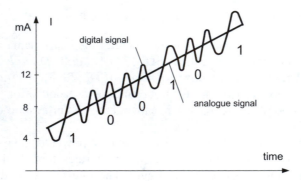

Figure 6.65. The analogue signal modulated by digital signal in HART interface

However, the digital interface can be more flexible; enabling for example easy data read and send on the external command, change of the measured range, performing diagnostics, etc. Therefore, in 1989 Rosemount introduced HART (*Highway Addressable Remote Transducer*). This interface is a hybrid joining analogue and digital data sending using practically the same infrastructure. In this way factories could add the digital data transfer without modification of the cabling.

HART interface uses the *FSK* (*Frequency Shift Keying technique*) to modulate the DC current output by AC signal. In this AC signal zero (*0*) corresponds to a *2200 Hz* sine wave and one (*1*) corresponds to a *1200 Hz* sine wave (Fig. 6.65).

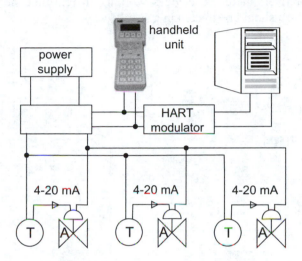

Figure 6.66. The HART interface used to control of the set of transmitters T and actuators A.

Since the digital signal has an average value equal to zero it does not disturb the analogue system. Usually, in the HART interface there is one master device (it can be a computer) and several slave devices (up to 15). Each slave device obtains a unique address and master unit can send to this device commands, for example: read variable and units, read current output and percent of range, calibrate, perform self-test, etc. The data frame usually consists of 1 start bit, 8 data bits, 1 odd parity bit and 1 stop bit. Apart from the central unit operating as a master device it is also possible to use a special handheld terminal (Fig. 6.66) that can also send the commands or receive the data.

6.4.11. Industrial communication standards – Fieldbus, Profibus, SCADA

Fig. 6.59 presents the control network used in modern cars. In the case of a control system in a plant situation is more complex – Fig. 6.67 presents the example of network organization in a factory.

In the typical industrial communication there are various "levels" of communication and related to these levels requirements. In the first level there are simple sensors and actuators, next are transducers, drive units, valves and operator terminals; at the next level there are controllers, then at the higher level is management. Al of these levels should cooperate, but requirements of the communication interfaces on different levels are different. Moreover, there are devices working in real-time or dangerous conditions, which should be treated in a special way.

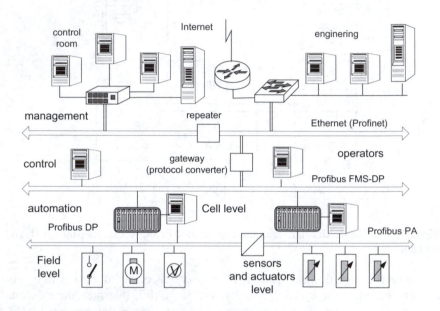

Figure 6.67 The example of the network organization in plant

All parts must be compatible and the whole network should operate in well organized way. It is recommended that the devices are suitable to communication profiles, and the devices provided by various manufacturers should be replaceable.

Fieldbus is the communication system designed for organization of the optimal conditions of data transmission as well as engineering and managing of the manufacturing process and technology. There are many various Fielbus systems: AS-i, DeviceNet, SDS, Interbus-S, Foundation Fieldbus. The Profibus system is one of the most widely used, particularly in Europe (Bender 1993, Mitchell 2003,Weigman 2004).

Table 6.3 presents the internationally accepted model of communication system. It is expected that the communication standard describes requirements related to these levels.

Table 6.3. The ISO/OSI communications reference model

Level	Designation	Function
7	Application	Application oriented commands
6	Presentation	Representation of data for analysis
5	Session	Stations connections, synchronization
4	Transport	Controlling data transmission, transport errors
3	Network	Connection, methods of avoiding collisions
2	Data-link	Bus access protocol (Medium Access Control MAC)
1	Physical	Hardware, coding and speed of data transmission

The Profibus (*PROcess Field BUS*) introduced in 1987 by 21 German companies was first described by German DIN standard and since 1999 is described by IEC Standard 61158 (*Digital Data Communication for Measurement and Control – Fieldbus for Use in Industrial Control Systems*). This standard describes mainly 1, 2 and 7 OSI levels (Table 6.3). The main goal of Profibus system is well illustrated in Fig. 6.68, taken from a Profibus brochure (Profibus 2005).

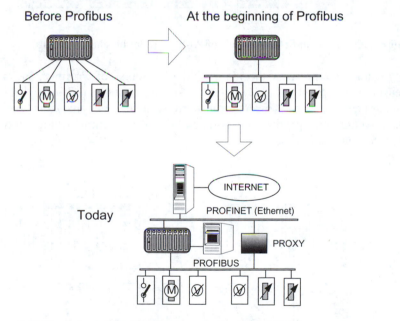

Figure 6.68. Evolution of the Profibus system (Profibus 2005)

Profibus is a set of standards and regulations describing a precisely physical layer (cabling, sensors, actuators) and internal interfaces (RS-485, MPB-IS, HART), network organization, communication protocols. Currently it is observed that the Profibus is divided into two parts: *Profibus* (*Automation Technology*) and *ProfiNet* (*Technology and Application*). The

Profibus infrastructure uses a communication standard described by IEC 61 158 Standard, the ProfiNet uses the Ethernet technology. Profibus system is integrated in Profinet using a proxy technique. Proxy is a network service that allows clients to make indirect connections to other network services.

market area	Factory automation	Motion Control	Process automation	Safety
designation	PROFIBUS DP	PROFIdrive	PROFIBUS PA	PROFIsafe
application profile		PROFIdrive	PA devices	PROFsafe
communication	DP Protocol	DP Protocol	DP Protocol	DP Protocol
transmission	RS485	RS485	RS485 MBP-IS	RS485 MBP-IS

Figure 6.69 The main field of interest of Profibus (after brochure Profibus)

Figure 6.69 presents the main field of interest of Profibus. From the measurement point of view the most important is *Profibus PA*. It standardizes devices (sensors, transmitters, actuators, analyzers, etc.) to ensure vendor-independent operation and describes communication between the controllers and these devices.

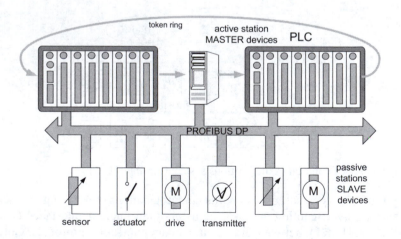

Figure 6.70. Profibus configuration in field/control level

In development of Profibus system PLC devices play important role. The *PLC (Programmable Logic Controller)* is a programmable microcontroller that substitutes hundreds of relays or timers used in the past. The PLC devices manage the actuators and sensors and according to the program stored in the memory are able to perform various controlling operations. The PLC devices are also capable of autonomously executing simple logic processes without involving the master computer. Figure 6.70 presents the Profibus configuration in the field/control level.

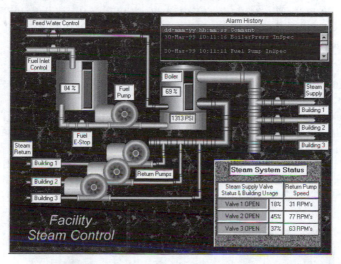

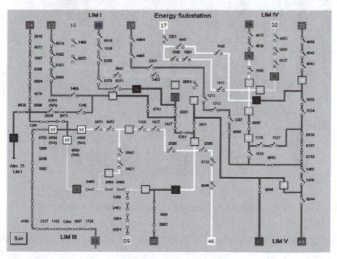

Figure 6.71. The two examples of the HMI (*Human-Machine Interface*) screen of Elipse SCADA system (Elipse 2005)

There have been developed computer systems (software) helping in the gathering, analyzing and control of the data flow in the Fieldbus interface. The most widely used are *DCS (Distributed Control System)* and *SCADA (Supervisory Control and Data Acquisition)*. In the SCADA system it is possible to monitor directly with the *computer control console* the state of all factory devices and also perform the operations in the case of alarm state (Bailey D. 2003, Boyer 2004).

Modern SCADA systems apply a user friendly computer interface called also *HMI (Human-Machine Interface)*. Two examples of such an interfaces – one used to control the steam flow, the second to control the electrical energy distribution are presented in Fig. 6.71.

6.4.12. Modular systems –VXI, PXI

The modular systems represent another philosophy of interfacing. In such systems all components are prepared in a form of standard format cards (for example *Eurocard*), which can be plugged together into a standardized chassis. This chassis is equipped with a series of slots into which the cards are slid – similarly to books on a bookshelf. An example of a modular system PXI device is presented in Fig. 6.72.

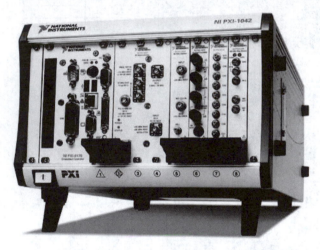

Figure 6.72. The example of PXI type system of National Instruments (NI 2005)

The important advantage of the modular system is that all connection lines are stable and short. Therefore the reliability and speed of data transfer is enhanced. Typically there are used separate 32-bit bus for data and separate for clock, synchronization and triggering. In 32-bit connectors the parallel data transmission is possible with a speed of *132 MBps*, and in 64-bit system even up to *264 MBps* (in PXI modular system).

The oldest modular system was the *CAMAC* system (*Computer Aided Measurement and Control*) introduced in 1960s as a reliable system in the nuclear industry. Currently, this system is gradually substituted by modern modular systems *VXIbus* (*VME Extension for Instrumentation*) and *PXI (PCI Extension for Instrumentation)*. Both standards use Eurocards. Eurocards have standardized sizes based on the "U" unit. The U is *44.45 mm* and for example 3U Eurocard is *100 mm × 160 mm*.

The VXI system is based on the *VMEbus (Versamodule Eurocard Bus)* introduced in 1981 by Motorola. VME standard separates both 32-bit data and address buses. In order to control the bus special nine lines known as arbitration bus are used.

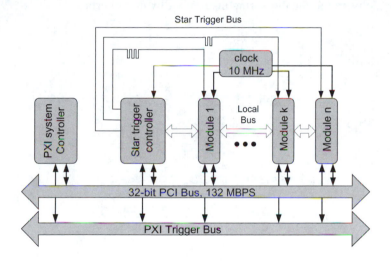

Figure 6.73. PXI timing and triggering buses

The PXI system (introduced in 1997 by National Instruments) uses a PCI bus, the same that is currently used in most personal computers. Therefore, the PC based cards are very similar to the PXI modules; they use the same drivers and software. The PXI system operates in "plug and play" technology – theoretically it is sufficient to plug the module into the chassis to start working with the application.

Figure 6.73 presents the organization of the communication in the PXI system. The whole system is controlled by the PXI controller module, which is practically an additional computer. To manage the system an external computer can be used instead of the internal controller. To obtain the best speed and synchronization a special Star Trigger Bus in form of a star is applied. In PXI standard there are available not only typical devices as data acquisition boards, but also whole measuring instruments – such as multimeters or oscilloscopes.

6.4.13. Standard Commands for Measuring Devices - SCPI

Lack of common commands for measuring instruments led a group of manufacturers to develop a standard command set (language) facilitating the users to control the measuring process. On the basis of GPIB.2 interface commands in 1990 a standard of commands called *SCPI – Standard Commands for Programmable Instruments* was proposed. It was assumed that this standard should be user friendly (written in the text mode) and can be delivered to the measuring instrument by any interface.

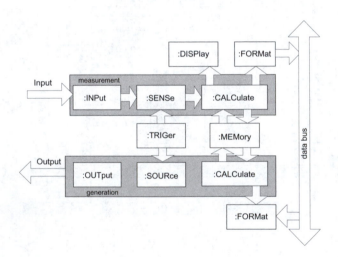

Figure 6.74. SCPI instrument model

The SCPI model of measuring instrument is presented in Fig. 6.74. The inscriptions in each block describe most important operations:

- *INPut* command controls the signal conditioning: input impedance (:IMPedance), gain (:GAIN), filtering, attenuation or other input operations;
- *SENSe* command describes range, resolution, type of the signal;
- *CALCulate* describes processing of measured data, for example calculation of the *rms* value;
- *OUTput* command defines the parameters of the output signal;

- *SOURce* describes signal parameters, such as frequency, power, modulation;
- *TRIGger* determines the method of synchronization;
- *MEMory* manages the saving procedures;
- *DISPlay* describes the method of presentation of the data.

The structure of commands is hierarchical – the command is followed by the more detailed subcommand, as presented in Fig. 6.75.

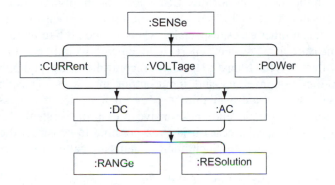

Figure 6.75. The example of hierarchical command tree

The commands are separated by colons. The blank space separates the parameter from the command and a semicolon separates commands within the same subsystem. For example the command

SENS:VOLT:AC:RANG 5

means the measurement of the AC voltage in the measuring range *5V*. In the case of

CONF:VOLT:DC 100,0.01; READ?

the voltmeter is configured as the *100V* DC voltmeter with resolution *0.01V* and the result of the measurement is transferred to the buffer.

The braces enclose the parameter choice while the vertical bar separates multiple parameter choices, for example

VOLT:DC:RANG {<range > |MIN|MAX}

Most available measuring instruments are capable of receiving and processing the SCPI commands and detailed command syntax is described in the *User Guides* for each particular device.

6. 5. MEASURING SYSTEMS BASED ON SIGNAL PROCESSORS

6.5.1. Microcontrollers and signal processors in measurements technique

So far, when the measuring systems were described above, it was assumed that the system is controlled by a typical PC type computer. Each functional block (interface, data acquisition board, etc.) can be connected to the computer by the internal PCI slots, by external PCIMCIA slots in the case of a laptop computer or by USB port. Such organization of the measuring system is rational if we want to perform the measurement in the laboratory. But in other cases, especially when we would like to construct a portable measurement system more convenient would be to use a microcontroller device. Many of the microcontrollers are equipped with analogue-to-digital converters and input/output interfaces. Thus, because currently such devices are available at a reasonable price it is possible to construct relatively simple and inexpensive measuring systems based on microcontrollers. In special cases it is even possible to construct the whole measuring system as one chip. Such microsystems are called *SoC* (*System on Chip*) systems.

The *microcontroller* (Fig. 6.76) is the microprocessor circuit supplemented by additional periphery circuits, such as memory, input/output circuit, interruption system. Such a device is able to work as a fully autonomous computer.

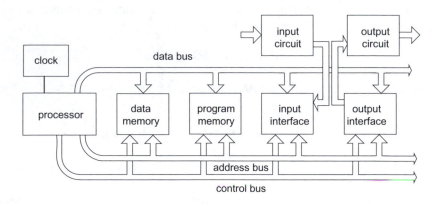

Figure 6.76. The architecture of a typical microcontroller

Microcontrollers are designed mainly as control devices. For signal processing more useful are specialized microcontrollers called *Digital Signal Processors DSP* (Lapsley 1997, Parhi 1998, Wahanhamer 1999). The DSP circuits are optimized taking into account the speed of mathematical operations, especially those related to signal processing. Since digital signal processors have internal fast multiply-accumulate module, then many digital operations are much faster in comparison with the microcontroller case (the same operations performed by DSP in one cycle need about forty cycles in the case of the microcontroller). The DSP operates in faster memory access mode and uses a special addressing mode. Therefore, in digital processing DSP for such operations such as FFT calculation, digital filter operations are faster not only in comparison to microcontrollers, but also even in comparison to Pentium microprocessor computers.

The first computer architecture was proposed in the 1930s by Howard Aiken from Harvard University. It was proposed to use two additional memories – one for data and one for programs. This kind of architecture is called as *Harvard architecture* (Fig. 6.77b). In the 1940s at University of Pennsylvania other computer architecture was proposed and realized. In this computer one common memory module was used. Such architecture is called *von Neumann* architecture (one of designers of the computer in Pennsylvania) (Fig. 6.77a). The von Neumann architecture was dominant for many years due to its simplicity.

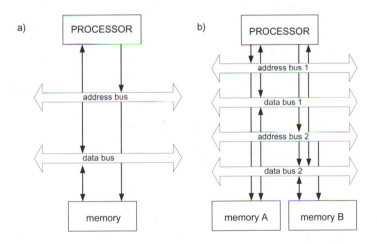

Figure 6.77. Architectures of microcomputer – von Neumann (a) and Harvard (b)

In the 1970s designers of signal processors returned to Harvard architecture, with two memories and two sets of buses. Although the cost of such processors was greater, their performances were significantly better. Today, both types of architecture are in use, but in microcontrollers more often von Neumann architecture is used, while in DSP mainly Harvard architecture is applied. Figure 6.78 presents a comparison of both architectures.

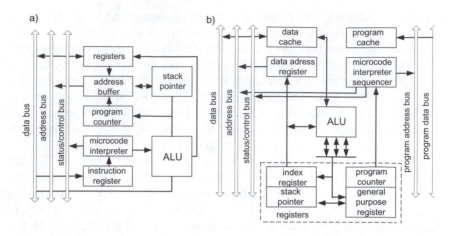

Figure 6.78. Comparison of architectures of microcontroller (a) and DSO (b) (Kularatna 2003)

Fig. 6.79 presents the block diagram of one of the most popular signal processor – TMS320C5x.

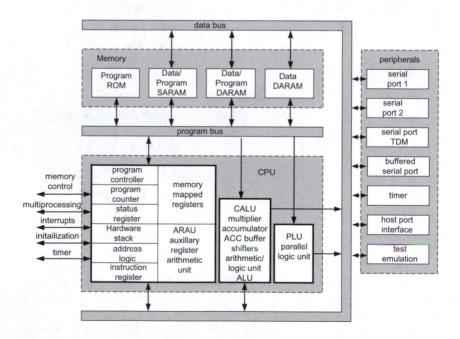

Figure 6.79. Block diagram of the TMS320C5x digital signal processor of Texas Instruments (TI 2005)

Various functions and performances of microcontroller and DSP means that in some measuring systems it is necessary to use two processors – one (microcontroller) for measurement control, and the other (DSP) for signal processing. Therefore, Infineon (Siemens) developed a new architecture called *TriCore* that is a hybrid DSP/microcontroller and combines the advantages of both types of processors (Infineon 2005).

In digital signal processor beside main arithmetic unit ALU (arithmetic-logic unit) the module MAC (multiplier/accumulator) is added enabling to perform the operation of multiplication and addition in one cycle

$$A = BC + D$$

This set of operations is often used in digital signal processing, among other in digital filter realization. The operation of multiplication and addition is performed in the arithmetic-logic unit called CALU (Fig. 6.80). It consists of 16×16 bits multiplier producing 32-bit product, conventional arithmetic-logic unit ALU, 32-bit accumulator, accumulator buffer and shifters.

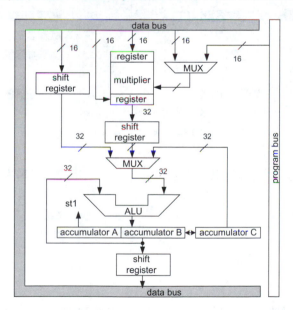

Figure 6.80. Central Arithmetic-Logic Unit CALU

In contrary to the processors used in personal computers, where new models substitute older ones, on the microcontroller market there are a lot of models. For example, one of the oldest but very popular 8-bit 8051 microcontroller is still in use. There were introduced special parameters for comparison of the microcontrollers and DSP. One of them is *MIPS – Million Instructions per Second*. In the case of floating point microcontrollers is used also the *MFLOPS – Million Floating Point Operations per Second*. MIPS is information about efficiency – for comparison the first IBM computer XT had MIPS = ¼ while a Pentium computer has MIPS of about 3000. Designers of microcontroller based devices question the meaning of the MIPS parameter (jokingly this acronym is translated as: *Meaningless Indicator of Performance*). In some cases more valuable would be determination of the time necessary to perform a typical calculation, as for example the FFT or digital filtering – such a comparison is presented in Fig. 6.81. Indeed, we can see from this comparison that between MIPS and the time used for calculations the relation is not very obvious – the Pentium processor with MIPS = 3400 needed for filter operation requires significantly more time then a signal processor with MIPS = 1800.

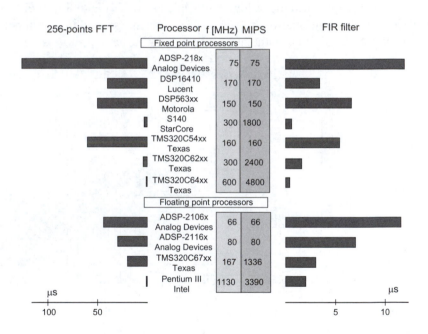

Figure 6.81. The benchmark of the main important processors performed by Berkley Design Technology Inc. (Eyere 2001)

Berkley Design Technology performs comprehensive analysis of the performance of digital signal processors. Such processors are tested through the realization of the most typical signal processing, as FFT analysis, FIR and IIR filter operation, demodulation, etc. Not only the time required for operation is analyzed, but also the memory usage, power consumption, cost of operation, etc. After such benchmark the BDTI (*Berkley Design Technology Inc.*) parameter is determined. Figures 6.81 and 6.82 present examples of such analysis.

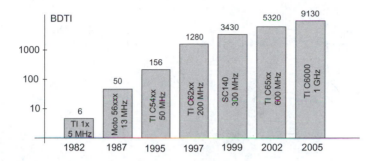

Figure 6.82. The progress in the DSP efficiency determined by Berkley Design Technology

In design of a DSP application for measurements other parameters can be as important as the memory and peripherals. For data acquisition it is important to use the DSP with internal analogue-to-digital converter. Simple microcontrollers use *UART* (*Universal Asynchronous Receiver/Transmitter*) or *SPI (Serial Peripheral Interface)* as output interface. But many devices are equipped with CAN, I^2C or USB interfaces.

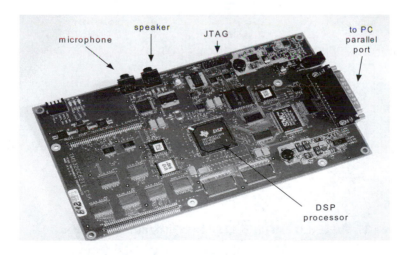

Figure 6.83. The example of the DSP Starter Kit "Composer Studio of Texas Instruments

It is reasonable for design purposes to use a digital signal processor in the form of *Development Board* (or Starter Kit, or Evaluation Board) (Chen 1999). Such a device is usually equipped with standard interfaces: RS-232C, USB or parallel interface. The example of a development board is presented in Fig. 6.83.

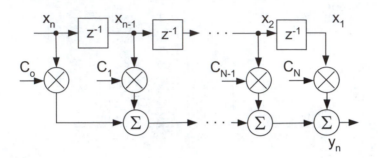

Figure 6.84. The mathematical operations in digital filtering process

There are described many applications of DSP for signal processing (Chasaing 2002, Dahnoun 2000, Kehtarnavaz 2000, Kehtarnavaz 2004, Singh 2003). One of a typical DSP operation is FIR filter. According to the circuit presented in Fig. 6.84 we should calculate

$$y_n = (x_n \times C_0) + (x_{n-1} \times C_1) + ... + (x_2 \times C_{N-1}) + (x_1 \times C_N) \tag{6.1}$$

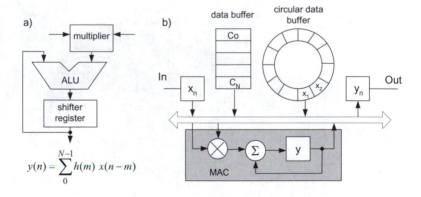

Figure 6.85. The CALU unit (a) as a tool for digital filtering operation (b)

For the 20-order filter it is necessary to perform 20 multiplications and 20 summations for every sample. The previous samples are registered in the temporary memory. As a result of filtering operation, a series of samples in the input after many times of multiplication by appropriate coefficients and by summation a new filtered series of samples is generated at the output.

Fig. 6.85 presents the CALU module performing MAC operation (multiplication and summation) in one clock cycle. To perform such a filtering operation it is convenient to use a circular buffer in which newly acquired sample replaces the oldest sample in the memory.

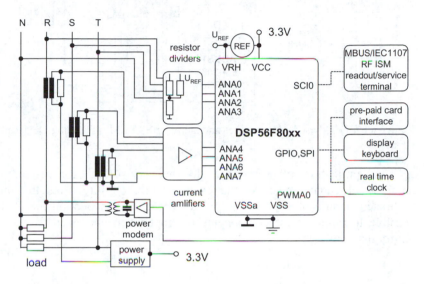

Figure 6.86. DSP based electric energy meter and analyzer designed by Freescale Semiconductor (Freescale 2005)

A digital signal processor can be used as the real-time power energy analyzer. Fig. 6.86 presents the design proposed by Freescale Semiconductors (Freescale 2005). DSP based meter can compute: active, reactive and apparent power and energy, peak power, multiple tariff counter, *rms* values and phase shift, frequency and higher harmonics signal. The DSP56F80x device from Motorola has 2×4 channel 12-bit ADC, general purpose port GPI, timer, FLASH memory. Included modem circuit transmits the PWM value through a buffer to one power phase.

A similar technique was applied by Analog Devices in the development of a DSP-based energy meter called SALEM (Fig. 6.87). In SALEM all three-phase or single phase calculations are performed in energy meter dedicated processor ADSST-EM-3035, which consists of 6-channel 16-bit ADC and DSP type processor. Logical operations as LCD display and RS-232C interface are controlled by additional microcontroller connected with DSP chipset via SPI interface. SALEM energy meter can operate as energy counter and energy analyzer computing power, energy, phase shift, *rms* calculation, frequency measurement and harmonic content analysis.

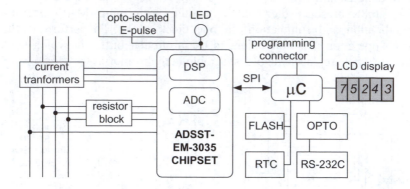

Figure 6.87. DSP based energy meter and analyzer SALEM of Analog Devices (AD 2005)

The application of microcontroller is very useful for routine and complicated calculations. As the example Fig. 6.88 presents the three-voltmeter method of impedance analysis (described in Section 5.5.1). It is possible to compute all components of impedance and even display the phasor diagram.

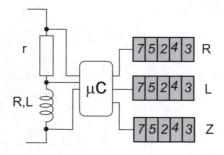

Figure 6.88. Microcontroller supported three-voltmeter methods of impedance analysis

6.5.2. Microinterfaces – SPI and UART

Microcontrollers are often equipped with a USB or CAN interface. These interfaces are described in Section 6.4.9. To connect an external device such as ADC, memory, real time clock, often a simple *SPI* interface (*Serial Peripheral Interface)* is used (Fig. 6.89).

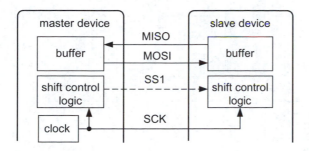

Figure 6.89. Connection of the devices via SPI interface

The SPI interface is a clocked serial stream of bits transmitted between one device established as master and another device set as slave. For connection of both devices three lines are used: *MISO* (*Master Input Slave Output*), *MOSI* (*Master Output Slave Input*) and *SCK* (*Serial Clock*). When there is more than one slave device additional line *SS* (*Slave Select*) is used. Data are transferred as 1 byte frames with a start and stop bit and then are transferred to the buffer and register. Fig. 6.90 presents the connection of microcomputer and digital-to-analogue converter.

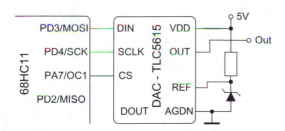

Figure 6.90. Connection of the DAC to microcontroller via SPI interface (ones and zeros 2005)

The *UART* (*Universal Asynchronous Receiver/Transmitter*) is an interface which transfers the data from the data bus (usually in parallel format) into a serial frame very similar to that used in the RS-232 interface. Therefore, the UART can be used directly as the interface to a computer (or microcontroller) or as the intermediate step to other standard interfaces (Fig. 6.91). It is also possible to use a similar interface for synchronous transmission (in such a case it is called *USRT* – *Universal Synchronous Receiver/Transmitter*). Both interfaces are also called USART.

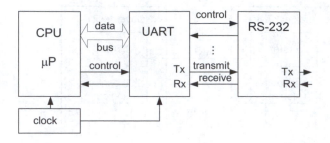

Figure 6.91. UART as the converter from parallel data to serial input/output

Fig. 6.92 presents the structure of UART transmitter and receiver. In the UART transmitter the data are loaded into a register. Between the transmitter holding register and transmitter shift register (which directly transmits the serial data) is the module SDU (serial data unit, which forms packet in a form: start, data, parity, stop bits).

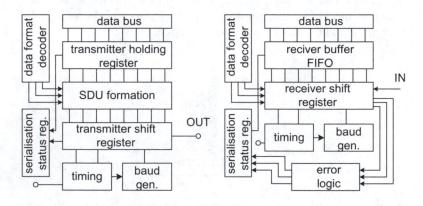

Figure 6.92. Block diagram of UART transmitter and receiver (Park 2003)

The UART receiver is waiting for the start bit, and when the packet of data appears the receiver shift registers form data as the parallel format word. This word is then transferred to FIFO (*first in first out*) memory. If the error logic shows that transmission was correct the data can be next transferred to the data bus.

The UART 8250 introduced by Motorola could transfer 1 byte packet of data with speed *19 200 bps*. Currently, instead of old UART 8250 the high speed UART 16550 is often used, which can send the 16-byte packets with a speed up to *115 kbps*. Figure 6.93 presents the main components UART interface used in personal computer serial port.

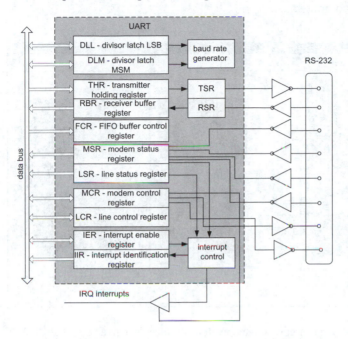

Figure 6.93. The main components of UART interface in PC serial port (James 2000)

6. 6. VIRTUAL MEASURING SYSTEMS

6.5.1. What is the virtual measuring device?

Most computer measuring systems use the graphical interface called *HMI (Human Machine Interface)* as the communicator between the computer and the operator. On such a screen, there can be presented digital displays very similar to those used in the real instruments. Fig. 6.94 presents an example of the user interface designed in LabVIEW program. Operator can change the parameters with virtual sliders, knobs or switches similarly to in the real instruments. Today most measuring instruments offer software with virtual HMI.

The virtual world surrounds us in various ways. We can buy various things in virtual shops (although the goods are real). In films, the expensive scenes are imitated by computer created graphics. In many computer games we can take actions as in real life. There are well known and popular games called "simulators". In such games the player can manipulate all navigation instruments and the computer simulates the effects in "real" situations and the results of bad decisions. It is possible to create the virtual world with the power of computers.

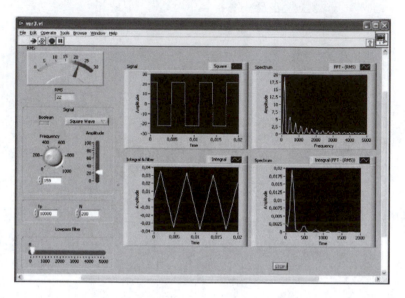

Figure 6.94. An example of the operator interface designed using LabVIEW

So, why not use this power in measurements? We can design the measuring instrument with all necessary "navigation" elements, with indicators, displays and graphs. In comparison with conventional instruments we can simply obtain the measuring results as documents, the print or transmit them. We can save the human environment because one computer can substitute many hardware instruments, with housings, supply parts and without unreliable mechanical parts. But as in a virtual shop the "goods" should be real – we should be able to measure the real signals using real data acquisition part of the instrument. Without real signals (real world in the background as it is presented in Fig. 6.95) we can use the simulator device and not a virtual measuring device.

The virtual instrument presented on the computer screen can by modified by the operator. This is a very important advantage of a virtual instrument, because we can have the measuring instrument exactly as we want it and not as it is available in the catalogues. And what is also very exciting is that we can design and create this instrument. Today, with the presence of good quality data acquisition boards we can design the measuring device and the result depends only on our knowledge, intellectual abilities and creative power.

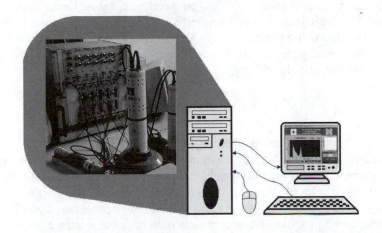

Figure 6.95 The real world in the background of virtual instrument (Tlaczala 2005)

The term *Virtual Measuring Device* sometimes can cause misunderstanding, because many electrical instruments communicate with the operator by an HMI type screen, but not all these instruments are virtual. As *virtual measuring instrument* we understand the computer measuring system consisting of a real part (data acquisition board, real measuring instruments controlled by interface – hardware) and a virtual part where measuring procedures (signal processing, result visualization – software) are realized only by computer. Thus in virtual instrument part of measuring procedures (earlier realized by equipment - hardware) is realized by software. Important role in the virtual measuring instruments plays the human-machine interface.

We can distinguish two types of virtual measuring instruments. The first one is built on the basis of the data acquisition board and practically the whole measuring procedure is performed by the computer. The second one can be created by standard measuring devices, such as multimeters or generators connected to the computer through interfaces. In such a system the measuring instruments operate as a tool (periphery) of the computer virtual instrument, where the results are processed and visualized.

Currently, the design and creation of virtual instruments is facilitated by several computer programs enabling the operator to realize such instrument with little knowledge of programming details. Usually such programs offer rich graphical tools for construction of sophisticated and even artistically attractive work.

National Instruments with *LabVIEW* program is a leader of graphical user friendly programming of virtual measuring instruments. But there are several other similar programs such as VEE Pro developed by Agilent, *TestPoint* developed by CapitalEquipment Corporation, *DasyLab* of Dasytec. Useful tools for data acquisition are also available in MATLAB *Data Acquisition Toolbox*. There is also free available graphical programming extension to MS Visual Studio.Net called *SoftWire* .

6.6.2. TestPoint

The *TestPoint* program was developed by Capital Equipment Corp. and was delivered as the support of measuring equipment produced by Kethley. At the moment of writing this book the future of Testpoint is not clear because it was sold to National Instruments – the creator of the most frequently used (and competing with TestPoint) program LabVIEW. Nevertheless, this program is worth presentation because it is an example of a very useful, simple and well designed program for data acquisition.

Let us start the presentation of TestPoint with a simple example. Assume that we intend to receive the signal (using a typical data acquisition board DAQ installed in the computer), then we want to observe the waveform on the oscilloscope-like screen and additionally we want to determine the *rms* value of this signal. After starting the program we see three main panels: *Stock* panel with tools, main panel (front panel of our virtual instrument) and the field with chosen tools called *Objects* (Fig. 6.96).

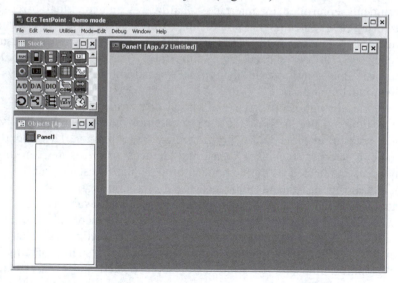

Figure 6.96. The screen of the computer after starting of the TestPoint

In the Stock panel (Fig. 6.97) we can see the tools like: pushbutton, switch, selector, slider, data-entry, indicator, display, bar indicator, table, graph display, ADC, DAC, digital I/O, RS232, GPIB, loop, conditional, case, math function, timer, etc. There are a number of tools, each of which has several variants. To design our measuring instrument assumed above we need to select: pushbutton, ADC, digital display, graph display and mathematical function tool to calculate the *rms* value.

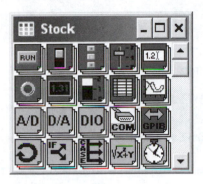

Figure 6.97. The stock panel of TestPoint

After selection these tools appear on the main panel and in the object panel as presented in Fig. 6.98. Additionally, a new file appears (called *Action List*) where we can construct the program as a sequence of operations.

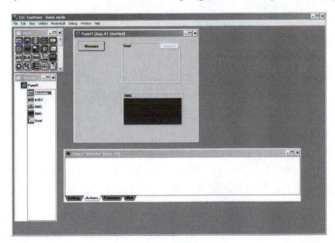

Figure 6.98. The computer screen after selection of the tools

Fig. 6.99 presents the action list of our program. As a first step we acquire the signal with set parameters (number of samples, sampling rate). Then, we calculate the *rms*, transfer the sampled data to the graph display and calculated value to the digital display. We can fill the grey field practically without writing, only by dragging the selected parameters by mouse. After this, we can put the pushbutton *Measure*, which will enable to start the program.

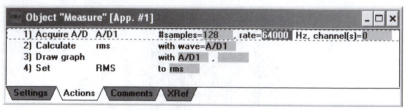

Figure 6.99. The Action List with the program as the sequence of operations

As a result we obtain the panel with measured data as presented in Fig. 6.100.

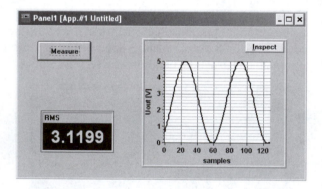

Figure 6.100. The panel of the virtual instrument after performed measurement

Thus, in a few minutes we were able to design and start to operate a complete measuring instrument – a digital voltmeter and a kind of simple oscilloscope.

In the TestPoint the function Inspect is very useful. After clicking this field the display is changed to a graphic file. We can now design a style of letters, colors and dimensions of fields, etc. After starting the copy operation this picture is transferred to the clipboard in two forms: as the graphical picture (Fig. 6.101a) and as the numerical set of samples (Fig.6.101b). So we can transfer the result as a figure (graph) to the document and using for example Excel we can convert digital data to the graph or we can perform additional mathematical operations if required.

a)

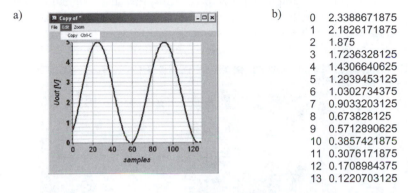

b)

0	2.3388671875
1	2.1826171875
2	1.875
3	1.7236328125
4	1.4306640625
5	1.2939453125
6	1.0302734375
7	0.9033203125
8	0.673828125
9	0.5712890625
10	0.3857421875
11	0.3076171875
12	0.1708984375
13	0.1220703125

Figure 6.101. Results of measurement in graphical form (a) and in digital form (b)

The example presented above is very simple. However, we can design much more complex measuring instruments. Figure 6.102 presents the example of the front panel of the measuring system for testing of the magnetic materials.

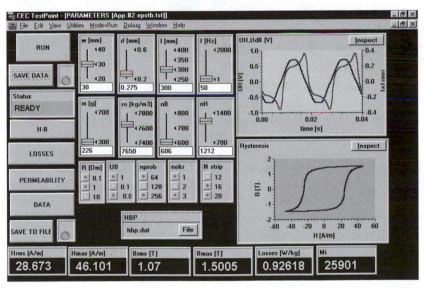

Figure 6.102. The user interface of the measuring system for testing of magnetic materials

In the central part of the user interface (Fig. 6.102) there are elements for setting the sample parameters (width w, thickness d, mass m, etc.) On the right hand side there are graph displays presenting measured signals and a hysteresis loop. In the bottom part we can see digital displays of measured values. These measured values can be saved in the form of a table (option Save data in Fig. 6.102) or as a file (option Save to file).

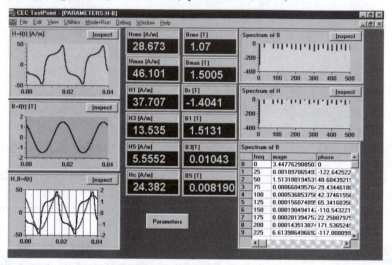

Figure 6.103. Sub-panel (of the panel presented in Fig. 5.101) with results of FFT analysis.

In the design of virtual elements the number of presented elements is practically unlimited. But when there are too many elements then the whole panel becomes cluttered and illegible – like in the pilot's cockpit. Therefore, it is reasonable to divide the main panel into several sub-panels. At the left hand side of the panel presented in Fig. 6.102 there are several buttons, which start appropriate sub-panels. Fig. 6.103 presents the sub-panel entitled H-B, in which there are additional, more detailed fields with measured results. For example we can see the results of FFT analysis. In this case, the return to the main panel is always possible by using the button "Parameters".

6.6.3. Agilent VEE Pro

VEE Pro (*Visual Engineering Environment*) is a graphical software environment designed for measuring instrument programming. In the program there are included thousands of drivers of various measuring instruments. It is possible to take complete control over each external device and it is possible to connect measuring devices into the measuring system with various interfaces.

Figure 6.104 presents the screen representing multimeter 34401 with the control panel simulating all functions of the device and the object diagram for programming. Real and virtual instruments, tools and signal processing functions are available as objects collected in menus. The objects on the work area of the program are represented by icons, which can be connected with virtual wires. For example, at the upper part of Fig. 6.104 we can see the icon of multimeter, which is connected to the display and to a repeater causing continuous processing of measurements. In the bottom part we can see the graphical interface of driver of this instrument with the possibility to control all functions of the multimeter by computer.

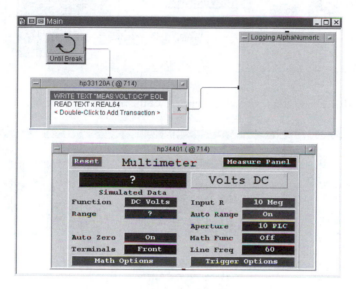

Figure 6.104. Graphical representation of digital multimeter in VEE program

Figure 6.105 presents the block diagram of the measuring system, which performs integration of the signal and computing the *rms* value. The results are presented on the graph and analogue/digital displays. Two virtual generators (signal generator and noises one) are connected as the source of the signal. The parameters of the signals (amplitude, frequency, level of noises) can be changed by the virtual sliders or knobs. All objects are connected with virtual wires similarly as in real circuits.

Selected objects (which could be any of the control devices, displays and indicators) can be moved to the user interface. Figure 6.106 presents the example of the user interface of the system (presented in Fig. 6.105). The operator can change the generator parameters and on the displays there are presented signals: input from the generator and output after integration. Additional instruments indicate the peak and *rms* values of the output signal.

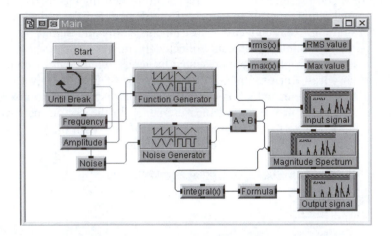

Figure 6.105. The example of a block diagram of the measuring system

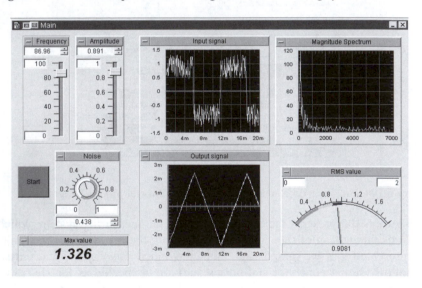

Figure 6.106. Example of the operator panel in the VEE of the measuring system presented in Fig. 6.105

Although VEE is designed for managing of measuring instruments it is also useful for design and organization of the measuring system controlled by computer. It is also possible to perform complex signal processing especially because MATLAB scripts and Signal Processing Toolbox with hundreds of functions are embedded in the program.

6.6.4. LabVIEW of National Instruments

LabVIEW is a programming system developed for data acquisition and processing. It is the most popular program, partly due to the importance of National Instruments, but also because it was the first such software environment. The LabVIEW designers developed special graphical programming language (*G programming language*) enabling users to work with such programs, even by not experienced in programming using higher level language, as for example C programming language. Every new edition of LabVIEW (currently, in 2005 version 7 is in use) introduced new possibilities and today the whole package is a powerful and sophisticated programming platform (Beyon 2000,Canway 2003, Chugani 1998, Essick 1998, Johnson 2001, Ritter 2001, Sokoloff 2003).

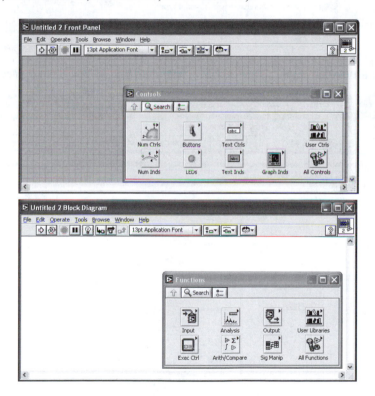

Figure 6.107. Two main windows in LabVIEW

Two main windows appear after starting a new LabVIEW program, these are: *Block Diagram* for graphical programming and *Front panel* for design the user interface (Fig. 6.107). In both windows there is palette of tools: in the block diagram there are icons representing various *Functions* (analysis, input, output, arithmetic, waveform etc), while in the Front Panel there are *Controls* (pushbuttons, displays, graphs, indicators) It is possible to design new objects (subprograms or libraries) and also it is possible to insert icons representing real instruments (in form of the drivers).

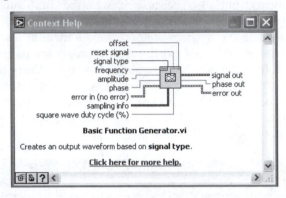

Figure 6.108. The example of context help with short information about connectors

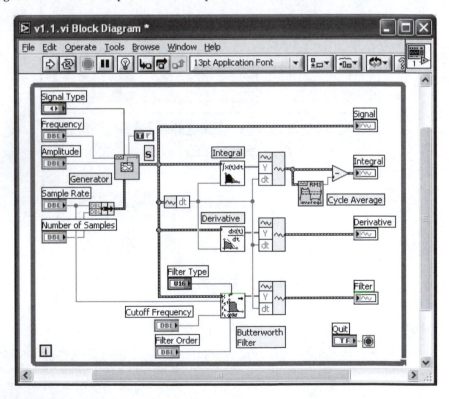

Figure 6.109. The example of the designed block diagram

All icons have various connectors (input and output ones) and the context help information are easy available to assist in the design process (Fig. 6.108).

During graphical programming the icons representing objects are connected with virtual wires. There are various wires with different thicknesses and colors. Thin wires represent single values, while thick are reserved for data sets (arrays of values). Various colors are used for real values, logic data, for strings, etc.

Figures 6.109 and 6.110 present the design of the data processing virtual device[1], similar to that presented in the previous section devoted to VEE. The signal from the virtual generator is integrated, differentiated or filtered by the Butterworth filter. The results are presented in the operator panel similarly as in the real measuring device (Fig. 6.94) or in the design more related to the laboratory equipment (Fig. 6.110).

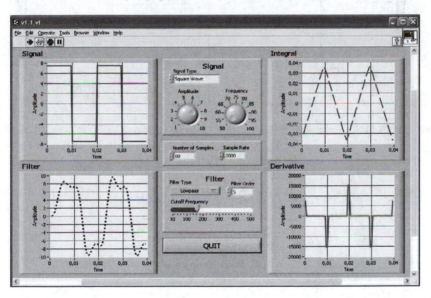

Figure 6.110 The example of operator interface designed for the signal processing device presented in Fig. 6.108

Fig. 6.111 presents another, frequently used application of LabVIEW – the spectrum analysis. This application consists of two parts – the virtual generator of signals with noises (*simulation*) and part of the spectral analyzer

---

[1] All examples of LabVIEW applications presented in this section have been designed by Slawomir Baranowski.

(*measurements*). Two *subVIs*[1] (which are analogous to subroutines) have been used as the generator: *Uniform White Noise Waveform.vi* and *Basic Function Generator.vi*.

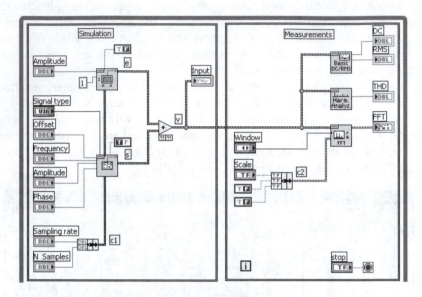

Figure 6.111. The example of the spectral analyzer

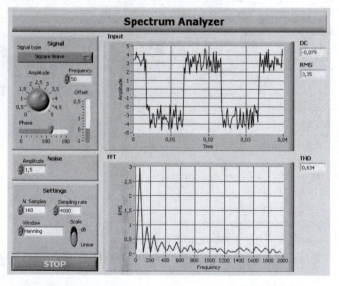

Figure 6.112. The operator interface of the spectrum analyzer

---

[1] The programs realised in LabVIEW are called VI.

The input signal of spectral analyzer (and output signal of generators) is presented in graph display. Computation of DC component and *rms* value of the signal is performed using subVI *Basic Averaged DC-RMS.vi*. The *Harmonic Distortion Analyser.vi* enables us to compute the THD coefficient. The subVI *FFT Spectrum (Mag-Phase)* can be used for spectrum analysis. As a result of this analysis the spectrum with *rms* values of harmonics is presented.

The operator in the front panel can set parameters of analyzed signal (amplitude, frequency, noise level) and parameters of data acquisition. As result obtains graph of the wave, FFT line spectrum, THD coefficient.

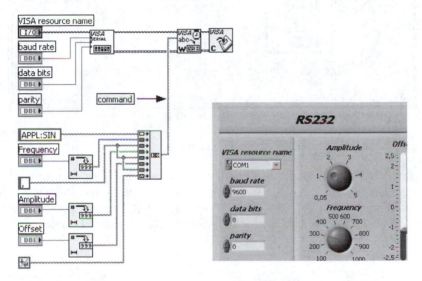

Figure 6.113. The example of remote control of generator via serial interface

Fig. 6.113 presents another important function of LabVIEW – control of the measuring device via an interface such as: GPIB, USB or RS-232C. Fig. 6.113 illustrates the example of the control of a generator using serial interface. As the first step, it is necessary to configure the connection. The serial port is configured by the tool *VISA Configure Serial Port.vi* where we can set the port identification (*VISA Resource Name*) and the serial data transfer parameters.

Next, we can send the command in the in the SCPI language in the form of the ASCII text, for example: *APPL: SIN <frequency>, <amplitude>, <offset>*. This command is converted to the string by the functions *Number to Decimal String* and *Concatenate String*. The prepared series of bytes is then sent through the serial port by the *VISA Write* function. The function

VISA Close is for closing the connection (after which the serial port can be used by another program).

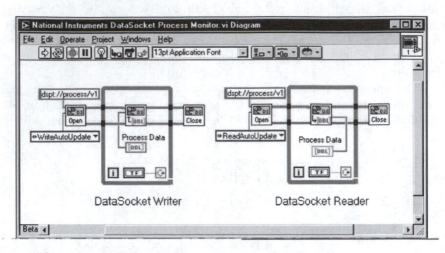

Figure 6.114. The DataSocket tool for communication and data transfer

LabVIEW offers also some simple tools for sending the data through the network. This tool is called *DataSocket* and it simplifies the data exchange between application and FTP or HTTP servers. It helps in sending the data to the server (*DataSocket Write*) and in receiving data (*DataSocket Read*).

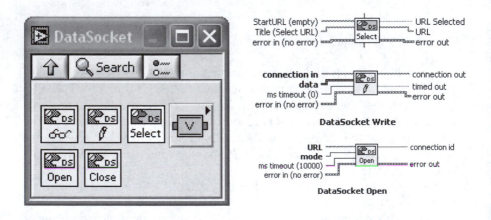

Figure 6.115. The functions of DataSocket: DS Read, DS Write, DS Open, DS Close and DS Select subVIs

Usually, to realize the data transfer between two computers through the network it is necessary to write appropriate programming code. Using DataSocket this is a relatively simple task, not more complicated than using a browser to read Web pages on the Internet. Protocol of DataSocket - DSTP (*DataSocket Transfer Protocol*) based on TCP/IP enables the data to be transferred between DataSocket clients (known as *DataSocket Reader* and *DataSocket Writer*) using *DataSocket Server*.

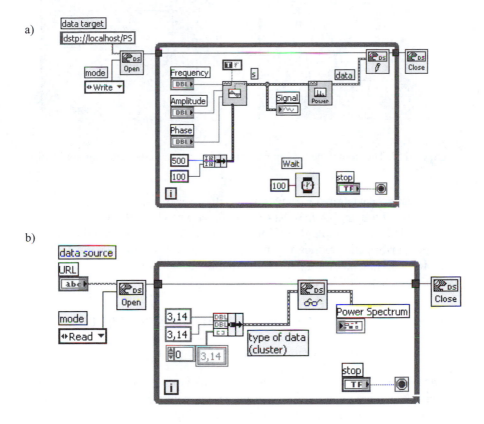

Figure 6.116. Block diagrams of write (a) and read (b) used in data transfer by DataSocket

Figures 6.116 and 6.117 present the application performing the following task: there is a network connection between two applications; in the first application the signal from a virtual generator is analyzed by *FFT Power Spetrum.vi* and the results are sent to the *DataSocket Write;* in the second application the data are read with function *Data Socket Open/Read* and displayed on the user interface panel.

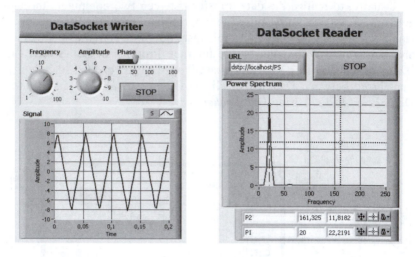

Figure 6.117. The front panels of operator interfaces used in data transfer application

To realize data transfer via network it is necessary first to connect the applications by using *DataSocket Open* function, where the URL addresses are specified (for example dstp://localhost/PS). After connection the analysis can be performed practically in on-line mode – every change of the waveform in one laboratory results in display of new results of FFT analysis in the remote, second laboratory.

6.7 THE EXAMPLES OF COMPUTER MEASURING SYSTEMS

6.7.1. The measuring system for testing of magnetic materials

At the end of this book the author would like to present several measuring systems which are practical illustrations of the material presented above. One of them is a measuring system designed for testing the properties of electrical steel. Electrical steels are tested by applying three main methods: Epstein frame, SST (*Single Sheet Tester*) device and SStT (*Single Strip Tester*) device. All these methods determine similar magnetic parameters (flux density B for defined values of magnetic field strength H, permeability μ, losses P, anisotropy δP). To determine of all these parameters it is necessary to ensure appropriate magnetizing conditions (see Section 6.7.2) and measure magnetic field strength H and flux density B.

There are two main methods of measurement of the magnetic field strength in the investigated steel sample. The first one (indirect method), enables calculation of magnetic field strength in the sample (of the length of magnetic path l) from the magnetizing current I in the primary winding n_1 according to the Ampere's law

$$Hl = I n_1 \qquad (6.2)$$

The second method uses the direct magnetic field measurement by means of a thin, tangential coil sensor, according to the rule that the tangential field component near the specimen surface is the same as the inside of the sample.

The flux density can be determined from the voltage e_2 induced in the secondary winding n_2 wound around the sample[1] with the cross-sectional area A, according to the Faraday's law

$$e_2 = -n_2 A \frac{dB}{dt} \qquad (6.3)$$

Thus, to determine the parameters of electrical steel we can measure the voltage drop across the resistor R (in this way we determine the current I and subsequently the value of magnetic field strength H) and the voltage induced in secondary winding e_2. For these measurements we can use a typical data acquisition board. In the case of processing the signal proportional to the H the problem is that the signal exhibits wide dynamic range (from several A/m to several kA/m). In the case of processing the signal proportional to B it is necessary to integrate the signal.

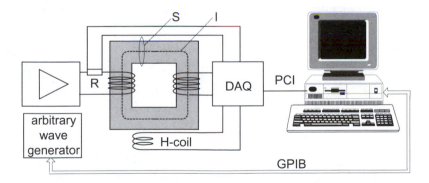

Figure 6.118. The main circuit of the measuring system for testing the magnetic materials

---

[1] Sometimes the induced voltage can be measured not in secondary winding, but between two needles electrically contacted with the specimen (Fiorillo 2004)

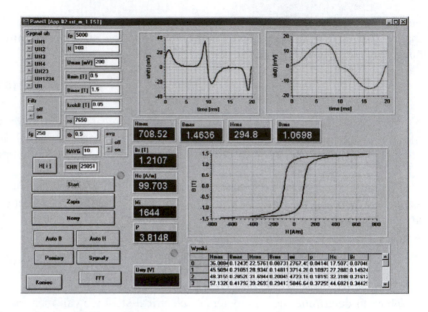

Figure 6.119 The example of the user interface for control of measured parameters

The operator sets the sample geometrical parameters and range and changes the interval of variables. The signals proportional to *B* and *H* after processing (for example integration of the coil signal) were used for computation of the magnetic parameters of the electrical steel. Fig. 6.120 presents the example of the operator interface.

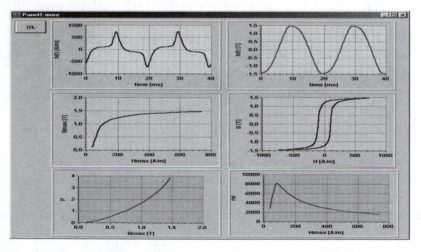

Figure 6.120. The operator interface with determined parameters

In automatic mode the operator set the range of variable, for example $B = 0 - 1.9\ T$ with interval $\Delta B = 0.1\ T$. The system measures B and via interface (RS-232C or GPIB) sets an appropriate value in the generator. However, due to the nonlinear dependence $B=f(H)$ and $B=f(U_{gen})$ special algorithm is necessary to set the B value iteratively. One of the simplest methods – bisection iteration is presented in Fig. 6.121.

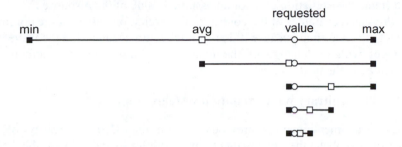

Figure 6.121. Bisection algorithm for searching of the requested value

In the bisection algorithm we assume that the requested value is inside some initial range *<min,max>*. Next, the average value of this range is calculated and compared with requested value. After this comparison the next range *<min,avg>* or *<avg,max>* is selected, respectively. The process is repeated until the difference (*avg-requested*) value reaches a satisfactory limit.

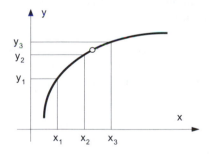

Figure 6.122. The polynomial-based algorithm for searching of requested value

The bisection algorithm is relatively slow. More effective (faster and more accurate) appears to be the polynomial-based algorithm. It is possible to determine coefficients of the third order polynomial from three points $P_0(x_0,y_0)$, $P_1(x_1,y_1)$, $P_2\ (x_2,y_2)$ (two points measured earlier and one point to be measured)

$$
\begin{bmatrix} a_0 \\ a_1 \\ a_2 \end{bmatrix} = \begin{bmatrix} y_0 & y_0^2 & y_0^3 \\ y_1 & y_1^2 & y_1^3 \\ y_2 & y_2^2 & y_2^3 \end{bmatrix}^{-1} \cdot \begin{bmatrix} x_0 \\ x_1 \\ x_2 \end{bmatrix} \tag{6.4}
$$

Knowing the dependence $y=f(x)$ ($U_{gen}=f(B)$ in our case) it is possible to determine the excitation U_{gen} for an assumed value of the response B.

Presented examples of the control of the excitation values of the nonlinear object are simplified because it does not take into account that the waveform changes with the change of the magnitude. More complex methods are presented in the next section.

6.7.2. The arbitrary wave excitation systems

As was mentioned in previous section due to non-linearity of the investigated object the same magnetizing conditions need to be established in order to compare results of investigations of electrical steel. As such reference condition the sine shape of flux density waveform was chosen by the standards. There are various methods of obtaining such condition, Fig. 6.123 presents the application of the neural perceptron network.

The signal X is connected to the input of the neural network from a virtual generator. Such a signal represented by N samples is processed by the neural network. The output signal Y from the network is converted to the analogue form by DAC device in the data acquisition/generation board. The analogue signal after amplification (power amplifier) is used as the signal in primary winding of the magnetized sample. The secondary voltage after integration is compared to the reference flux density. The difference (error) is used to control the neural network.

In the first step the weights of perceptron layers $W1$ and $W2$ are randomly initialized. The input signal of the first layer of neurons is computed by the $U1$ element of LabVIEW program (Fig. 6.124)

$$
u_i^{(1)} = \sum_{j=0}^{N} w_{ij}^{(1)} x_j \tag{6.5}
$$

The output signal of the first layer is computed by $V1$ element

$$
v_i = f(u_i^{(1)}) \tag{6.6}
$$

Respectively the input signal of the second layer is computed by $U2$ element

$$u_i^{(2)} = \sum_{j=0}^{N} w_{ij}^{(2)} v_j \tag{6.7}$$

and output signal of the second layer is computed by element Y

$$y_i = f(u_i^{(2)}) \tag{6.8}$$

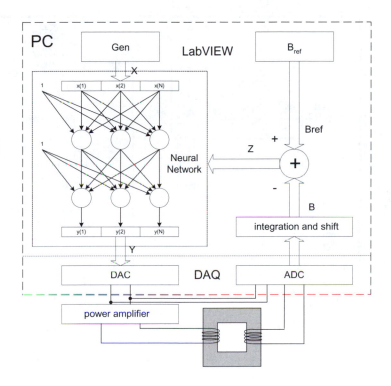

Figure 6.123. The operation principle of the system of arbitrary wave generation

Elements $f'(U2)$ and $f'(U1)$ compute the derivative of the activation function of first and second layer

$$\frac{df\left(u_i^{(1)}\right)}{du_i^{(1)}}, \quad \frac{df\left(u_i^{(2)}\right)}{du_i^{(2)}}$$

Element OBJ connects the output signal of neural network to DAQ input, measures the output signal of the magnetized element and determines the flux density.

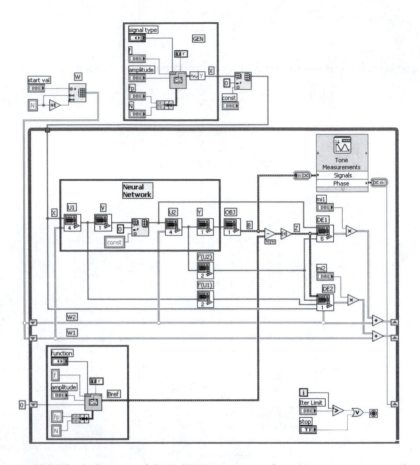

Figure 6.124. Block diagram of the LabVIEW program for arbitrary wave generation

The *DE1* and *DE2* elements compute the gradient coefficients according to the principle of the back propagation of errors

$$\frac{\partial E}{\partial w_{ij}^{(1)}} = \sum_{k=1}^{N} \left(b_k - b_{refk} \right) \frac{df(u_k^{(2)})}{du_k^{(2)}} w_{ki}^{(2)} \frac{df(u_i^{(1)})}{du_i^{(1)}} x_j \tag{6.9}$$

and

$$\frac{\partial E}{\partial w_{ij}^{(2)}} = \left(b_k - b_{refk} \right) \frac{df(u_1^{(2)})}{di_i^{(2)}} v_j \tag{6.10}$$

where $E(w)$ is the objective function

$$E(w) = \frac{1}{2}\sum_{k=1}^{N}\left(b_k - b_{refk}\right)^2 \qquad (6.11)$$

The adaptation of the weights is realized according to the relation: $w(k+1)=w(k)+\Delta w$, where $w(k)$ are the weights in the k-step, $\Delta w=-\mu\cdot\nabla E(w)$. Additionally, to improve the speed of iteration the phase of B signal is determined and respectively the phase of B_{ref} is modified.

a)

b)

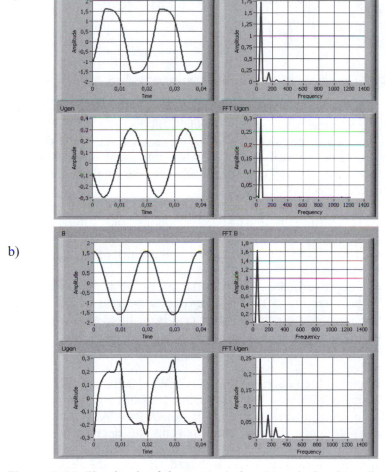

Figure 6.125. The signals of the system before (a) and after (b) realization of the algorithm

Fig. 6.125 presents the initial signals and final signals after realization of the algorithm. In the initial state the generator signal is a sine wave but the flux density is disturbed. After several iterations the flux density is a sine wave while the generator signal is suitably changed.

Other, competitive solution of the same problem was developed in Wolfson Centre of Magnetics Technology in Cardiff (Zurek 2004). The special "control" in the form of a feedback circuit has to be employed in order to ensure the sinusoidal magnetizing conditions. In the past, the negative feedback circuits were based on analogue electronics (Fig. 6.126).

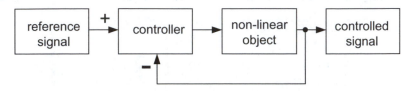

Figure 6.126. A simplified concept of negative analogue feedback

Although the analogue electronic circuits were working in true real-time mode, their performance was limited. Especially, in the cases of high level of magnetization or highly non-linear materials the required sinusoidal shape of flux density could not be easily achieved. Also, only a sine wave shape was relatively easy to obtain – any other shape of waveform would require much more sophisticated electronics.

However, recently the power of digital technology can be used. If the computer is equipped with a data acquisition and an arbitrary signal generation board (often combined into one device) then the whole "control" procedure can be achieved in the software, which is much more flexible than the electronic circuits.

The idea of negative feedback remains the same – the acquired waveform (controlled signal –see Fig. 6.126) is compared to the reference waveform and the output waveform generated by the card is modified according to the difference between them. The process has to be repeated many times in order to bring the difference between the controlled and reference waveforms to the required level.

The program has been developed in LabVIEW, which provides excellent signal processing possibilities. The main digital feedback algorithm program consists of four main parts, as shown in Fig. 6.127. Part "A" is a measuring thread responsible only for the measurements of appropriate values. Therefore, the raw, as measured data is fed straight to that part without any additional filtering or calculations. The thread "B" uses the same data as thread "A", but some additional signal processing (low-pass filtering, precise software triggering, etc.) is carried out. The data/waveforms in the thread

"B" are fed to a proportional controller (part "C" in Fig. 6.127), which compares the controlled and reference waveforms and modifies the output waveform. However, before the generation there is another part of the digital feedback. Part "D" compares the phases for each harmonic in the controlled signal (flux density in this case) and the magnetizing current, decomposes with Fourier Transform the output waveform, introduces the calculated phase shifts for all harmonics, composes back the output waveform with an inverse Fourier Transform (see element "Correct phase" in part "D" in Fig. 6.127), and only then the output waveform is generated. The calculations performed in part "D" introduce a correction for the shift in phase characteristics due to non-linearity of any devices between the voltage generation stage (the output of the card) and the controlled waveform (flux density); these devices could be for example: low-pass filter, power amplifier, separating transformer, magnetizing yoke. Therefore, the whole feedback algorithm becomes fully adaptive, since all these non-linearities are automatically taken into account during the magnetization process. For that reason the part "D" is called "adaptation module".

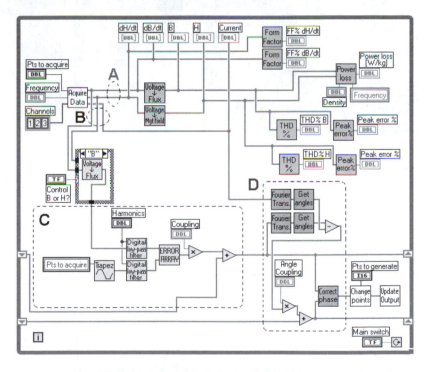

Figure 6.127. Simplified adaptive digital feedback algorithm written in LabVIEW (not all objects are shown): A – measurement thread, B – control thread, C - proportional controller, D – adaptation module

The adaptive digital feedback algorithm presented in Fig. 6.127 is so versatile that it can be used without any modification with various magnetizing systems and samples (ring sample, single sheet tester, Epstein frame) without any modification of the software. If more channels are added (i.e. the parts "A-D" are copied), then such a program is capable of controlling a more complex systems (rotational, arbitrary two-dimensional and three-phase magnetizing yokes).

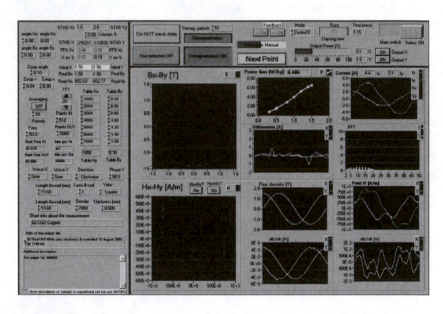

Figure 6.128. The front panel of the LabVIEW program for two-dimensional and rotational measurements under controlled arbitrary flux density and magnetic field conditions

Obviously, such an approach is not real-time anymore, but the required shape of the controlled waveform is achieved within a few seconds. However, the adaptive iterative feedback can automatically measure and save the full output data (power loss, permeability, flux density and magnetic field waveforms) for selected magnetizing parameters (frequency, sample, flux density). The stability and accuracy of the control is very good: for controlled sinusoidal flux density the errors of peak value and the FF (form factor) are below *0.05%*. Also, the magnetic field waveform can be controlled (Zurek 2004). Moreover, the shape of the waveforms can be arbitrary: sine, triangle, trapezoid, or even PWM. The control of sinusoidal shape of flux density waveform (which is still the most important in the industry) can be achieved for any frequency between *0.5 Hz* and *50 kHz*.

6.7.3. The scanning device for magnetic field imaging

The system for magnetic field scanning, called *Magnetovision* has been designed for testing of magnetic field distribution in three-dimensional volume (Tumanski 1998, Tumanski 2002). After scanning the system processes data and the magnetic field distribution is presented in a form of a color map (Fig. 6.129).

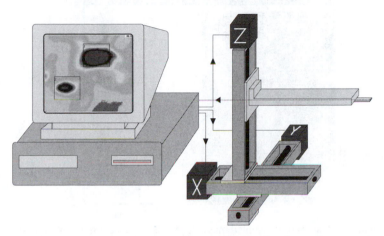

Figure 6.129. The system for scanning and visualization of magnetic field distribution

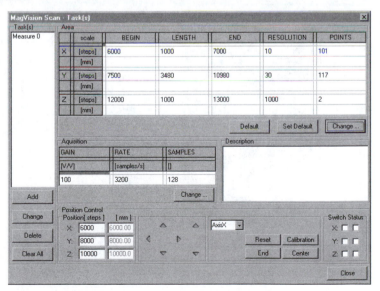

Figure 6.130 The operator panel of the scanning system

The three-axis positioning device consists of three micrometer screws driven by stepper motors. The movement of each of the stepper motor is controlled by computer. The front panel of the scanning subroutine is presented in Fig. 6.130. The results of scanning can be presented graphically in on-line mode during movement of the sensor. Fig. 6.131 presents the screen picture during the scanning process.

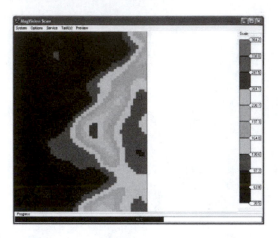

Figure 6.131. The map of scanned magnetic field presented in on-line mode

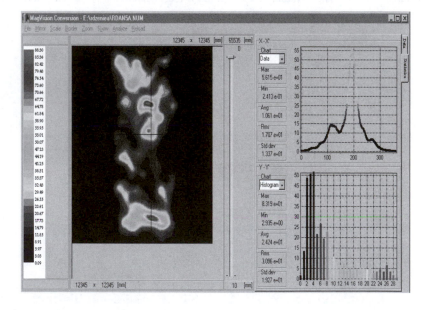

Figure 6.132. The presentation of scanning result

After finishing the movement, the registered data can be processed to present the map as well as the histograms of magnetic field distribution in a selected line. Figure 6.132 presents the example of such a presentation.

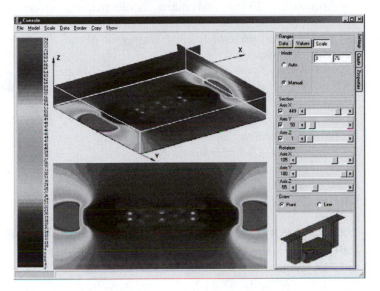

Figure 6.133. The presentation of magnetic field distribution in 3D volume

Fig. 6.133 presents the post-processed presentation of a three-dimensional distribution of a magnetic field. It is possible to obtain the results in the form of a color map of arbitrary selected cross-section of the magnetic field. Fig. 6.133 presents the example of investigation of magnetic field distribution around the C-shaped yoke with magnetizing winding.

The three examples of the computer measuring system presented above illustrate that the possibilities of such systems today are practically unlimited, mostly due to the power of computers. However in the background of the computer technology still the most important is the designer, with his knowledge and experience. The author hopes that this book can help in widening the knowledge about measurement principles.

REFERENCES

AD 2005 Analog Devices, www.analog.com

Advanced Information 2005 Advanced Information Networks, www.advancedinformation.net

Anderson D., Dzatko D., Mindshare I. 2001 *Universal Serial Bus System Architecture*, Addison-Wesley

Anderson D. 1998 *FireWire System Architecture*, Addison-Wesley

Armson M. 2003 *The Status of IEEE 1451.4 Plug and Play*, Honeywell Sensotec

Axelson J.L. 1999 *Serial Bus Complete*, Lakeview Research

Axelson J.L. 2005 *USB Complete*, Lakeview Research

Bailey D. 2003 *Practical SCADA in Industry*, Newnes

Bekkers R. 2001 *Mobile Telecommunications Standards: UMTS, GSM, TETRA, ERMES*, Artech House

Bender K., Katz M. 1993 *Profibus – the Fieldbus for Industry Automation*, Prentice Hall

Beyon J.V. 2000 *LabVIEW Programming, Data Acquisition and Analysis*, Prentice Hall

BDTI 2005 Berkley Design Technology Inc., www.bdti.com

Bluetooth 2005 www.bluetooth.com

Boyer S.A. 2004 *SCADA*, Instrumentation Systems

Burns A. 2001 *Real-time Systems and Programming Languages*, Addison Wesley

Callaway E.H. 2003 *Wireless Sensor Networks*, CRC Press

Caristi A.J. 1989 *IEEE-488 General Purpose Instrumentation-Bus Manual*, Academic Press

Castro J.P. 2001 *The UMTS Network and Radio Access Technology*, Wiley & Sons

Cisco 2005 Cisco Systems, www.cisco.com

Chasaing R. 2002 *DSP Application using C and TMS320C6X DSK*, Wiley & Sons

Chen A. Wong J. 1999 *Designing the TMS320C548/9 DSP Development Board*, Texas Instruments Technical Brief No. 465

Chugani M.L., Samant A.R., Cerna M. 1998 *LabVIEW Signal Processing*, Pearson Education

Conway J. 2003 *A Software Engineering Approach to LabVIEW*, Prentice Hall

Cooling J.E. 2002 *Software Engineering for Real-time Systems*, Addison Wesley

Dahnoun N. 2000 *Digital Signal Processing Implementation*, Prentice Hall

Dataq 2005 Dataq Instruments, www.dataq.com

DataTaker 2005 DataTaker www.datataker.com

Essick J. 1998 *Advanced LabVIEW*, Prentice Hall
Eyere J. 2001 *The Digital Signal Processors Derby*, IEEE Spectrum 6
Fiorillo 2004 *Measurement and Characterization of Magnetic Materials*, Elsevier
Firewire 2005, www.firewire.org
Freescale 2005, Freescale Semiconductors, www.freescale.com
Garney J., Solari E., Jaff K., Hosler B., Clallahab S. 1998 *USB Hardware and Software*, Independent Pub Group
Gook M. 2004 *PC Hardware Interfaces*, A-list
Held G. 2000 *Data over Wireless Networks: Bluetooth, WAP, WLAN*, McGraw-Hill
Held GT. 2002 *Ethernet Networks*, Wiley & Sons
Held G. 2004 *Virtual Private Networking*, Wiley & Sons
Honeywell 2005 Honeywell – Solid State Electronics Center, www.ssec.honeywell.com
IEEE P1451 www.ieee1451.nist.gov
Ilyas M. 2004 *Handbook of Sensor Networks*, CRC Press
Infineon 2005 Infineon Techn ologies AG, www.infineon.com
IrDA 2005 www.irda.org
Johnson G.W., Jennings R. 2001 *LabVIEW Graphical Programming*, McGraw-Hill
Kehtarnavaz N., Keremet M. 2000 *DSP System Design*, Prentice Hall
Kehtarnavaz N. 2004 *Real-time Signal Processing*, Elsevier
Kosiur D. 2001 *Build and Managing of Virtual Private Network*, Wiley & Sons
Lawanz W. 1997 *CAN System Engineering*, Springer
Liu J.W. 2000 *Real-time Systems*, Prentice Hall
Mitchell R.W. 2003 *Profibus*, Instrumentation Systems
Morrow R. 2002 *Bluetooth Operation and Use*, McGraw-Hill
Muller N.J. 2000 *Bluetooth Demystified*, McGraw-Hill
Nawrocki W. 2005 *Mesurement Systems and Sensors*, Artech House
NI 2005 National Instruments www.ni.com
Omega 2005 Omega Engineering Inc. www.omega.com
Ones and zeros 2005, www.onz.com/uci
Ontrak 2005 Ontrak Control Systems, www.ontrak.net
Paret D. Fenger C. 1997 I^2C *Bus*, Wiley & Sons
Park J., Mackay S., Wright E. 2003a *Practical Data Communications for Instrumentation and Control*, Newnes
Park J., Mackay S. 2003b *Practical Data Acquisition for Instrumentation and Control Systems*, Newnes
Pfeiffer.R., Ayre A., Keydal C. 2003 *Embedded Networking with CAN and CANopen*, RTC Books
Profibus 2005 Profibus International, www.profibus.com

Putten 2003 A.V. Electronic Measurement Systems – Theory and Practice, IOP Publ.

Radiometrix 20005, Radiometrix, www.radiometrix.co.uk

Reynders D., Wroght E. *Practical TCP/IP Ethernet Networking for Industry*, Elsevier

Ritter D.J. 2001 *LabVIEW GUI* , McGraw-Hill

Rob O. 1999 *Low-power Wireless Infrared Communication*, Kluwer

Roshan P. 2003 *Wireless Local Area Networks Fundamentals*, Cisco

Satel 2005 Satel, www.satel.fi

Schneeman R.D. 1999 *Implementing of Standard-based Distributed Measurement and Control Application on the Internet*, NIST

SDI-12 2005, www.sdi-12.org

Shay W. 2003 *Understanding Data Communications and Networks*, Thompson

Singh A., Srinivasan S. 2003 *Digital Signal Processing Implementation*, Brooks Cole

Sokoloff L. 2003 *Applications in LabVIEW*, Prentice Hall

Spurgeon C.E. 2000 *Ethernet – Definitive Guide,* O'Reilly

Superlogics 2005 Superlogics www.superlogics.com

TI 2005 Texas Instruments , www.ti.com

Tlaczala W. 2005 LabVIEW – *Integrated Programming Environment,* Przeglad Elektrotechniczny (Electrotechnical Review), 81, pp. 1-6

Tumanski S. 1999 *Magnetovision*, McGraw-Hill Yearbook of Science and Technology, pp. 242-244

Tumanski A. Liszka A. 2002 *The Methods and Devices for Scanning of magnetic Fields*, Journal of Magnetism and Magnetic Materials, 242, pp.1253-1256

USB 2005 www.usb.org

Weigman J. Killian G. 2004 *Decentralization with Profibus*, Wiley & Sons

Winiecki W. 2004 *Mesurement Systems*, Chapter 7 in Modern Metrology (in Polish), WNT

Wooi Ming Tan 1999 *Developing USB Peripherals*, Annabooks

Zurek S., Marketos P., Meydan T., 2004 *Control of Arbitrary Waveforms by Means of Adaptive Digital Feedback Algorithm*, Przeglad Elektrotechniczny (Electrotechnical Review), 80, pp. 122-125

Symbols used in the Book

A – area [m²]
B – flux density [T]
C – capacitance [F]
E – energy [Ws]
F – force [N]
G – weight [N]
H – magnetic field strength [A/m]
I – current [A]
J – current density [A/mm²]
L – inductance [H]
M – mutual inductance [H]
P – power [W]
Q – reactive power [VAr]
R –resistance [Ω]
S – apparent power [VA]
T – period [s]
T – temperature [°C or K]
U – voltage [V]
X – reactance [Ω]
Z – impedance [Ω]

b - damping factor (degree of damping)
f – frequency [Hz]
l – length [m]
m – mass [g]
n – number (of turns, of measurement etc)
t – time [s]
u – uncertainty

α – temperature coefficient of resistance
σ – standard deviation
ε – relative change of parameter $\varepsilon = \Delta X / X$
φ, $\cos\varphi$ - phase shift
$\eta = \omega/\omega_o$ – relative frequency
ρ – resistivity [Ωm]
ω – angular frequency [Hz]

δX – relative error, relative uncertainty
ΔX – absolute error, absolute uncertainty
$f(x)$ – density of probability
$F(x)$ – distribution function
K_u – amplification factor [V/V]
$h(n)$ – impulse response
$h(n) * x(n)$ – convolution
Pr – probability
S_x - sensitivity
$X(j\omega)$ – Fourier transform
$X(s)$ – Laplace transform
$X(z)$ – z-transform

$x(t)$ – instantaneous value
$x(n)$ – discrete value
X or X_{rms} or $\hat{X}$ – rms value
X_m – magnitude
X_{pp} – peak to peak value
X_{AVG} – average rectified value
$\bar{x}$ - mean value
$\mathbf{X}$ – vector value

c – speed of light $c = 2.99792\ 10^8\ m/s$
ε_o – permittivity in vacuum $\varepsilon_o = 8,8541810^{-12}\ F/m$
g – Earth's acceleration factor $g = 9.8\ m/s^2$
h – Planck's constant $h=6.62617\ 10^{-34}\ Js$
k – Boltzman constant, $k = 1.38\ 10^{-23}\ J/K$
μ_o – permeability in vacuum $\mu_o = 1.25663\ 10^{-6}\ H/m$
q – the electric charge $q = 1.6\ 10^{-19}\ C$

Abbreviations used in the Book

AC – alternating current
ADC – analogue to digital converter
ALU – arithmetic–logic unit
AWG – arbitrary Wave Generator
BDA – Bluetooth Device Address
BIPM – Bureau International Poids at Mesures
bit – binary digit
bps – bits per second
Bps – bytes per second
BTS – Base Transceiver Station
CALU – central arithmetic–logic unit
CAMAC – Computer Aided Measurement and Control
CAN – Controller Area Network
CD – compact disc
CF – Crest Factor
CMR – Common Mode Rejection
CMRR – Common Mode Rejection Ratio
CRO – cathode ray oscilloscope
CRC – Cyclic Redundancy Check
CRT – cathode ray tube
CSD – Circuit Switched Data
CSMA/CD – Carrier Sense Multiple Access Collision Detect
DAC – digital to analogue converter
DAQ – data acquisition board
DDS – Direct Digital Synthesis
dB – decibel
DC – direct current
DCC – Direct Current Comparator
DCE – Data Communication Equipment
DFS – Discrete Fourier Series
DFT – Discrete Fourier Transform
DNL – Differential nonlinearity
DSD – Direct Stream Digital
DSP – Digital Signal Processor
DTE – Data terminal equipment
EDGE – Enhanced Data rates for Global Evolution
ENOB – Effective Number of Bits
FCS – Frame Check Sequence
FDDI – Fiber Distributed Data Interface
FF – Form Factor
FFT – Fast Fourier Transform
FIR – Finite Response Filter
FPBW – Full Power Bandwidth
FS – full scale

FSK – Frequency Shift Keying
FSR – Full Scale Range
GPIB – General Purpose Interface Bus
GPRS – General Packet Radio Services
GPS – Global Positioning System
GSM – Global System of Mobile Communication
HART – Highway Addressable Remote Transducer
HLDLC – High Level Data Link Control
HMI – Human–machine Interface
ICA – Independent Component Analysis
IDFT – Inverse Discrete Fourier Transform
IEC – International Electrotechnical Commission
IEE – Institute of Electrical Engineers
IEEE – Institute of Electrical and Electronics Engineers
IFFT – Inverse Fast Fourier Transform
IIR – infinite response filter
IML – intermodulation distrotion
IMRR – Isolation Mode Rejection Ratio
INL – integral nonlinearity
IrDA – Infrared Data Association
ISM – Industrial, Scientific and Medical
ISO – International Standard Organization
ISO VIM – International Vocabulary of Basic and General Terms in Metrology
I2C – Inter-Integrated Circuit
LAN – Local Area Networks
LCD – liquid crystal display
LMS – Least mean square
LSB – Least Significant Bit
LTI – Linear time invariant system
MAC – multiplier/accumulator
MASH – Multistage Noise Shaping
MFB – multifeedback filters
MFLOPS – Million Floating Point Operations per Second
MIPS – Million Instructions per Second
MSPS – mega samples per second
NCAP – Network Capable Application Processor
NI – null indicator
NIST – National Institute of Standards and Technology USA
OBDII – On Board Diagnostics
OpAmp – operational amplifier
PCM – Pulse Code Modulation
PFD – phase frequency detector
PLC – programmable logic controller
PLL – phase locked loop
ppm – parts per million – 10^{-6}
PWM – Pulse Width Modulation
rdg – reading
RFI – radio frequency interference

rms – Root Mean Square
RS232c – Recommended Standard 232
RxD – data receiving
SACD – Super Audio Compact Disc
SAR – Successive Approximation Register
SCADA – Supervisory Control and Data Acquisition
SCPI – Standard Command for Programmable Instruments
SDI-12 – Serial Digital Standard for Microprocessor Based Sensors
SDU – Serial Data Unit
SFDR – Spurious Free Dynamic Range
SH – sample and hold circuit
SHA – sample and hold amplifier
SI – System Internationale
SIM – Subscriber Identity Module
SINAD – signal to noise and distortion ratio
SMS – Short Message Service
SNR – signal to noise ratio
SOFM – Self Organising Feature Maps
SPI – Serial Peripheral Interface
SPS – samples per second
STFT – Short Time Fourier Transform
STIM – Smart Transduce Interface Module
TCPIP – Transmission Control Protocol/Internet Protocol
TDM – Time Division Multiplier
TEDS – Transducer Electronic Data Sheet
THD – Total Harmonic Distortion
TII – Transducer Independent Interface
TSH – track and hold circuit
TTCAN – Time Triggered Communication on CAN
TxD – Data Transmission
UART – Universal Ssynchronous Receiver Transmitter
UMTS – Universal Mobile Telecommunication System
USB – Universal Serial Bus
VEE – Visual Engineering Environment
VPN – Virtual Private Network
WAMS – Wide Area Measuring System
WAN – Wide Area Network
WAP – Wireless Application Protocol
WDF – wave digital filters
WLAN – Wireless Local Area Network
WUSB – Wireless Universal Serial Bus
ZOH – Zero Order Hold

Index

Absolute error, 14
AC bridge circuits, 99-104
AC/DC conversion, 131-141
Accuracy, 2,14
Acquisition (signals), 122,362-366
Activation function, 333
Adaptive filters, 327-331
    Kalman filter, 330-331
    Wiener filter, 328-329
Adaptive measuring methods, 326
Adder (amplifier), 168
Agilent VEE Pro, 428-430
Aliasing, 208,320
Alternate display (oscilloscope), 92
Ammeter, 78
Ampere (unit), 14,15,16
Ampere's law, 439
Amplifications (signals), 143-178
Amplifier
    adder, 168
    antilog, 163-164
    auto-zero, 152-154
    charge, 159-161
    chopper, 150-152
    differential, 143-144
    differentiating, 162
    dynamics, 143
    electrometers, 159-161
    exponential, 163-164
    frequency bandwidth, 143
    function, 161-168
    instrumentation, 146-147
    integrating, 161
    isolation, 147-150
    large input resistance, 159-161
    linear distortion, 143
    logarithmic, 163-164
    lock-in, 154-159
    non-linear distortion, 143
    operational, 145-146
    small DC signals, 150-154
    small AC signals, 154-159

Analogue filters, 191-201
Analogue measuring technique, 18
Analogue signal processing, 18,121-203
Analogue to digital converter (ADC), 205-238
    1 bit, 225-230
    delta-sigma, 225-230
    SAR, 220-222
    flash, 222-223
    half-flash, 223
    integrating (dual slope), 230-234
    MASH, 230
    multibit delta sigma, 230
    pipeline, 223-224
    second order delta-sigma, 229
Anderson loop, 112-114
Antialias filter, 209
Antilog (exponential) amplifier, 163-164
Aperture time, 218
Arbitrary Wave Generator (AWG), 302
arithmetic-logic unit (ALU), 413
Aron method, 83-84
Artificial intelligence, 327
Artificial model of measurement, 30
Artificial neural network, 331-340
    Activation function, 333
    Back propagation algorithm, 335-
    Delta rule, 336
    Hebb's theory, 338
    Hopfield network, 335
    Kohonen neural network, 339
    Perceptron, 333-334
    Self-organizing network, 338
Asynchronous data transmission, 370
Asynchronous Fourier analysis, 264,295
Atomic clock, 61-62
Autocorrelation, 253
Auto zero, 45,152-154,234
Auto zero amplifier, 152-154
Average rectified value (signal), 123
Averaging of signal, 303
Averaging in the frequency domain, 304

Axon, 332
Back propagation algorithm, 335-337
Balanced type bridge circuit, 24,96-99
Ballistic galvanometer, 77
Bandwidth (frequency), 32,44,48
Base quantities, 16
Base transceiver station (BTS), 386
Baud rate, 370-371
Berkley Design Technology, 414
Bessel filter, 196
Bifilar winding, 66
Bilinear transform, 282
Binary code, 215-216
    fixed pint number, 215-216
    floating point number, 215,217
    offset binary, 216
    sign and magnitude, 216
    two's complement, 216
    unsigned integer, 215
Bureau International Poids at Mesures (BIPM), 3,14,16,57
Bisection algorithm, 441
Bit, 18
Bit per second (bps), 369
Bitstream converter, 225-230
Blackman window, 279
Blind deconvolution, 312
Blind signal separation, 339
Bluetooth interface, 383-384
Bluetooth Device Address (BDA), 384
Boolean logic, 340
Boxcar averager, 308
Boxcar integrator, 308
Bridge circuit, 7,19,23-25,46,94-112
    AC bridge circuit, 99-104
    balanced, 24,96-99
    defection type, 24,107-112
    equilibrium condition, 95-96
    Kelvin, 98-99
    linearisation, 109-110
    Maxwell-Wien, 102
    null type, 24,96-99
    RLC, 104
    Sauty-Wien, 101-102
    Schering, 101
    sensitivity, 111-112
    transformer, 104-106
    unbalanced, 24,94,107-112

Wien, 100-101
    Wheatstone, 97-98
Burst mode of acquisition, 364
Butterworth filter, 196
Calculable standards, 63
Calibrator, 21,57,69-70
Calibration, 57
Candela (unit), 16
Capacitance converter, 131
Capacitive interferences (coupling), 180
Capacity measurement, 314-315
Carrier Sense Multiple Access Collision Detect CSMA/CD, 392,396
Casual system, 250
Cauer (elliptic) filter, 196
Celsius (unit), 15
Central Arithmetic-Logic Unit (CALU), 413
Central Limit Theorem, 40
Centronics interface, 381
Cesium atomic clock, 3,16,61-62
CF (crest factor), 49
Charge amplifiers, 159-161
Chebyshew filter, 197
Circuit switched data (CSD), 386
Circular convolution, 255
Chopper amplifier, 150-152
Chopper display (oscilloscope), 92
Cladding, 185
Class of accuracy, 40,42
Coarse approximation, 273
Coherent averaging, 303
Common mode rejection (CMR), 144
Common mode rejection ratio (CMRR), 144
Colpitts oscillator, 131
Comb filter, 283
Combined uncertainty, 55
Comparator (amplifier), 168
Comparator (measuring instrument), 22,116-118
Comparative method, 22,116-118
Compensation method, 21,114-115
Competitive learning, 338
Computational intelligence, 327
Computer Aided Measurement and Control (CAMAC), 407
Computer measuring system, 1,349-454

Conditioning (signals), 8,121,126-131, 353-354
Conductive interferences, 182
Continuous time, 205
Converter of capacitance, 131
Convolution, 251,252
Controlled Area Interface (CAN), 396-397
Cooley and Tukey, 266
Correction, 28-29
Correlation, 55,253,309
    Autocorrelation, 253
    Cross-correlation, 253
Cost of measurement, 31
Covariance, 55
Coverage factor, 32
Cathode ray oscilloscope (CRO), 88
Critical resistance of galvanometer, 77
Cross correlation, 253
Cathode ray tube(CRT), 88
Cumulative histogram, 36
Cumulative distribution function, 36
Current loop, 372,400
Current standard, 58
Current steering converter, 243
Current to voltage converter, 169
Current transformer, 117
Current weight, 18,58,176-177
Cyclic redundancy check (CRC), 370
Damping factor, 50
Data acquisition board (DAQ), ,362-366
Data logger, 357-359
Data communication equipment (DCE), 369
DataSocket, 436_437
Data terminal equipment (DTE), 369
Data transmission
    full-duplex, 372
    half-duplex, 372
    asynchronous, 371
    simplex, 372
    synchronous, 371
Data transmission frame, 371
Data transmitter, 356-357
Daubechies wavelet, 271,273
DC bridge circuits, 96-99
Decade, 192
Decibel dB, 44,291

Decimation filter, 245
Decimation in frequency, 268
Decimation in time, 268
Decomposition, 250,274
Deconvolution, 253,309-311
Deflection type bridge circuits, 24,107-112,127-130
Deglitch filter, 248
Degree of correlation, 55
Delta-sigma converter, 225-230
Delta rule, 336
Dendrite, 332
Density of probability, 36
Derivative quantities, 16
Detail information, 273
Deterministic signals, 123
Differential amplifier, 143-144
Differential measurement method, 45-46
Differential nonlinearity (DNL), 235-236,247
Differential sensors, 46-47,107
Differentiating amplifier, 162
Digital energy meters,323-326
Digital filters, 275-287
    comb, 283
    finite response FIR, 276,278-280
    infinite response IIR, 276, 280-281
    non-recursive filters, 276, 278-280
    recursive (IIR) filters, 276, 280-281
    wave, 284-285
Digital filter averager, 307
Digital measuring instruments, 312-326
    digital multimeter, 312-317
    digital oscilloscope, 318-323
    frequency meter, 317
    Scopemeter, 315
    wattmeter, 323-326
Digital measuring technique, 18
Digital oscilloscope, 318-323
Digital signal processing, 205-348
Digital Signal Processor (DSP), 282,410-415
    arithmetic-logic unit (ALU), 413
    Berkley Design Technology, 414
    (CALU), 413
    Harvard architecture, 411
    TriCore, 413
    (MIPS), 414

(MFLOPS), 414
multiplier/accumulator (MAC), 413
von Neumann architecture, 411
Digital signals, 18,122-123
Digital signal synthesis, 297-303
Digital Sound Processor, 297
Digital to analogue converter (DAC),
                238-248
   current steering,243
   R-2R, 243
   string (segmented),243
   weighted resistors, 242
Diode rectifier, 132
Dirac (delta) function, 249
Direct current comparator (DCC),
                22,115-116
Direct Digital Synthesis (DDS), 300-301
Direct measurement method, 18
Direct Stream Digital (DSD), 245
Dirichlet kernel, 262
Discrete Fourier Transform (DFT), 254,
254-256, 259-268
Discrete Fourier Series (DFS), 253
Discrete time, 205
Distributed measuring system, 362
Distribution function, 36
   normal (Gaussian), 36-38,54
   rectangular, 39,54
   Student's, 39
   triangular, 39,54
   uniform , 39,54
Dithering, 240
Dual slope converter, 230-234
Duplex data transmission, 372
Dyadic sampling, 272
Dynamics, 17,32,219
Effective number of bits (ENOB), 237-
Electrodynamic instrument, 7,82
Electromagnetic interferences, 45
Electromechanical instruments, 73
Electrometer, 159-161
Elliptic (Cauer) filter, 196,209
Energy meter, 86-87,323-326
Enhanced Data rates for Global
Evolution (EDGE), 387
Epstein frame, 30,438
Error, 14,15,28
   absolute, 14

differential nonlinearity, 235-247
of dynamics, 50-52
hysteresis, 43,47
integral nonlinearity, 235-236,247
of linearity 28,44
of monotonicity, 247
nonlinearity, 24,43,44
quantization, 213
random, 27
relative, 14
resolution, 43-44
rise time (oscilloscope), 321
sensitivity, 43
systematic, 27
zero drift, 43
Estimation, 2,27
Estimated value, 28
Estimate of variance, 52
Etalon, 57
Ethernet, 4,7,392-395
   hub, 394
   repeater, 394
   router, 394
   switch, 394
   Transmission Control
      Protocol/Internet Protocol 395
   Virtual Private Network VPN, 395
Ethernet blue book, 392
Eurocard, 406-407
Evanohm, 66
Expanded uncertainty, 32,53
Expected value (expectation), 37-38
Experimental standard deviation, 52
Exponential (antilog) amplifier, 163-164
Exponential averaging, 304
Faraday's law, 17,439
Fast Fourier Transform (FFT),
                266-268,289-291
Fast Wavelet Transform, 273
Feedback, 19,21,24,169-178
Ferrari's system, 86
Feussner potentiometer, 114
FF (form factor), 49
Fiber-optic cable, 184-185
Fiber Distributed Data Interface (FDDI),
                392
Fielbus, 402
Filter coefficients, 277

Filter kernel, 277
Filters,
   active, 193-195
   analogue, 191-201
   antialias, 209
   Bessel, 196
   Butterworth, 196
   Cauer (elliptic) 196,209
   Chebyshev, 196
   cut-off frequency, 192
   decimation, 245
   digital, 275-287
   interpolation, 246
   multifeedback (MFB), 195
   passive, 193
   passband, 191
   polynomials, 197-198
   Sallen-Key, 194
   state variable, 200
   stoppband, 191
   switched capacitor, 199-201
Finite response filter FIR, 276,278-280
FireWire interface, 376-377
FireWire socket, 376
First order inertia, 50
Fixed point number, 215-216
   offset binary,216
   sign and magnitude, 216
   two's complement, 216
   unsigned integer, 215
Flash converter, 222-223
Flicker noise, 180
Floating ground, 187
Floating point number, 215,217
Fluxmeter, 77
Force to current converter, 176-177
Fountain Cesium Atomic Clock, 62
Four wire connection, 188
Fourier series, 124-125
Frame check sequence (FCS), 371
Frequency hopping spread spectrum
      (FHSS), 384
Frequency bandwidth, 32,47
Frequency meters, 317
Frequency of oscillations, 50
Frequency Shift Keying (FSK), 401
Full power bandwidth (FPBW), 237
Full scale range (FSR), 42

Full scale(FS), 41,213
Function amplifiers, 161-168
Fuzzy logic, 3,340-344
Fuzzy logic inference, 342
Fuzzy logic operations
   concentration, 341
   dilation, 341
   fusion, 341
   intersection, 341
Fuzzy rule set, 342
Galvanic isolation (separation), 147-150
Galvanometer, 22,77
   ballistic, 77
Gated integrator, 308
Gateway GPRS Support Node (GGSN),
      386
Gaussian (normal) distribution , 36-37
General packet radio Services (GPRS),
      86-388
General Purpose Interface Bus (GPIB)
      9,377-382
   bus, 379
   connector, 378-379
   expander, 381
   extender, 381
   handshaking lines, 380
Gibbs phenomenon, 279
Gilbert multiplier, 164
Glitch, 248
Global System Mobile Communication
      (GSM), 385,391-394
Gradient sensor, 46
Graphical programming language, 431
Grounding, 45,186-187
GSM interface
   base transceiver station (BTS), 386
   circuit switched data (CSD), 386
   EDGE, 386
   (GPRS), 386-388
   Short Message Service (SMS), 387
   (WAP), 388
Guard, 187-188
Guide on uncertainty, 14,27
Haar wavelet, 275
Half-duplex data transmission, 372
Half-flash converter, 223
Hamming window, 264
Hanning window, 264,295

Harvard architecture, 411
Hartley oscillator, 131
Hebb's theorem, 338
Hebbian learning, 338
Heisenberg uncertainty, 15,270
Hermitian symmetry, 255
High level data link control (HLDLC),
	371
High speed handshake protocol
	(HS488), 381
High Speed Circuit Switched Data
	(HSCSD), 387
Highway Addressable Remote
	Transducer (HART), 400-401
Histogram, 34-36,
	cumulative , 36
Hopfield neural network, 335
Hewlett Packard Interface Bus (HPIB),
	377
Hub, 374,394
Human-machine Interface (HMI), 421
Hysteresis error, 43,47
I-button, 355
International Electrotechnical
	Commission (IEC), 16
IEC 625 interface, 377-482
IEC Standard 61158, 408
IEEE P1451 standard, 359-362
IEEE 1284 interface, 381
IEEE 1394 interface, 376
IEEE 488 interface, 377-482
Ilink interface, 377
Impedance to frequency converter, 178
Impulse response, 251,277
Independent component analysis (ICA),
	339
Indicating measuring instrument, 73-85
	electrodynamic, 82
	moving coil, 74-80
	moving iron, 81
Indirect method of measurement, 18
Inductive watt-hour meter, 86-87
Inductive interferences (coupling), 182
Industrial, Scientific and Medical (ISM)
	band, 389
Inertia, 50
Infinite response filters IIR, 276,280-
	281

Information, 3,13
Infrared data association (IrDA), 382
In phase (quadrature) component, 294
Instrumentation, 13
Instrumentation amplifier, 146-147
Integral nonlinearity (INL), 235-236,247
Integrating amplifier (integrator), 161
Integrating converter, 230-234
Intelligent data analysis, 326-344
	adaptive filters, 327-331
	artificial neural networks, 331-341
	fuzzy logic, 340-344
Intelligent sensors, 3,7,354-356
Interface, 9,367
	bus, 367
	protocol, 367
	serial, 367
	parallel, 367
Interfaces, 367-401
	1-wire, 399
	4-20 mA, 400
	Bluetooth, 383-384
	CAN, 4,396-397
	Ethernet, 4,7,392-395
	FireWire, 376-377
	GPIB, 9,377-382
	GSM, 385-389
	HART, 400-401
	I2C, 398-399
	IrDa, 382-383
	radio, 389-391
	MicroLan, 399
	RS-232C, 4,7,368-373
	RS-485, 372-373
	SDI, 399
	SPI, 418
	UART, 419
	USB, 7,373-377
	UMTS, 389
	WUSB, 385
Interferences, 44,180-182
	capacitive, 180
	conductive, 182
	inductive, 182,185
	magnetic, 185
	radio frequency interference, 182
Inter-Integrated Circuit (I2C), 398-399
Intermodulation distortion (IML), 237

Internet, 395
Interpolation filter, 246
Inverse Discrete Fourier Transform
        (IDFT), 254
Inverse Fast Fourier Transform (IFFT),
        268
International Standard Organization
        (ISO), 3,13,16
ISO VIM, 13
ISO Guide, 14,27
Isolation amplifier, 147-150
Isolation mode rejection ratio (IMRR),
        152
Isolation
    transformer, 148
    capacitive, 149
    light, 149
    magnetic, 150
Isolatiom Mode Rejection Ratio
        (IMRR), 148
Jitter, 248
Johnson (thermal) noise, 180
Josephson effect, 59
Kalman filter, 330-331
Kelvin (unit), 16
Kelvin bridge circuit, 98-99
Kernel, 251,310
Kilogram (unit), 16
Klitzing (von) constant, 60
Kohonen neural network, 339
LabVIEW, 10,431-438
Lack of linearity, 44
Latch (register) circuit, 241
Latency, 352
Liquid cristal display (LCD), 89-90
Leakage, 260
Least mean square LMS, 327
Least significant bit (LSB), 42,213
Level of confidence, 28
Line Printer Terminal (LPT), 381
Linear time invariant system (LTI), 250
Linearity, 24,28
Linearization, 24
Linguistic variables, 340
Local Area Network LAN, 392
Lock-in amplifiers, 154-159
Logarithmic amplifier, 163-164
Log/antilog multiplier, 169

Losses, 30
Magnetoresistor, 46,150,173
Magnetovision, 449-451
Mallat transform, 273
Manganin, 66,78
Mask, 310
Mathematical model, 17
Maximal limiting uncertainty, 56
Maxwell-Wien bridge circuit, 101
McCulloch-Pitts model, 332
Mean value, 38,123
Measurand, 13,14
Measurement
    definition, 2,3,11,13,15
    range, 20
    scale, 15
Measurement methods, 18-26
    comparative, 21
    compensation, 21
    differential, 19
    direct, 18
    indirect, 18,21
    null, 19
    null-differential, 20
    substitution, 25-26
Measurement of
    capacity, 314-315
    current, 19
    electric power, 86-90,323-326
    frequency, 317
    hardness, 15
    impedance, 318
    period, 317
    reactive power, 82-86
    resistance, 3
    temperature, 15
Measuring
    electric signal, 18
    procedure, 15
Measuring microsystem, 352
Median filter, 306
Memberhsip function, 341
Meter (unit), 16
Method of measurement, 15
Metrology, 3,13
Microammeter, 77
Microcontroller, 410-412
MicroLan interface, 399

Microohmmeter, 160
Millivoltmeter, 78
Million Instructions per Second (MIPS), 414
Million Floating Point Operations per Second (MFLOPS), 414
Missing code, 236
Mistakes, 30
Modus Ponens, 342
Mohs scale, 15
Mole (unit), 16
Monotonicity, 247
Morlet wavelet, 271
Most significant bit (MSB), 215
Movement of the pointer, 76
Moving average filter, 305
Moving coil instrument, 6,74-80
Moving window method (MWM), 269
Moving iron instrument, 7,81
Multibit delta sigma converter, 230
Multilayer perceptron, 334
Multimeter, 69-70,312-317
multifeedback filters (MFB), 195
Multiple FFT analysis, 294
Multiple point random sampling, 320
Multiplier, 114,164-167,324
Multiplier/accumulator (MAC), 413
Multistage Noise Shaping (MASH), 230
National Institute of Technology (NIST), 22,61,116,359
Network capable application processor (NCAP), 360
Neural network, 331-340
Noise, 44,179-180
    1/f , 180
    flicker, 180
    matching, 190
    pink, 180
    white, 179
    shot, 180
    thermal (Johnson), 180-181
Noise rejection, 232,328
Noise shaping, 228
Nominal conditions, 43
Nonlinearity, 24,43,44
Non-recursive (FIR) filters, 276,278-280
Normal (Gaussian) distribution, 36-37
Null indicator (NI), 22

Null measurement method, 19
Null-differential method, 20
Null-modem connection, 368
Null type bridge circuit, 24,96-99
Nyquist frequency, 209
On Board Diagnostics OBDII, 5
Octave, 192
Offset binary number, 216
Ohm's law, 16,127
Operating conditions, 43
Operational amplifier, 145-146
Optical coupler (OC), 185
Oscillations, 50
Oscilloscope, 88-93,318-323
Oversampling, 210,228,244
Paced mode of acquisition, 364
Panel meters, 80
Parseval theorem, 51,256
PCI Extension for Instrumentation (PXI), 407
Peak value (signals), 123
Peak value converter, 137
Peak-to-peak value (signal), 138
Perceptron, 333-334,442-443
Performance Spectrum Analyzer (PSA), 293
Phase circle, 300
Phase frequency detector (PFD), 298
Phase locked loop (PLL), 298-299
Phase-sensitive detector (rectifier), 130, 139-141,155-156
Phosphor coating (oscilloscope), 89
Piconet, 383
Pink noise, 180
Pipeline converter, 223-224
Plug-in acquisition board, 362-363
Poles of the transmittance, 257
Potentiometer, 22,114-115
ppm (parts per million), 41
Probability, 28,37-39
    density, 36-37
    distribution , 36
    normalized function, 37
Profibus, 403-404
Profinet, 404
Programmable logic controller (PLC), 405
Propagation of uncertainty, 55

Pulse Code Modulation (PCM), 244
Pulse Width Modulation (PWM), 244
Quadrature (in-phase) component, 294
Quadrature sampling, 294
Quantization, 206,211,
Quantisation error, 213
Quntum Hall effect, 60
Quantum Josephson effect, 59
Quartz oscillator, 69
Radio data Transmission, 389-391
Radio frequency interference (RFI), 182
Radio modem, 389
Radio transceiver, 391
Random errors, 28
Random uncertainty, 30
Range of measuring device, 15,20
rdg (reading), 41
Reactive power meter, 85-86
Real time measuring system, 352-353
Recording measuring instruments, 88-94
Recorders, 93-94
Recovery (signal), 44,122
Rectangular distribution, 39,54
Rectifier, 132
Recursive (IIR) filters, 276, 280-281
Reference multimeter, 69
Reference value, 15
Register (latch) circuit, 241
Relative error, 14
Relative uncertainty, 41,43
Reliability of measurement, 32
Repeater, 394
Resistance to current converter, 172-173
Resistor decade box, 66
Resolution, 20,42,43
Resolution of digital instrument, 41,42
R-2R converter, 243
RLC bridge circuit, 104
RLC to frequency converter, 178
rms averaging, 304
rms (root mean square) value, 49,123
rms converters, 133-137
Router, 394
RS-232C, 4,7,368-373
RS-485, 372-373
Ryleigh current weight, 58-59
SACD technique, 33,210
Sallen-Key filter, 194

Sample and hold amplifier (SHA), 218
Sample and hold circuit (SH), 217-218
Sample and hold oscilloscope, 320-321
Samples per second (SPS), 206
Sampling, 206
    burst mode, 364
    frequency, 206
    Nyquist frequency, 209
    oversampling, 210
    paced mode, 364
    random, 319
    sequential, 319
    undersampling, 210
Sawtooth signal in oscilloscope, 88
Sauty (de)-Wien bridge circuit, 101-102
Scalling process, 21
Scientific intrumentation, 13
Schering bridge circuit, 101
Scopemeter, 315
Screen (oscilloscope), 89
Second (unit), 16
Segmented (string) converter, 243
Self-organising neural network, 338
Self organising feature maps (SOFM),
        338
Sensitivity, 20,43
Sensitivity error, 43
Sensor, 4
Sensors,
    coil sensor, 17
    Hall sensor, 7
    strain gauge sensor, 7
    thermocouple sensor, 7
    thermoresistive, 28
    thermistor sensor, 7
Separation
    transformer, 148
    capacitive, 149
    light, 149
    magnetic, 150
Sequential sampling, 319
Serial data unit (SDU), 370
Serial-Digital Interface Standard for
        Microprocessor-Based Sensors
        (SDI), 399
Serial Peripheral Interface (SPI), 418-
Series resistor, 77
Settling time, 241

Shannon theorem, 208
Shielding, 45,183
    electrostatic, 183
    magnetic, 183
Short Message Service (SMS), 387-388
Short Time Fourier Transform (STFT),
            268-270
Shot noise, 180
Shottky relation, 180
Shunt resistor, 77
Sigma-delta converter, 225-230
Sign and magnitude number, 216
Signal
    acquisition, 122
    amplification, 122,143-178
    analogue processing, 18,121-203
    average rectified, 123
    conditioning, 121,126-131
    conversion, 122
    deterministic, 123
    digital processing, 18,205-348
    filtering, 122
    harmonization, 122
    instantaneous, 123
    magnitude, 123
    mean value, 123
    modulation, 122
    peak value, 123
    peak to pik, 123
    recovery, 122
    rms, 123,133-138
    separation, 122,147-150
    stochastic, 123
Signal recovery, 45,122,303-312
    averaging method, 303
    coherent averaging, 303
    correlation, 309
    deconvolution, 309-310
    exponential averaging, 304
    rms averaging, 304
    vector averaging, 304
Signal to noise ratio (SNR), 44,180,237,
signal to noise and distortion ratio
            (SINAD), 237
Simplex data transmission, 372
SI system, 3,16-17,
Smart digital magnetometer, 355
Smart sensor, 7,359

Smart Transduce Interface Module
        (STIM), 360-362
Spectral analysis, 125, 287-297
    digital, 297
    multiple filter, 291
    performance spectrum analyser, 293
    tuned filter, 292
    superheterodyne, 292
Spectrum analyzers, 288-297,434
Spectral density noises, 180
Spectrum lines, 125
Spectrogram, 297
Spurious free dynamic range (SFDR),
            237
Squarer, 166
Staircase reconstruction, 239
Stand-alone data loggers, 357
Standards, 3,14,57-69
    calculable, 63
    definition, 57
    of current, 58
    of capacitance, 62,67-68
    of frequency, 61-62,69
    of inductance, 68-69
    of length, 57-58
    resistance (quantum), 60
    of resistance, 65-66
    of voltage (quantum), 59
    of voltage, 63-65
Standard Command for Programmable
        Instruments (SCPI), 381,408-409
Standard deviation, 37-39,53
Standard reference, 13
Standard unit, 2,3,16
Standard uncertainty, 32
State variable filter, 200
Stationary signals, 269
Stochastic signals, 123
Strain gauge sensor, 7,46
Stain-gauge bridge, 156-157
String (segmented) converter, 243
Student's distribution, 39
Subband coding, 273
Subranging converter, 228-229
Subscriber identity module (SIM), 385
Substitution measurement method, 25
Successive Approximation Register
        (SAR), 220-222

Super Audio Compact Disc (SACD), 210
Superconducting effect, 59
Superheterodyne, 292
Superposition principle, 250
Supervisory Control and Data Acquisition (SCADA), 406
Switch, 394
Switched capacitor (filter), 199-201
Synapse, 332
Synchronous data transmission, 370
Synchronous Fourier analysis, 260,264
Synchronisation (oscilloscope), 89-90
Synthesis, 250
System on Chip, 410
Systematic errors, 27
Termocouple sensor, 7
TestPoint, 10,424-428
Thompson-Lampard capacitor, 62
Time constant T, 50
Time division multiplier (TDM), 324
Time invariant (stationary) system, 250
Time Trigerred Communication on CAN (TTCAN), 397
Thermal (Johnson) noise, 180
Thermistor, 7
Three-fold-slope converter, 233
Three phase wattmeter, 85-86
Three voltmeter method, 317,418
Token Ring, 392
Total harmonic distortion (THD), 48-49,237
Traceability, 3
Track and hold circuit (TSH), 218
Transconductance multiplier, 164
Transducer electronic data sheet (TEDS), 360-361
Transducer independent interface (TII), 360
Transfer function, 257
Transmission Control Protocol/Internet Protocol (TCPIP), 395
Transformer bridge circuits, 104-106
Transformer separation (isolation), 148
Transient response, 237
Transmitter, 360,355-357
Triangular distribution, 39,54
Triggering (oscilloscope), 90-91

TriCore, 413
True *rms*, 49
True *rms* converters, 133-137
True value, 28
Tunning word, 299
Twiddle factor, 267
Twisted wire, 184
Two's complement number, 216
Two wire transmitter, 360
Type A of uncertainty, 31,52
Type B of uncertainty, 31,54
Unbalanced bridge circuits, 24,107-112,128-130
Uncertainty, 2,11,13,14,26-57
    absolute, 43
    combined, 55
    expanded, 32,54
    maximal limiting, 56
    of measuring device, 30
    random, 30
    relative, 41,43
    standard, 32
    type A, 31,52
    type B, 31,54
Undersampling, 211
Unit of measurement, 15
Universal asynchronous receiver transmitter (UART), 358,419-421
Universal Mobile Telecommunication System (UMTS), 385, 389
Universal RLC bridge circuit, 104
Universal serial bus (USB) interface, 7,373-377
    USB bulk transfer, 375
    USB connector, 375
    USB isochronous transfer, 375
    USB host controller, 374
    USB hub, 374
    USB interrupt transfer, 375
    USB data packet, 375
Universal shunt resistor, 79
Unsigned integer, 216
Useful storage bandwidth (USB), 322
Validation, 31,58
Variance, 38,53
Varicap diode, 299
Vector averagins, 304
Virtual instrument, 10, 421-438

Virtual measuring system, 421-438
  Agilent VEE Pro, 428-430
  LabVIEW, 10,431-438
  TestPoint, 424-428
Virtual Private Network VPN, 395
Visual Engineering Environment (VEE), 428-430
VME Extension for Instrumentation (VXI), 407
Voltage controlled oscillator (VCO), 298
Voltage Controlled Source (VCS) filters, 195
Voltage divider, 126-127
Voltage to current converter, 169
Voltage to frequency converter, 131-132,142-143
Voltmeter, 78
Voltammeter, 79
von Neumann architecture, 411
Wagner earth (ground), 99-100
Warshawski bridge, 118
Watt-hour meter, 90-92,323-326
Wattmeter (electrodynamic), 82-83
Wave adaptor, 285
Wave digital filter, 284-285
Wavelet transform, 271-275
Wavelet
  approximation, 273
  Daubechies, 271,273
  decomposition, 274
  detail, 273
  Haar, 275
  Morlet, 271

Weather Transmitter, 356-357
Weighted resistors converter, 242
Weston cell, 63-64
Wheatstone bridge circuit, 97-98
Wideband Code Division Multiple Access (WCDMA), 389
Wide area measuring system (WAMS), 351
Wide Area Network WAN, 392
Widespread measuring system, 351
Wien bridge circuit, 100-101
Wiener filter, 328-329
Window function, 261
  Blackman, 279
  Hamming, 264
  Hanning, 264,295
Winner take all, 338
Wireless Application Protocol (WAP), 388
Wireless interfaces, 382-385
Wireless Local Area Network (WLAN), 395
Wireless Markup Language (WML), 388
Wireless Universal Serial Bus (WUSB), 385
XOR gate, 299
XY mode (oscilloscope), 93
XY recorder, 94
Zero drift error, 44,45
Zero order hold (ZOH) reconstruction, 239
Z-transform, 256-257